VOLUME FIVE HUNDRED AND FIFTY

METHODS IN ENZYMOLOGY

Riboswitches as Targets and Tools

METHODS IN ENZYMOLOGY

VOLUME FIVE HUNDRED AND FIFTY

METHODS IN ENZYMOLOGY

Riboswitches as Targets and Tools

Edited by

DONALD H. BURKE-AGUERO

Department of Molecular Microbiology & Immunology and Department of Biochemistry, University of Missouri, USA

ELSEVIER

AMSTERDAM • BOSTON • HEIDELBERG • LONDON
NEW YORK • OXFORD • PARIS • SAN DIEGO
SAN FRANCISCO • SINGAPORE • SYDNEY • TOKYO

Academic Press is an imprint of Elsevier

Academic Press is an imprint of Elsevier
225 Wyman Street, Waltham, MA 02451, USA
525 B Street, Suite 1800, San Diego, CA 92101-4495, USA
125 London Wall, London, EC2Y 5AS, UK
The Boulevard, Langford Lane, Kidlington, Oxford OX5 1GB, UK

First edition 2015

Notices

Knowledge and best practice in this field are constantly changing. As new research and experience broaden our understanding, changes in research methods, professional practices, or medical treatment may become necessary.

Practitioners and researchers must always rely on their own experience and knowledge in evaluating and using any information, methods, compounds, or experiments described herein. In using such information or methods they should be mindful of their own safety and the safety of others, including parties for whom they have a professional responsibility.

To the fullest extent of the law, neither the Publisher nor the authors, contributors, or editors, assume any liability for any injury and/or damage to persons or property as a matter of products liability, negligence or otherwise, or from any use or operation of any methods, products, instructions, or ideas contained in the material herein.

ISBN: 978-0-12-801123-2
ISSN: 0076-6879

For information on all Academic Press publications
visit our website at store.elsevier.com

CONTENTS

CONTRIBUTORS

Robert T. Batey
Department of Chemistry and Biochemistry, University of Colorado at Boulder, Boulder, Colorado, USA

Sanchita Bhadra
Department of Chemistry and Biochemistry, Institute for Cellular and Molecular Biology, Center for Systems and Synthetic Biology, University of Texas at Austin, Austin, Texas, USA

Alicia Soler Canton
Department of Bionanoscience, Kavli Institute of Nanoscience, Delft University of Technology, Delft, The Netherlands

James M. Carothers
Department of Chemical Engineering, Molecular Engineering and Sciences Institute, Center for Synthetic Biology, University of Washington, Seattle, WA, USA

Yaroslav G. Chushak
711th Human Performance Wing, Air Force Research Laboratory, Wright-Patterson Air Force Base, Dayton, Ohio, USA

Rodrigo A. Correa-Rojas
Department of Chemical Engineering, Molecular Engineering and Sciences Institute, Center for Synthetic Biology, University of Washington, Seattle, WA, USA

Christophe Danelon
Department of Bionanoscience, Kavli Institute of Nanoscience, Delft University of Technology, Delft, The Netherlands

Andrew D. Ellington
Department of Chemistry and Biochemistry, Institute for Cellular and Molecular Biology, Center for Systems and Synthetic Biology, University of Texas at Austin, Austin, Texas, USA

Tom Ellis
Centre for Synthetic Biology and Innovation, and Department of Bioengineering, Imperial College London, South Kensington Campus, London, United Kingdom

Sven Findeiß
Research Group Bioinformatics and Computational Biology, Faculty of Computer Science, and Institute for Theoretical Chemistry, University of Vienna, Vienna, Austria

Michael S. Goodson
711th Human Performance Wing, Air Force Research Laboratory, Wright-Patterson Air Force Base, Dayton, Ohio, USA

Ming C. Hammond
Department of Chemistry, and Department of Molecular & Cell Biology, University of California, Berkeley, California, USA

Svetlana V. Harbaugh
711th Human Performance Wing, Air Force Research Laboratory, Wright-Patterson Air Force Base, Dayton, Ohio, USA

Jörg S. Hartig
Department of Chemistry, Konstanz Research School Chemical Biology, University of Konstanz, Konstanz, Germany

J.V. Hines
Department of Chemistry & Biochemistry, Ohio University, Athens, Ohio, USA

Norikazu Ichihashi
Graduate School of Information Science and Technology, Osaka University, and Exploratory Research for Advanced Technology, Japan Science and Technology Agency, Suita, Osaka, Japan

Samie R. Jaffrey
Department of Pharmacology, Weill Medical College, Cornell University, New York, New York, USA

Sungho Jang
Department of Chemical Engineering, Pohang University of Science and Technology, Pohang, Gyeongbuk, Republic of Korea

Gerald F. Joyce
Department of Chemistry, The Skaggs Institute for Chemical Biology, The Scripps Research Institute, La Jolla, California, USA

Gyoo Yeol Jung
Department of Chemical Engineering, and School of Interdisciplinary Bioscience and Bioengineering, Pohang University of Science and Technology, Pohang, Gyeongbuk, Republic of Korea

Colleen A. Kellenberger
Department of Chemistry, University of California, Berkeley, California, USA

Nancy Kelley-Loughnane
711th Human Performance Wing, Air Force Research Laboratory, Wright-Patterson Air Force Base, Dayton, Ohio, USA

Benedikt Klauser
Department of Chemistry, Konstanz Research School Chemical Biology, University of Konstanz, Konstanz, Germany

Shungo Kobori
Graduate School of Information Science and Technology, Osaka University, Suita, Osaka, Japan

J. Liu
Department of Chemistry & Biochemistry, Ohio University, Athens, Ohio, USA

J.A. Means
Department of Chemistry & Biochemistry, Ohio University, Athens, Ohio, USA

Mario Mörl
Institute for Biochemistry, University of Leipzig, Leipzig, Germany

Zohreh Nourian
Department of Bionanoscience, Kavli Institute of Nanoscience, Delft University of Technology, Delft, The Netherlands

Atsushi Ogawa
Proteo-Science Center, Ehime University, Matsuyama, Japan

Charles Olea Jr.
Department of Chemistry, The Skaggs Institute for Chemical Biology, The Scripps Research Institute, La Jolla, California, USA

René C.L. Olsthoorn
Department of Molecular Genetics, Leiden Institute of Chemistry, Leiden University, Leiden, The Netherlands

Georgios Pothoulakis
Centre for Synthetic Biology and Innovation, and Department of Bioengineering, Imperial College London, South Kensington Campus, London, United Kingdom

Silvia Ramundo
Departments of Molecular Biology and Plant Biology, University of Geneva, Geneva, Switzerland

Charlotte Rehm
Department of Chemistry, Konstanz Research School Chemical Biology, University of Konstanz, Konstanz, Germany

Jean-David Rochaix
Departments of Molecular Biology and Plant Biology, University of Geneva, Geneva, Switzerland

Martin M. Rudolph
Fachbereich Biologie, Synthetische Biologie, Technische Universität Darmstadt, Darmstadt, Germany

Jessica C. Seeliger
Department of Pharmacological Sciences, Stony Brook University, Stony Brook, New York, USA

Sang Woo Seo
Department of Chemical Engineering, Pohang University of Science and Technology, Pohang, Gyeongbuk, Republic of Korea

David Sparkman-Yager
Department of Chemical Engineering, Molecular Engineering and Sciences Institute, Center for Synthetic Biology, University of Washington, Seattle, WA, USA

Peter F. Stadler
Institute for Theoretical Chemistry, University of Vienna, Vienna, Austria; Bioinformatics Group, Department of Computer Science and the Interdisciplinary Center for

Bioinformatic, University of Leipzig, Leipzig, Germany; Center for RNA in Technology and Health, University of Copenhagen, Frederiksberg, Denmark; Max Planck Institute for Mathematics in the Sciences; Fraunhofer Institute for Cell Therapy and Immunology, Leipzig, Germany, and Santa Fe Institute, Santa Fe, New Mexico, USA

Rita L. Strack
Department of Pharmacology, Weill Medical College, Cornell University, New York, New York, USA

Beatrix Suess
Fachbereich Biologie, Synthetische Biologie, Technische Universität Darmstadt, Darmstadt, Germany

Daniel Summerer
Department of Chemistry, Konstanz Research School Chemical Biology, University of Konstanz, Konstanz, Germany

Jeremiah J. Trausch
Department of Chemistry and Biochemistry, University of Colorado at Boulder, Boulder, Colorado, USA

Pauline van Nies
Department of Bionanoscience, Kavli Institute of Nanoscience, Delft University of Technology, Delft, The Netherlands

Erik R. Van Vlack
Department of Chemistry, Stony Brook University, Stony Brook, New York, USA

Michael-Paul Vockenhuber
Fachbereich Biologie, Synthetische Biologie, Technische Universität Darmstadt, Darmstadt, Germany

Manja Wachsmuth
Institute for Biochemistry, University of Leipzig, Leipzig, Germany

Jina Yang
Department of Chemical Engineering, Pohang University of Science and Technology, Pohang, Gyeongbuk, Republic of Korea

Tetsuya Yomo
Graduate School of Information Science and Technology; Graduate School of Frontier Biosciences, Osaka University; Exploratory Research for Advanced Technology, Japan Science and Technology Agency, Suita, Osaka, and Earth-Life Science Institute, Tokyo Institute of Technology, Meguro, Tokyo, Japan

Chien-Hung Yu
Department of Molecular Genetics, Leiden Institute of Chemistry, Leiden University, Leiden, The Netherlands

C. Zeng
Department of Chemistry & Biochemistry, Ohio University, Athens, Ohio, USA

S. Zhou
Department of Chemistry & Biochemistry, Ohio University, Athens, Ohio, USA

PREFACE

The early years of the twenty-first century have seen an explosion of interest in the diverse capabilities of RNA. Riboswitches capture the excitement and promise of this field. They are structurally dynamic, they sense and respond to specific molecular partners, their occupancy states governs gene regulatory decisions, and they can be engineered to reprogram gene regulatory circuitry. Importantly, many of the experimental and theoretical tools that have been used to study riboswitches can also be applied to other RNAs, and tools developed for studies of other RNAs can be applied to riboswitches.

These two volumes (*Methods in Enzymology* 549 and 550) include 40 contributions that outline cutting-edge methods representing a wide spectrum of research questions and scientific themes. The first volume emphasizes natural riboswitches, from their discovery to assessment of their structures and functions. The second volume shifts the focus to applying riboswitches as tools for a variety of applications and as targets for inhibition by potential new antibacterial compounds. A third volume (*Methods in Enzymology* 553) will appear shortly after these two focusing on computational methods for predicting and evaluating dynamic RNA structures. Although the chapters are organized into discrete themes, many of them cut across thematic boundaries by weaving together methodological solutions to multiple issues, and several of the chapters could fit comfortably into more than one section.

VOLUME 1

Riboswitch discovery. In the early days of the riboswitch field, new riboswitches were discovered at a frenetic pace, often by comparing large sets of bacterial genomes. While that approach continues to identify new members of known riboswitch families, the interval between discoveries of new riboswitch families widens. The series begins with two chapters outlining new methods that utilize informatics approaches in combination either with RNASeq and genome-wide methods (the Martin-Verstraete) or with *in vitro* selection (Lupták) to discover new natural riboswitches.

Sample preparation. Any effort to characterize purified, functional RNAs will only be as good as the corresponding sample preparations. Therefore, the next five chapters are dedicated to methods for the synthesis and preparation of large RNAs. Three groups exploit specialty nucleic acids with functionalities of their own. The first chapter in this set describes the use of cotranscribed aptamer affinity tags that are removed by activatable self-cleavage (Legault). This is followed by methods for using catalytic deoxyribozyme ligases to assemble large RNAs from synthetic fragments, some of which carry site-specific spin labels for electron paired resonance studies (Höbarter). The third chapter in this set describes the combined use of aminoacyl transferase ribozymes and chemical protection to generate charged tRNAs on a large scale (Ferré-D'Amaré). These are followed by two chapters that integrate organic chemical methods with improved enzymology to produce photocleavable biotinylated guanosine that incorporates at the 5′ end of *in vitro* transcripts (Sintim) and large quantities of selectively $^{13}C/^{15}N$-labeled RNA in previously unattainable labeling patterns for improved spectroscopic analysis (Dayie).

Structure and function. The biochemical functions of riboswitches are inextricably linked with their three-dimensional structures. The next several chapters, therefore, provide methods for evaluating riboswitch structure and function. Updated protocols are provided for widely utilized SHAPE method of structural probing, along with details of how to implement new software for data interpretation (Weeks). It is well recognized that structural context can perturb pK_a values within RNA and DNA; hence, the next chapter details how to measure them without falling into traps of oversimplifying the underlying molecular processes (Bevilacqua). The next chapter provides methods for obtaining appropriate crystals for ligand–RNA complexes, with emphasis on fragment-bound TPP riboswitches (Ferré-D'Amaré). This section ends with a detailed description of experimental and analytical methods for using small-angle X-ray scattering to define RNA conformations in solution (Rambo).

Conformational dynamics. Spectroscopic methods are ideal for following riboswitch conformational dynamics in real time. The first two chapters of this section describe site-specific incorporation of spectroscopic labels and their use in addressing specific question, first with ^{19}F NMR to probe conformational exchange (Greenbaum) and then with spin-label probes for electron paramagnetic resonance spectroscopy of large RNAs (Fanucci). Single-molecule methods such as smFRET have become a staple of modern biophysical analysis. Three chapters provide detailed guidance on many facets of smFRET, from sample preparation, data acquisition, and analysis to explorations of folding landscapes (Penedo, Walter, and Cornish). The last chapter of this section describes how to integrate surface plasmon resonance (SPR), isothermal titration calorimetry, and circular dichroism to examine tertiary docking (Hoogstraten).

Ligand interactions. One of the most important characteristics of riboswitches is their ability to sense the presence of specific metabolites by forming bound molecular complexes. Isothermal titration calorimetry is one of the most powerful methods for evaluating the energetics of RNA–small molecule complexes (Wedekind). SPR is another powerful tool for characterizing aptamer kinetic and equilibrium binding properties and is detailed in two chapters (Smolke and Sigel). Finally, an innovative and relatively new technique known as DRaCALA is described in the last chapter of the first volume (Lee).

VOLUME 2

The second volume in this series takes a different perspective on riboswitches. Specifically, now that nature has shown us that RNA modules can sense metabolites and report on them, how can we take advantage of that ability to engineer new properties into cells and biochemical systems? Necessarily, this volume takes a much broader view of riboswitches than those found in nature, encompassing ligand-responsive transcriptional and translational modules, ribozymes, sensors, and modules that induce fluorescence in a fluorophore upon formation of the bound complex. It encompasses Synthetic Biology applications as tools to understand normal biological processes, and as tools to reprogram metabolite flux in workhorse organisms. Finally, it comes full circle by screening small-molecule libraries for inhibitors of natural riboswitches.

In short, this second volume details methods at the cutting edge of the translational science of riboswitches.

Artificial riboswitches. The first six chapters of the second volume provide methods for several approaches to construct and optimize artificial riboswitches. There has been substantial progress toward designing artificial riboswitches from scratch, especially when guided by experimental validation (Mörl). A contrasting approach uses *in vitro* selection/evolution to obtain ligand-responsive ligase ribozymes from highly diverse starting populations (Joyce), or to reshape and reprogram the ligand-binding and expression platforms of natural riboswitches (Batey). The next chapter presents methods for optimizing signal transduction (Kelley-Loughnane), since regulation sometimes benefits from maximizing suppression of basal expression in the OFF state and sometimes from maximizing expression in the ON state. The next two chapters address optimization in two very different cell-free systems, first using coupled transcription–translation to optimize a ligand-responsive self-cleaving ribozyme, or "aptazyme" (Yomo), and then taking advantage of a eukaryotic mechanism by which ribosomes "shunt" past certain secondary structures, which can be stabilized to increase shunting efficiency by binding to the analyte ligand (Ogawa).

Ligand-responsive fluorescent sensors. There has been longstanding interest in coupling the binding of ligands to RNA with the emission of light. One such system is that of the recently described Spinach (and Spinach2) aptamer mimics of green fluorescent protein, which are the focus of the next five chapters, each in a different system. The first chapter in this section, from the lab that discovered and first described the Spinach system, presents methods for using it to image intracellular RNA in mammalian cells (Jaffrey). The next two chapters describe how to use these modules in bacterial cells, first as intracellular sensors of intracellular cyclic dinucleotide levels (Hammond) and then for simultaneous and independent monitoring of mRNA and protein levels (Ellis). The next chapter takes this same question into solution and into vesicle-based artificial cells (Danelon). The fifth chapter in this section couples sensing of oligonucleotide "ligands" with Spinach2 output in real time for sequence-specific target quantitation and potential point-of-care applications (Ellington).

Synthetic biology: Conditional control of gene expression. The third section of this volume lays out several methods for using artificial or natural riboswitches to study gene function. This has proven to be a powerful tool in organisms for which limited genetic tools are available, such as the intracellular pathogen Mycobacteria (Seeliger), as well as in more readily manipulated, nonpathogenic bacteria such as *Streptomyces coelicolor* (Suess). Eukaryotes can be similarly studied. A clever variation on this approach is to make the expression of query genes to be dependent upon a regulatory protein whose expression is controlled by a natural riboswitch, as demonstrated here for the unicellular alga *Chlamydomonas reinhardtii* vitamin-repressible riboswitch (Rochaix). This concept is extended to a self-cleaving aptazyme by identifying variants that respond to various ligands to regulate gene expression in the yeast *Saccharomyces cerevisiae* (Hartig).

Synthetic biology: Pathway optimization. The fourth section provides methods that illustrate two examples of using riboswitches as tools to optimize metabolic pathways for Synthetic Biology applications. The first chapter details a computational approach focused on kinetic folding with experimental validation to build aptazymes that respond cotranscriptionally to the presence of metabolites, and the experimental validation of those devices (Carothers). The second chapter describes a method for using riboswitches to impose selective growth advantages on cells that optimally channel their metabolic output into production of a desired compound (Jung).

Antiriboswitches drug screens. The final section reverses the perspective, treating riboswitches as targets for antibacterial drug development and ligand-binding specificity as the basis for identifying antibiotic candidates. The first chapter lays out a sensitive, fluorescence-based screening cascade for identifying compounds that target the T-box riboswitch antiterminator element (Hines). The second chapter describes screening platform that uses cell-free lysates to monitor translational read-through of a mammalian frameshift signal that is under the control of the preQ1-I riboswitch (Olsthoorn).

I first encountered riboswitches in a conference on RNA-Based Life in 2001 in separate presentations from Miranda-Ríos and Breaker. At the time of this writing (October 2014), a PubMed search turns up 763 hits for the term "riboswitch," and it will be well over 800 by the time of the publication of these volumes. The field is moving fast and in many directions. A great number of talented people with diverse expertise have contributed to these volumes, and all of us hope that they will serve as a useful resource to advance RNA research both within the riboswitch field and beyond.

DONALD H. BURKE-AGUERO

CHAPTER ONE

Design of Transcription Regulating Riboswitches

Sven Findeiß*[,†], Manja Wachsmuth[‡], Mario Mörl[‡,1], Peter F. Stadler[†,¶,§,#,||,]**

*Research Group Bioinformatics and Computational Biology, Faculty of Computer Science, University of Vienna, Vienna, Austria
[†]Institute for Theoretical Chemistry, University of Vienna, Vienna, Austria
[‡]Institute for Biochemistry, University of Leipzig, Leipzig, Germany
[¶]Bioinformatics Group, Department of Computer Science and the Interdisciplinary Center for Bioinformatic, University of Leipzig, Leipzig, Germany
[§]Center for RNA in Technology and Health, University of Copenhagen, Frederiksberg, Denmark
[#]Max Planck Institute for Mathematics in the Sciences, Leipzig, Germany
[||]Fraunhofer Institute for Cell Therapy and Immunology, Leipzig, Germany
**Santa Fe Institute, Santa Fe, New Mexico, USA
[1]Corresponding author: e-mail address: moerl@uni-leipzig.de

Contents

Abstract

In this chapter, we review both computational and experimental aspects of *de novo* RNA sequence design. We give an overview of currently available design software and their limitations, and discuss the necessary setup to experimentally validate proper function *in vitro* and *in vivo*. We focus on transcription-regulating riboswitches, a task that has just recently lead to first successful designs of such RNA elements.

Methods in Enzymology, Volume 550
ISSN 0076-6879
http://dx.doi.org/10.1016/bs.mie.2014.10.029

1. INTRODUCTION

Directly encoded in their host transcript, riboswitches represent one of the most efficient mechanisms to regulate expression on the transcriptional and translational level. The term is composed of two parts: one refers to the *ribo*nucleic acid building blocks and the second indicates the implemented structural *switch*ing mechanism which is triggered by an external stimulus. On single-molecule scale, the latter can often not be understood as a back and forth switch. In particular, if transcription is regulated using an intrinsic terminator structure, there is (currently) no mechanism known that recruits the RNA polymerase back to terminated transcripts. Consequently, if the riboregulator is in its OFF state, it cannot be switched on again. However, with respect to a population of mRNA transcripts over time and depending on the presence or absence of the trigger, the two states can alternate.

Based on that description, diverse riboregulators could be classified as riboswitches. For a detailed review with a more experimental perspective, we refer to Suess and Weigand (2008). Riboswitches combine a sensor domain that detects an external stimulus and an effector domain that exerts a regulatory function conditioned on the state of the sensor. The sensor is often an aptamer, i.e., a short nucleic acid sequence that specifically recognizes and binds a small molecule. An adaptive binding process ensures the aptamer's high specificity that may discriminate even between a hydrogen atom and a methyl group in the bound molecule. These minimal structural elements are known to regulate a wide variety of processes, including translation, transcriptional elongation, splicing, and virus replication, in a ligand-dependent manner (Suess & Weigand, 2008).

Given a particular small-molecule ligand, it is currently not possible to computationally design an aptamer for it *de novo* because we lack sufficiently detailed information how nucleic acids interact with other molecule types. A design task in this context is, therefore, limited to sequence optimization in order to increase specificity and functionality. This is, however, hampered by the fact that we lack a detailed picture of how RNA binding pockets look. Time-consuming SELEX (Systematic Evolution of Ligands by EXponential enrichment) experiments starting from a random pool of RNA sequences, therefore, are still the experimental method of choice to raise aptamers against almost any target (Klussmann, 2006).

Based on a PubMed search the term riboswitch appeared in 2002 for the first time in literature (Winkler, Nahvi, & Breaker, 2002). It typically refers to a two-component structural RNA element that consists of an aptamer which represents the sensor element and an expression platform that functions as actuator. To tightly link the sensor and the actuator, both components overlap and molecule binding in the first triggers a reaction within the latter. To design a riboswitch in its classical meaning, a preselected aptamer sequence and the regulation mechanism to be implemented are typically taken as input. The design task here focuses on the implementation of the communication between sensor and actuator, and the triggered effect, i.e., terminator formation or ribosome binding site sequestration.

A slightly more exotic triggered effect is implemented in aptazymes. These elements are composed of an *apta*mer and a self-cleaving RNA element, called ribo*zyme* (Ogawa & Maeda, 2007). In a trigger-dependent way, the cleavage process is regulated. Similar to aptamers, little is known about requirements such as sequence and structure composition of ribozymes. A larger set than the commonly referred glmS and hammerhead ribozymes and detailed simulations of the cleavage process are mandatory to make *de novo* design of such elements possible in the future.

The most common approach in synthetic biology is to construct "gadgets" such as riboswitches by combining well-defined components, in analogy with electrical or mechanical engineering. Biopolymers, however, cannot be strictly partitioned into independent modules so that interfacing "prefabricated" components is a serious and difficult practical problem. Breaker and coworkers, for example, engineered ribozymes that respond to the bacterial second messenger *c*yclic *di*guanosyl 5′-*m*ono*p*hosphate (Gu, Furukawa, & Breaker, 2012). In order to connect a hammerhead ribozyme and a c-di-GMP-binding aptamer, a SELEX approach starting from random sequences of different lengths was used. The use of SELEX-style experimental techniques to obtain functional sequences from pools of candidates is, in fact, a quite common step in riboswitch engineering (Fowler, Brown, & Li, 2008).

In Wachsmuth, Findeiß, Weissheimer, Stadler, and Mörl (2013), we used a more general approach, striving to construct as much of the sequence as possible from scratch using a mechanistic understanding – and thus a computational model – of the regulatory principle. Only those parts of the RNA for which a mechanism-based design is not feasible at present (the aptamer's interaction with the ligand) were introduced as prefabricated components. A good example of a *de novo* design is the construction of "multi-state RNA devices," which makes use of the quite well-known rules governing RNA–RNA interactions and design goals prescribing

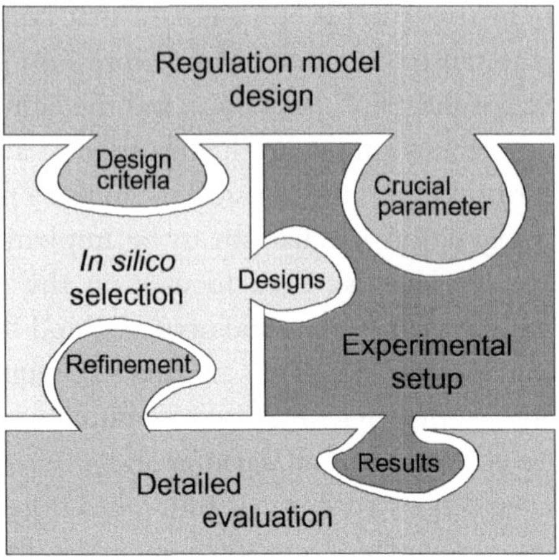

Figure 1 The molecule design puzzle starts with the description of the *regulation model* that will be implemented. *Design criteria* describing the system and *crucial parameters* that need to be experimentally measured need to be drafted. The first are used to come up with a suitable *in silico selection* process. The modeling of external triggers and the selection of appropriate tools, e.g., to solve the inverse folding problem for multi-stable RNAs, are important to generate promising riboswitch *designs*. The *experimental setup* encompasses ideally *in vitro* and *in vivo* experiments, e.g., cloning strategy and suitable reporter gene assays, to show that the system is working as intended. The first design attempt is often not successful and *detailed evaluation* of positive and negative results has to be done to *refine* the *in silico selection* process. Experts in computational (yellow (light gray in the print version)) and experimental (blue (dark gray in the print version)) biology have to join forces to come up with new models and evaluate the results (green (gray in the print version)).

interactions to generate candidate sequences (Rodrigo, Landrain, Majer, Darós, & Jaramillo, 2013).

The rational *de novo* approach is more demanding on our knowledge of the system in question since we need to be able to explicitly model the function given just sequence information (Fig. 1). To incorporate prefabricated components into the model, detailed knowledge of their structure and function is essential. We found that even for the well-described theophyline aptamer, integration into a *de novo* RNA design is not straightforward. Experimentally determined values, in our case the aptamer-metabolite binding constant, and those estimated from predicted structures can be significantly different. The model has to be adjusted accordingly. On the other hand, the *de novo* design approach is by far more flexible in the goals that it can achieve than any approach restricted to prefabricated components only. As we shall see in the following section, design goals have to be phrased in terms of structures that have to be attained together with thermodynamic constraints on these structures. In general, both positive design goals, which should be matched as close as possible, and negative design goals that specify situations to avoid, play important roles in successful RNA designs.

2. COMPUTATIONAL DESIGN OF RNA STRUCTURES

2.1. The inverse folding problem

Secondary structures, i.e., the Watson–Crick and wobble base pairs providing the scaffold of the three-dimensional structure, have turned out to be adequate to describe many aspects of RNA molecules in sufficient detail and accuracy for practical applications. This includes thermodynamic, kinetic, and evolutionary aspects, see, e.g., Gorodkin and Ruzzo (2014) for a series of recent reviews. At the same time, the simplicity of the secondary structure model, together with detailed experimentally determined energy parameters, enables structure predictions with acceptable accuracy within fractions of a second (Mathews & Turner, 2006; Lorenz et al., 2011) and thus sets the stage for computational approaches to RNA design. In particular, for every sequence x and every secondary structure θ of the same length, we can compute the energy $g(x,\theta)$.

In its simplest form, the RNA design problem is the inverse of the folding problem. Folding algorithms predict the ground-state secondary structure $\varphi(x)$ of a given RNA sequence x and its folding (free) energy $g(x) := \min_{\theta} g(x,\theta)$. Inverse folding, hence, asks for the set $S(\psi) := \{x|\varphi(x) = \psi\}$ of sequences whose ground-state structure is the given secondary structure ψ. In contrast to the secondary structure prediction problem $x \mapsto \varphi(x)$, there are no efficient exact algorithms to solve the inverse folding problem. Hence, one has to resort to heuristic approaches, typically by transforming it into the optimization problem of finding a sequence x whose ground-state structure $\varphi(x)$ minimizes the distance to the target structure ψ, formally $\mathrm{d}(x) := \mathrm{d}(\varphi(x),\psi) \rightarrow \min$, where of course the sequence x has the same length as the target structure ψ. Of course, we have $S(\psi) = \{x|\mathrm{d}(x) = 0\}$. Several measures of distances between RNA secondary structure have been devised; the simplest and most frequently used is the base pair distance $\mathrm{d}(\theta,\psi) = |(\theta\backslash\psi) \cup (\psi\backslash\theta)|$, which simply counts the number of base pairs unique to one of the two structures.

The logic of base pairing provides an easy *a priori* check whether a sequence x can possibly attain the secondary structure ψ: all base pairs of ψ must correspond to pairs of nucleotides in x that correspond to either

Watson–Crick or wobble pairs. If this is the case, we say that the sequence x is consistent with the structure ψ, in symbols $x \in C[\psi]$. Of course, it suffices to search for solutions of the inverse folding problem within $C[\psi]$.

The simplest inverse folding algorithms, therefore, start with a randomly selected initial structure $x_0 \in C[\psi]$ and seek to improve x_0 by point mutations or the exchange of types of base pairs. Such a mutant sequence x' is accepted if $d(x') < d(x_0)$; otherwise, another modification of x_0 is tested. Somewhat surprisingly, simple "adaptive walk" heuristics of this form are often successful even for moderately large sequences (Hofacker et al., 1994). This is a consequence of the peculiar structure of the sequence-to-structure map of RNAs (Gruener et al., 1996; Reidys, Stadler, & Schuster, 1997; Schuster, Fontana, Stadler, & Hofacker, 1994): the neutral network of ψ, i.e., the set of sequences that have ψ as the ground state is approximately uniformly (and for naturally occurring structures also densely) embedded in $C[\psi]$.

The adaptive walk approach is computationally rather inefficient since for each attempted improvement step the secondary structure of x' has to be computed. Although secondary structures are not completely modular, i.e., a single base exchange can lead to global refolding, a high degree of structural neutrality nevertheless allows us to gain from a divide-and-conquer strategy where the target structure ψ is carved up into a hierarchy of substructures for which candidate sequences are pre-optimized. For the technical details, we refer to the original publications on RNA inverse folding tools: `RNAinverse` (Hofacker et al., 1994), `RNA-SSD` (Andronescu, Fejes, Hutter, Hoos, & Condon, 2004), `INFO-RNA` (Busch & Backofen, 2006), and `DSS-Opt` (Matthies, Bienert, & Torda, 2012).

Applying local search starting from a seed sequence comes with the limitation that the above-mentioned approaches might get stuck in shallow local minima. This is mainly caused by the fact that promising candidate sequences cannot be reached using point-mutation steps during the optimization procedure. To a certain extent, this problem can be alleviated by using more sophisticated search algorithms such as evolutionary algorithms (Esmaili-Taheri, Ganjtabesh, & Mohammad-Noori, 2014). `RNA-ensign` (Levin et al., 2012) implements an efficient global sampling approach by interactively increasing the number of possible mutations on the candidate sequence until the target secondary structure is predicted for one of the mutants. Its successor, `IncaRNAtion` (Reinharz, Ponty, & Waldispuhl, 2013), implements a weighted sampling procedure that allows the user to control explicitly the GC content of the resulting sequences. This enables the user to adapt the designed sequences to the organism in which *in vivo* experiments will be conducted. CPdesign, a component of the RNAiFold framework (Garcia-Martin, Clote, & Dotu, 2013a, 2013b), implements an exact constraint programming approach and is thereby able to produce all possible sequences that adapt a certain structure. As this comes with a high computational demand, the authors also implemented a large neighborhood search heuristic in LNSdesign to analyze long sequences and structures. RNAexinv (Avihoo, Churkin, & Barash, 2011), finally, takes a shape instead of a fully specified structure as input. Structural motifs, i.e., stems and the different types of loops, are produced in the requested order and nesting, while their lengths and precise positioning are adjusted more flexibly.

In the simplest case, which is the one addressed by nearly all currently available software tools, ψ is a fully specified secondary structure (Fig. 2). Relaxed constraints, such as variable stem length, can be treated only by solving the inverse folding problem independently for all structures that satisfy the prescribed constraints. Loose constraints, thus, become computationally excessively expensive.

RNA sequences do not just form a single structure; instead, they spawn an ensemble $\Phi(x)$ of secondary structures in which every structure θ that is compatible with x, i.e., $x \in C[\theta]$, appears with relative frequency $p(x,\theta) = \exp(-g(x,\theta)/RT)/Z$, where $Z := \sum_\theta \exp(-g(x,\theta)/RT)$ is the partition function. This observation allows a refinement of the inverse folding heuristics. Instead of a simple optimization, we now have two *design principles*:

(+) enrichment of the desired structure ψ in the ensemble

(−) depletion of all other undesirable structures ξ from the ensemble.

The negative design principle can be interpreted, for instance, to discourage alternative structure ξ more stringently; the more they differ from ψ, the larger $d(\xi,\psi)$ becomes. Thus, considering the entire Boltzmann ensemble of sequence x, it is natural to minimize the "ensemble defect"

$$\nu(x,\psi) := \sum_\theta d(\theta,\psi)p(x,\theta) \tag{1}$$

which can be computed efficiently from the base pairing probability matrix of x and is implemented in NUPACK (Zadeh, Wolfe, & Pierce, 2011).

The choice of the design goals is far from obvious. An interesting way to evaluate the principles followed by human players is the Web-based game EteRNA devised by Rhiju Das and collaborators. The game asks human players without prior knowledge of the topic to solve inverse folding tasks as a kind of puzzle. After several rounds of design cycles, the users were asked

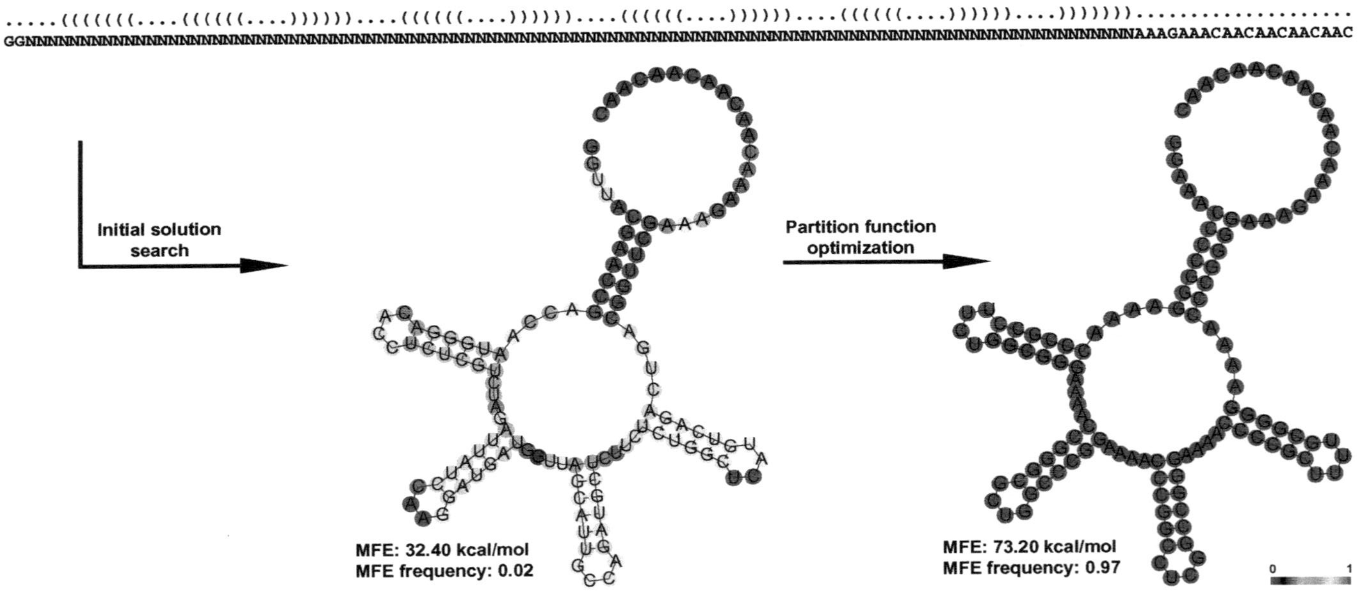

Figure 2 Example output of `RNAinverse` solving one task of the `EteRna` lab competition (Lee et al., 2014). The secondary structure (in dot–bracket notation) and the sequence constraint (nucleotides A, C, G, U, and N for any of those) strictly define the design target. `RNAinverse` initially searches for a sequence that adapts the target structure as ground state. Using the partition function option, it further optimizes the initial solution such that the energy is minimized and that the ground state dominates the structure ensemble. The color code indicates base pair probability. Note that structural probing experiments of this and other examples have shown that the prediction and the structure of the synthesized RNA molecule can differ significantly (Lee et al., 2014). (See the color plate.)

to formulate design rules they acquired while playing. Interesting ideas, such as a preferred direction of GC base pairs in multi-loops and a maximum number of AU base pairs per stem, have been suggested by users and were used as the basis of `EteRNABot`, which showed a performance superior to classical RNA inverse software (Lee et al., 2014). This work indicates that strategies that interpret RNA inverse folding as a multi-objective optimization problem, such as `MODENA` (Taneda, 2011), could be further improved by implementing additional design goals. The state of the art of design rules was recently summarized in Bida and Das (2012).

2.2. Designing multi-stable RNAs

So far, we have been concerned only with the inverse folding problem. More interesting design tasks, however, ask for sequences that can fold into two or more alternative, prescribed structures $\Theta = \{\theta_1, \theta_2, \ldots, \theta_K\}$. A simple mathematical result decides whether there is a sequence that can conceivably fold into each $\theta_i \in \Theta$.

Theorem 1. *(Flamm, Hofacker, Maurer-Stroh, Stadler, & Zehl, 2001) The set* $C[\Theta] := C[\theta_1] \cap C[\theta_2] \cap \cdots \cap C[\theta_K] \neq \emptyset$ *if and only if the graph with edge set* $\theta_1 \cup \theta_2 \cup \cdots \cup \theta_K$, *i.e., the union of the base pairs of the input structures, is bipartite.*

An immediate consequence of this so-called intersection theorem is that the design problem for $K = 2$ prescribed structures is always solvable (Reidys et al., 1997). A numerical survey (Hoener zu Siederdissen et al., 2013) shows that for $K = 3$ a solution still exists quite frequently, while for $K \geq 4$ it becomes quite unlikely that arbitrary K-tuples of input structures admit a sequences that can fold into all θ_i.

Usually, one is not so much interested in just any sequence $x \in C[\Theta]$. The interesting ones are those for which each of the structures θ_i has an energy close to the ground state and/or is separated by a high energy barrier from the alternatives. These additional constraints are naturally regarded as an optimization problem over $C[\Theta]$ with a suitably chosen cost function. A simple example is

$$\sum_{i=1}^{K} p(x, \theta_i) - \alpha \sum_{i<j} \left(p(x, \theta_i) - p(x, \theta_j)\right)^2, \qquad (2)$$

which tries to maximize the total weight of all design goals in the Boltzmann ensemble and penalizes large deviations of their relative frequencies.

Several software tools have become available for the design of such multi-stable (or "allosteric") RNAs, starting with the `PERL` script `switch.pl`

(Flamm et al., 2001). The more efficient implementation, `RNAdesign` (Hoener zu Siederdissen et al., 2013), uses enumerative techniques to ensure that the initial points in $C[\Theta]$ are sampled uniformly. `ARDesigner` (Shu, Liu, Chen, Bo, & Wang, 2010) and `Frnakenstein` (Lyngsø et al., 2012) implement similar ideas and use heuristics to generate initial sequences in $C[\Theta]$ and to modify sequences when optimizing their objective functions. The most complex issue in practice, which remains at least partially unresolved, is the translation of the user's design goal into a suitable objective function.

2.3. Modeling external triggers

From a computational point of view, the simplest environmental factor to which an RNA may react is a change in temperature. Since enthalpies and entropies are tabulated separately in the RNA energy parameter sets, temperature is an explicit parameter in the `ViennaRNA` package (Lorenz et al., 2011). Design functions of the form $\mathrm{d}(\varphi_{T'}(x),\psi_{T'}) + \mathrm{d}(\varphi_{T''}(x),\psi_{T''})$ thus can be readily used to find sequences x that switch from ground-state structure $\psi_{T'}$ to another ground-state $\psi_{T''}$ when the temperature is changed from T' to T''.

The *de novo* design of very simple RNA thermometers composed of only a single small stem-loop structure containing the ribosome binding site and the functionality of these synthetic RNA thermometers at physiological temperatures has been shown (Neupert & Bock, 2009; Neupert, Karcher, & Bock, 2008). In Waldminghaus, Kortmann, Gesing, and Narberhaus (2008), a combination of computer-based rational design and *in vivo* screening was used. After only two rounds of selection several RNA thermometers with efficiencies comparable to naturally evolved examples were obtained. The majority of mutations introduced during the selection process were located in stem regions which had to be destabilized in order to obtain functional thermosensors. This approach started from *de novo* designs produced with `switch.pl` (Flamm et al., 2001). A three-state thermoswitch proposed by `ARDesigner`, however, has never been experimentally validated (Shu et al., 2010). Naturally occurring building blocks, namely an RNA thermometer and the hammerhead ribozyme, have been recently combined to engineer a temperature-sensitive, self-cleaving RNA molecule (Saragliadis, Krajewski, Rehm, Narberhaus, & Hartig, 2013). The above-mentioned successful thermometer designs suggest that a thermozyme construction might also be possible by *de novo* approaches alone.

Conceptually, pH sensors (Nechooshtan, Elgrably-Weiss, Sheaffer, Westhof, & Altuvia, 2009) are very similar to RNA thermometers. So far, they have not become amenable to computational design strategies since no ready-to-use parameter set for the influence of pH on RNA folding energies is available.

An other computationally feasible approach is to implement switches that are triggered by nucleic acid sequences. In the context of switching RNA molecules, the naturally occurring rpoS-DsrA mRNA-sRNA pair (Majdalani, Cunning, Sledjeski, Elliott, & Gottesman, 1998) and the *de novo* designed RAJ11 system (Rodrigo, Landrain, & Jaramillo, 2012) are of interest. They have in common that the ribosome binding site of the target is sequestered in a local structural element. Upon binding of the trigger, a structural rearrangement enables ribosome binding and thereby activates translation. In a follow-up study (Rodrigo et al., 2013), the authors formulated a generally applicable mapping of truth tables which describe the system's states using Boolean encoding, i.e., presence and absence of molecules and the resulting output, to a cost function which is the essential ingredient for most design tools. The combination of RNA design and target prediction software, e.g., `RNAcofold` (Bernhart et al., 2006) and `RNAup` (Mückstein et al., 2006), can be used to implement nucleic acid-triggered switches. It is worth noting that the `NUPACK` suite (Zadeh et al., 2011) and the `RNAiFold` framework (Garcia-Martin et al., 2013a, 2013b) already combine both to design multi-strand switches. A similar strategy is used by `MultiSrch/StochSrchMulti` (Ramlan & Zauner, 2011). The beauty of using nucleic acid triggers is the possibility to model the complete system, i.e., structural states of the components and their most likely interaction, *in silico*.

In the riboswitch context, small-molecule triggers, usually metabolite ligands, are of interest. A simple remedy to our current inability of including ligand binding directly into RNA folding algorithms is to use structural constraints to model the ligand-bound state. In particular, RNA folding algorithms are capable of specifying a set α of paired and unpaired bases, the structure the ligand can bind to that is enforced during structure calculation. The most stable structure under this constraint $\varphi(x|\alpha)$ and its associated folding energy $g(x|\alpha)$ can be computed efficiently, e.g., with `RNAfold -C`. As usual, $\varphi(x)$ and $g(x)$ denote the (unconstrained) ground state and its energy. Importantly, $\varphi(x)$ must not display the intact aptamer structure. Ideally, we banish all aptamer base pairs from the unconstrained secondary structure, thus setting $\varphi(x) \cap \alpha = \emptyset$ as negative design goal. Natural riboswitches may not adhere to this ideal; nevertheless, it seems reasonable to go for large structural changes in an engineering setting. By design, we have $g(x) < g(x|\alpha)$. If the ligand is present, $\varphi(x|\alpha)$ is stabilized by the ligand binding energy $e_L < 0$, which has to be sufficiently high to make the ligand-bound state the ground state. For nucleic acid ligands, it is possible to compute e_L explicitly; in the case of small-molecule (metabolite) ligands, one has to rely on literature values or estimates for e_L. Any design in which a structural change is to be triggered by ligand binding must have the ligand-bound state as ground state; hence, the inequalities

$$g(x|\alpha) + e_L \leq g(x) \leq g(x|\alpha) \tag{3}$$

must hold. A generic way of encoding Eq. (3) in a design objective is to ask that $g(x)$ is roughly halfway between $g(x|\alpha) + e_L$ and $g(x|\alpha)$, i.e., to set $g(x) - g(x|\alpha) \approx e_L/2$.

Current design tools, however, do not yet make very systematic use of constrained ensembles of structures. `RNAdesign` (Hoener zu Siederdissen et al., 2013), for instance, instead favor energies of partial structures as building blocks for the objective function. Nevertheless, interesting design problems can already be solved, see Fig. 3 for an example.

The energy differences between the alternative states are directly related to the quality of the switch. The downside of the constraint folding framework is that all energetic contributions of ligand binding have to be integrated "by hand." We reported that in case of theophylline, the estimated ligand binding energy of -8.86 kcal/mol does not account for the predicted energy differences between bound and unbound state which range from -13.2 to -15.8 kcal/mol for functional RNA switches (Wachsmuth et al., 2013). The discrepancy of about -7 kcal/mol might be caused by several factors. First, different experimental conditions such as salt condition and pH value have been used for estimating the energy parameters of the secondary structure prediction software and for evaluating the riboswitch. Second, the structure of the RNA's ligand binding pocket usually not known in detail. Thus the exact binding mechanism and the exact nucleotides involved in ligand binding are typically uncertain. Finally, binding constants reported for independent experiments may not have been corrected for the energetic effects of structural changes outside the binding site itself.

2.4. Current limitation of design software

Despite the fact of several attempts to engineer (based on naturally occurring building blocks) and to design riboswitches *de novo*, no ready-to-use software is currently available. Here, we summarize problems we recognized while designing transcriptional ON switches, i.e., RNA elements that

A

Constraints:

```
ON:  (((((...((((((((.....)))))...)))...))))).....................
OFF: .......................(((((((((((((....)))))))))))))........
     NNNNNAUANNNNNNNNGUCUUNNNNNCCUNNNCAGNNNNNNNNNNNNNNNNNNUUUUUUUUU
```

Objectives:

(1) $g(x|\mathrm{ON}) + e_\mathrm{L} \leq g(x|\mathrm{OFF}) \leq g(x|\mathrm{ON})$

(2) $g(x|\mathrm{OFF}) - g(x|\mathrm{ON}) \approx e_\mathrm{L}/2$

$e_\mathrm{L} = -8.86$ kcal/mol

(3) $p(x|\mathrm{ON}) + p(x|\mathrm{OFF}) \approx 1$

B

GUCUCAUACCCAUAUAGUCUUUAUAUCCUGGGCAGGAGGCGUUUUGCUUGGGGUUUUUUUUU

ON: **OFF:**

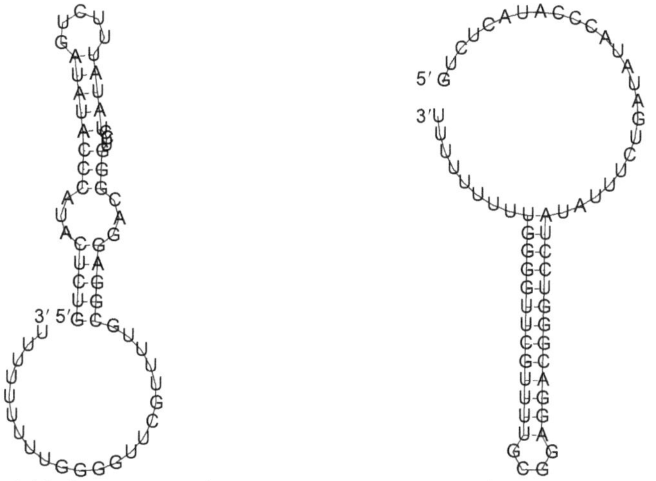

$g(x|ON)$: −11.10 kcal/mol
$p(x|ON)$: 0.31

$g(x|\mathrm{OFF})$ −14.30 kcal/mol
$p(x|\mathrm{OFF})$: 0.51

(1) $-19.96 \leq -14.30 \leq -8.86$ ✓

(2) $-3.2 \approx -4.43$ ✓

(3) 0.82✓

Figure 3 Example of a multi-stable structure design. (A) Input are two structural states (in dot–bracket notation) and a sequence constraint (nucleotides A, C, G, U, and N for any of those). The 5′ part of the sequence and the depicted ON state might correspond to an aptamer in its binding competent state. We assume that the nucleotides within the loops are important for the interaction with the trigger and are therefore fixed in the sequence constraint. In the OFF state, the RNA molecule should form a rho-independent terminator structure followed by a poly-U track at the 3′ end of the sequence. This state should be adapted if no trigger is bound and represents the ground state. Objectives (1) and (2) guarantee that the energy differences between the two states are optimized such that the binding energy of the trigger of − 8.86 kcal/mol enables switching between both states. The objective (3) ensures that both states should be fairly frequent in the ensemble. (B) Solution to the design problem satisfying input constraints and objectives computed with `RNAdesign`.

activate transcription upon ligand binding. These problems make the development of standard design approaches a challenging task.

A sequence design is successful if and only if all nucleotides adapt their desired state, i.e., paired or unpaired. In riboswitch design this is not true for the entire molecule. It is only mandatory for sub regions that have to fold into a well defined structure to be functional, e.g., rho-independent transcription terminators. The sensing component of a riboswitch is typically an aptamer of which sequence and structure are predefined. These "hard" sequence and structure constraints are implemented in existing design software. In addition, regions typically flexible in length, structure, and sequence which should not show certain features are important for a successful design. The ON state of transcription regulating riboswitches should, for instance, not contain stable substructures upstream of the poly-U track that might unintentionally terminate transcription or interfere with translation. These "soft" constraints are essential to implement uncertain/flexible elements in order to explore the candidate space. However, they are not easy to implement into an efficient algorithm.

With an increasing number of constraints and structural states, the possibility to design a solution decreases. For these cases, it would be of interest to have a framework to propose designs that are close but not exact, or at least one that identifies problems for the user and suggests possible solutions.

That is why we could not use an existing RNA design tool and came up with a two-step approach: (i) creation of plausible candidates and (ii) subsequent filtering and optimization of this set. The success of the latter depends on the knowledge about thermodynamic or kinetic features of functional riboswitch designs. Parts of the second design step could be migrated into the first one, provided detailed experimental knowledge becomes available for individual systems, either designed or naturally occurring. Results of Haller, Souliére, and Micura (2011) and Quarta, Sin, and Schlick (2012), who investigated dynamics and folding landscapes of naturally occurring riboswitches, might be helpful in this respect.

Another issue, especially important for riboswitches that should regulate transcription, is the detailed analysis of thermodynamic and kinetic effects. We modeled a transcriptional ON switch, i.e., an RNA element that activates transcription upon ligand binding, using thermodynamic features only. After experimental testing, we ended up with a 50% success rate of functional designs. Our recent analysis revealed that in addition to thermodynamic effects, kinetic effects during terminator formation also seem to have a great impact (unpublished data). We speculate that typically both thermodynamic and kinetic effects act in concert to facilitate RNA function.

As we investigated, riboswitch regulation during transcription timing and effects of cotranscriptional folding play an important role. To address this question, we used simple MFE (minimum free energy)-based folding simulations in our proof-of-concept design. Of course, using the ground-state structure only neglects the fact that RNA molecules do not adapt a single structure. Instead, an RNA can fold into a diverse ensemble of possible structures. Applying partition function algorithms should help to further improve prediction accuracy. Furthermore, a diverse set of cotranscriptional folding software, such as `Kinefold` (Xayaphoummine, Bucher, & Isambert, 2005), `Kinwalker` (Geis et al., 2008), and `BarMap` (Hofacker et al., 2010), is available to estimate more realistic folding simulations.

Secondary structures have proven to be a very useful level of description for many, but certainly not all, aspects of RNA structures. In particular, they are insufficient to model more subtle aspects of the 3D structure with loops. Especially, multi-branched structures – common to many natural riboswitches – are often insufficiently predicted by inverse folding software (Lee et al., 2014). A possible remedy, albeit not yet explored in the context of RNA design tasks, is to use the more fine-grained Leontis–Westhof representation (Leontis, Stombaugh, & Westhof, 2002; Leontis & Westhof, 2001), which also includes the many types of nonstandard base pairs and is more restrictive regarding isosteric replacements of base pairs. *GU* pairs are already fairly different from Watson–Crick pairs, and it may be a good idea not to use them interchangeably with Watson–Crick pairs at least in the vicinity of aptamer binding pockets. Constraints on *GU* pairs are rather straightforward to handle. A folding algorithm that attempts to predict Leontis–Westhof-style extended secondary structures is `RNAwolf` (Hoener zu Siederdissen, Berhart, Stadler, & Hofacker, 2011). Its parametrization, however, has rather limited accuracy, and it has not been used in an RNA design context so far.

3. EXPERIMENTAL EVALUATION OF DESIGNED RNA STRUCTURES

3.1. Considerations on candidate selection and cloning procedures

After the *in silico* design of putative RNA regulators, the functionality of the predicted riboswitch constructs has to be tested experimentally. In the best of all cases, this is done *in vitro* and subsequently *in vivo*. Experiments of the latter type are not necessarily successful even if functionality has been shown *in vitro* (Link et al., 2007; Weigand et al., 2008). So far, a definite prediction of functionality at the sequence level is not possible. Hence, a set of candidates need to be tested experimentally. This set should reflect diversity such that uncertain variables and features of the design are clarified.

Investigating designed transcriptional riboswitches, we found that functionality is likely if no mismatches are present in the terminator stem and when the loop is no longer than 4–5 bp. Furthermore, a stem length of about 14 nt seems to be advantageous. In particular, we recognized that a high stability of the terminator stem can interfere with formation of a stable aptamer–ligand complex, while low terminator stabilities seem to be detrimental to terminator formation (Wachsmuth et al., 2013). However, the terminator stability required for an efficient functionality of the switch strongly depends on the sequence composition of the individual aptamer, its structure, and its affinity to the ligand and might require further adaptation for each selection purpose.

For functional analysis, candidates need to be fused to a reporter construct. In principle, the riboswitches should work regardless of the chosen reporter system. As they are regulating at the level of transcription, RNA as well as protein-based reporter systems can be applied. The fastest way to differentiate between functional and nonfunctional candidates is a reporter gene assay with robust and well-characterized read out systems, such as β-galactosidase or GFP reporter genes (Ghim, Lee, Takayama, & Mitchell, 2010). For transcriptional riboswitches, however, it might be best to choose a reporter system that allows functional analysis at the transcriptional level. This may be done by *in vitro* transcription assays under synchronized conditions (Landick, Wang, & Chan, 1996). However, from such an *in vitro* analysis, one cannot conclude that the riboswitch is also functional within the cell. To allow *in vivo* investigations, the easiest way to prove a transcription-based regulation is the classical Northern blot analysis. For activity tests of riboswitches regulating at the transcriptional level, they are integrated into the 5′ untranslated region of a reporter construct. This can be done by site-directed mutagenesis or (if suitable restriction sites are present) using standard ligation protocols. To avoid interaction of riboswitch folding with upstream sequences, the riboswitch should be positioned close to the transcription start site. It is also necessary to consider that motifs upstream (Peters, Vangeloff, & Landick, 2011) and downstream (Carafa, Brody, & Thermes, 1990) of the terminator can have an impact on termination efficiency. Furthermore, it has been shown that uridine-rich sequences influence translation initiation (Zhang & Deutscher, 1992). As our riboswitches include eight uridine residues at the 3′-end of the terminator, this sequence should be kept distant from the Shine–Dalgarno sequence (Wachsmuth et al., 2013). Taken together, these examples clearly show that all sequences flanking the riboswitch must be selected according to the individual circumstances and adjusted to the chosen aptamer domains.

3.2. Further characterization and current limitations

So far, our synthetic transcription riboswitches based on an *in silico* selection have been shown to be functional in *E. coli*. While intrinsic terminators have not been detected in all bacterial species (Peters et al., 2011; Washio, Sasayama, & Tomita, 1998), they are a common instrument for transcriptional termination in most bacteria and their structural features are similar among different species (Hoon, Makita, Nakai, & Miyano, 2005). Hence, transcriptional riboswitches should be functional in different bacterial species. Nevertheless, it might be helpful to define species-specific terminator features prior to selection.

There are several possible reasons for a design approach to fail. Our algorithm determines the most stable structure (MFE) for a defined number of transcript intermediates of a riboswitch and assumes that a definite switch between aptamer and terminator fold happens after transcription of the terminator. This has been verified for a natural riboswitch showing differences in structure of the aptamer itself compared to the full-length riboswitch (Wickiser, Winkler, Breaker, & Crothers, 2005). Here, only the isolated aptamer is in a binding competent state. This is typical for a kinetic rather than a thermodynamic regulation. For analysis of binding properties as well as structures of transcriptional intermediates or full-length riboswitches, inline probing (Soukup & Breaker, 1999) or SHAPE experiments (Wilkinson, Merino, & Weeks, 2006) can be performed. These approaches give further information on the structural reality that consequently can be integrated into the selection process. In our first attempt, we used the naïve design hypothesis that the MFE structure of a defined sequence predicted *in silico* is present in nature. Although great effort is being taken in optimizing structure predictions, the reported accuracy varies between 20% and 60% for long and is close to 70% for short RNA molecules (Hofacker, 2014). Furthermore, RNA molecules adapt different structures with similar thermodynamic stabilities. Such a folding behavior might interfere with proper selection of functional riboswitches. Accordingly, detailed structural analysis of building blocks, e.g. the aptamer, is recommended prior to selection.

It is known that many intrinsic terminators comprise transcriptional pause sites that cause a reduction of transcription speed. This aids the folding of the RNA by allowing it more time to do so. This has been shown to be an important feature in several natural riboswitches (Lemay et al., 2011; Wickiser et al., 2005). Such transcriptional pausing can be detected by *in vitro* transcription assays (Landick et al., 1996). The incorporation of pause site elements into the design principles could improve the efficiency of synthetic transcriptional riboswitches. However, the exact sequence requirements for efficient pausing are still uncertain.

4. CONCLUDING REMARKS

Taken together, the combination of computer-based prediction of synthetic riboswitches combined with experimental validation and improvement is a valuable strategy for the development of new synthetic regulatory elements at the RNA level. The beauty of this system is the possibility that the feedback from experimental investigations allows to improve the predictive power of the algorithm, leading to a considerable optimization of artificial gene expression regulation at the transcriptional level.

In this review, we have emphasized the computational side of riboswitch design. Although recent work clearly demonstrates that the rational, modeling-based approach is not only feasible in principle but can effectively replace artificial selection, the available methods and tools are still in their infancy. The basic principles of inverse folding with multiple competing target structures are well understood. A systematic approach to incorporate the energetic constraints associated with external triggers, however, is still a subject of ongoing research. In the same vein, negative design goals are hard, and sometimes maybe impossible, to phrase in current software tools. A second limitation of current modeling capabilities comes from kinetic effects. Although it is possible to identify kinetic trap states, e.g., by simulating folding trajectories or using enumerative approaches to explore the energy landscape of a candidate sequence, these methods are computationally very demanding. So far, they are more effectively used as an *a posteriori* filter for candidate designs rather than an integral part of the design objective. Very recent advances in computational methods for RNA energy landscapes and kinetics (Kucharík, Hofacker, Stadler, & Qin, 2014;

Mann, Kucharík, Flamm, & Wolfinger, 2014) may be a first step toward changing this situation.

Rational RNA design holds great promise for RNA-based synthetic biology. Well-characterized sensor and ribozyme components can be integrated into single RNA molecules that can be designed in such a way that individual features are enabled or disabled in response to external triggering signals. Thus, it seems feasible to design complex logical functions into single RNAs. Such designs are presumably less prone to side effects and may operate at much shorter time scales than circuits comprising multiple steps of transcription factor-governed gene expression.

REFERENCES

Andronescu, M., Fejes, A. P., Hutter, F., Hoos, H. H., & Condon, A. (2004). A new algorithm for RNA secondary structure design. *Journal of Molecular Biology, 336*, 607–624.

Avihoo, A., Churkin, A., & Barash, D. (2011). RNAexinv: An extended inverse RNA folding from shape and physical attributes to sequences. *BMC Bioinformatics, 12*, 319.

Bernhart, S. H., Tafer, H., Mückstein, U., Flamm, C., Stadler, P. F., & Hofacker, I. L. (2006). Partition function and base pairing probabilities of RNA heterodimers. *Algorithms for Molecular Biology, 1*, 3.

Bida, J. P., & Das, R. (2012). Squaring theory with practice in RNA design. *Current Opinion in Structural Biology, 22*, 457–466.

Busch, A., & Backofen, R. (2006). INFO-RNA—A fast approach to inverse RNA folding. *Bioinformatics, 22*, 1823–1831.

Carafa, Y. d., Brody, E., & Thermes, C. (1990). Prediction of rho-independent Escherichia coli transcription terminators. *Journal of Molecular Biology, 216*, 835–858.

Esmaili-Taheri, A., Ganjtabesh, M., & Mohammad-Noori, M. (2014). Evolutionary solution for the RNA design problem. *Bioinformatics, 30*, 1250–1258.

Flamm, C., Hofacker, I. L., Maurer-Stroh, S., Stadler, P. F., & Zehl, M. (2001). Design of multistable RNA molecules. *RNA, 7*, 254–265.

Fowler, C. C., Brown, E. D., & Li, Y. (2008). A FACS-based approach to engineering artificial riboswitches. *Chembiochem, 9*, 1906–1911.

Garcia-Martin, J. A., Clote, P., & Dotu, I. (2013). RNAiFOLD: A constraint programming algorithm for RNA inverse folding and molecular design. *Journal of Bioinformatics and Computational Biology, 11*, 1350001.

Garcia-Martin, J. A., Clote, P., & Dotu, I. (2013). RNAiFold: A web server for RNA inverse folding and molecular design. *Nucleic Acids Research, 41*, W465–W470.

Geis, M., Flamm, C., Wolfinger, M. T., Tanzer, A., Hofacker, I. L., Middendorf, M., et al. (2008). Folding kinetics of large RNAs. *Journal of Molecular Biology, 379*, 160–173.

Ghim, C.-M., Lee, S. K., Takayama, S., & Mitchell, R. J. (2010). The art of reporter proteins in science: Past, present and future applications. *BMB Reports, 43*, 451–460.

Gorodkin, J., & Ruzzo, L. (Eds.), (2014). *Class-specific prediction of ncRNAs (Vol. 1097)*. New York: Humana Press.

Gruener, W., Giegerich, R., Strothmann, D., Reidys, C., Weber, J., Hofacker, I. L., et al. (1996). Analysis of RNA sequence structure maps by exhaustive enumeration. *II. Structures of neutral networks and shape space covering. Monatshefte für Chemie, 127*, 375–389.

Gu, H., Furukawa, K., & Breaker, R. R. (2012). Engineered allosteric ribozymes that sense the bacterial second messenger cyclic diguanosyl 5′-monophosphate. *Analytical Chemistry, 84*, 4935–4941.

Haller, A., Souliére, M. F., & Micura, R. (2011). The dynamic nature of RNA as key to understanding riboswitch mechanisms. *Accounts of Chemical Research*, *44*, 1339–1348.
Hoener zu Siederdissen, C., Berhart, S. H., Stadler, P. F., & Hofacker, I. L. (2011). A folding algorithm for extended RNA secondary structures. *Bioinformatics*, *27*, i129–i137, (ISMB).
Hoener zu Siederdissen, C., Hammer, S., Abfalter, I., Hofacker, I. L., Flamm, C., & Stadler, P. F. (2013). Computational design of RNAs with complex energy landscapes. *Biopolymers*, *99*, 1124–1136.
Hofacker, I. L. (2014). Energy-directed RNA structure prediction. *Methods in Molecular Biology*, *1097*, 71–84.
Hofacker, I. L., Flamm, C., Heine, C., Wolfinger, M. T., Scheuermann, G., & Stadler, P. F. (2010). BarMap: RNA folding on dynamic energy landscapes. *RNA*, *16*, 1308–1316.
Hofacker, I. L., Fontana, W., Stadler, P. F., Bonhoeffer, S., Tacker, M., & Schuster, P. (1994). Fast folding and comparison of RNA secondary structures. *Monatshefte für Chemie*, *125*, 167–188.
Hoon, M. J. L. d., Makita, Y., Nakai, K., & Miyano, S. (2005). Prediction of transcriptional terminators in Bacillus subtilis and related species. *PLoS Computational Biology*, *1*, e25.
Klussmann, S. (Ed.), (2006). *The aptamer handbook: Functional oligonucleotides and their applications*. Wiley-VCH.
Kucharík, M., Hofacker, I. L., Stadler, P. F., & Qin, J. (2014). Basin Hopping Graph: A computational framework to characterize RNA folding landscapes. *Bioinformatics*, *30*, 2009–2017. http://dx.doi.org/10.1093/bioinformatics/btu156.
Landick, R., Wang, D., & Chan, C. L. (1996). Quantitative analysis of transcriptional pausing by Escherichia coli RNA polymerase: His leader pause site as paradigm. *Methods in Enzymology*, *274*, 334–353.
Lee, J., Kladwang, W., Lee, M., Cantu, D., Azizyan, M., Kim, H., et al. (2014). RNA design rules from a massive open laboratory. *Proceedings of the National Academy of Sciences of the United States of America*, *111*, 2122–2127.
Lemay, J.-F., Desnoyers, G., Blouin, S., Heppell, B., Bastet, L., St-Pierre, P., et al. (2011). Comparative study between transcriptionally- and translationally-acting adenine riboswitches reveals key differences in riboswitch regulatory mechanisms. *PLoS Genetics*, 7, e1001278.
Leontis, N. B., Stombaugh, J., & Westhof, E. (2002). The non-Watson-Crick base pairs and their associated isostericity matrices. *Nucleic Acids Research*, *30*.
Leontis, N. B., & Westhof, E. (2001). Geometric nomenclature and classification of RNA base pairs. *RNA*, 7, 499–512.
Levin, A., Lis, M., Ponty, Y., O'Donnell, C. W., Devadas, S., Berger, B., et al. (2012). A global sampling approach to designing and reengineering RNA secondary structures. *Nucleic Acids Research*, *40*, 10041–10052.
Link, K. H., Guo, L., Ames, T. D., Yen, L., Mulligan, R. C., & Breaker, R. R. (2007). Engineering high-speed allosteric hammerhead ribozymes. *Biological Chemistry*, *388*, 779–786.
Lorenz, R., Bernhart, S. H., Siederdissen, C. H. Z., Tafer, H., Flamm, C., Stadler, P. F., et al. (2011). ViennaRNA Package 2.0. *Algorithms for Molecular Biology*, *6*, 26.
Lyngsø, R. B., Anderson, J. W. J., Sizikova, E., Badugu, A., Hyland, T., & Hein, J. (2012). Frnakenstein: Multiple target inverse RNA folding. *BMC Bioinformatics*, *13*, 260.
Majdalani, N., Cunning, C., Sledjeski, D., Elliott, T., & Gottesman, S. (1998). DsrA RNA regulates translation of RpoS message by an anti-antisense mechanism, independent of its action as an antisilencer of transcription. *Proceedings of the National Academy of Sciences of the United States of America*, *95*, 12462–12467.
Mann, M., Kucharík, M., Flamm, C., & Wolfinger, M. T. (2014). Memory efficient RNA energy landscape exploration. *Bioinformatics*, *30*, 2584–2591. http://dx.doi.org/10.1093/bioinformatics/btu337.

Mathews, D. H., & Turner, D. H. (2006). Prediction of RNA secondary structure by free energy minimization. *Current Opinion in Structural Biology, 16*, 270–278.

Matthies, M. C., Bienert, S., & Torda, A. E. (2012). Dynamics in sequence space for RNA secondary structure design. *Journal of Chemical Theory and Computation, 8*, 3663–3670.

Mückstein, U., Tafer, H., Hackermüller, J., Bernhart, S. H., Stadler, P. F., & Hofacker, I. L. (2006). Thermodynamics of RNA-RNA binding. *Bioinformatics, 22*, 1177–1182.

Nechooshtan, G., Elgrably-Weiss, M., Sheaffer, A., Westhof, E., & Altuvia, S. (2009). A pH-responsive riboregulator. *Genes & Development, 23*, 2650–2662.

Neupert, J., & Bock, R. (2009). Designing and using synthetic RNA thermometers for temperature-controlled gene expression in bacteria. *Nature Protocols, 4*, 1262–1273.

Neupert, J., Karcher, D., & Bock, R. (2008). Design of simple synthetic RNA thermometers for temperature-controlled gene expression in Escherichia coli. *Nucleic Acids Research, 36*, e124.

Ogawa, A., & Maeda, M. (2007). Aptazyme-based riboswitches as label-free and detector-free sensors for cofactors. *Bioorganic & Medicinal Chemistry Letters, 17*, 3156–3160.

Peters, J. M., Vangeloff, A. D., & Landick, R. (2011). Bacterial transcription terminators: The RNA 3′-end chronicles. *Journal of Molecular Biology, 412*, 793–813.

Quarta, G., Sin, K., & Schlick, T. (2012). Dynamic energy landscapes of riboswitches help interpret conformational rearrangements and function. *PLoS Computational Biology, 8*, e1002368.

Ramlan, E. I., & Zauner, K.-P. (2011). Design of interacting multi-stable nucleic acids for molecular information processing. *Biosystems, 105*, 14–24.

Reidys, C., Stadler, P. F., & Schuster, P. (1997). Generic properties of combinatory maps: Neutral networks of RNA secondary structures. *Bulletin of Mathematical Biology, 59*, 339–397, (SFI preprint 95–07–058).

Reinharz, V., Ponty, Y., & Waldispuhl, J. (2013). A weighted sampling algorithm for the design of RNA sequences with targeted secondary structure and nucleotide distribution. *Bioinformatics, 29*, i308–i315.

Rodrigo, G., Landrain, T. E., & Jaramillo, A. (2012). De novo automated design of small RNA circuits for engineering synthetic riboregulation in living cells. *Proceedings of the National Academy of Sciences of the United States of America, 109*, 15271–15276.

Rodrigo, G., Landrain, T. E., Majer, E., Darós, J.-A., & Jaramillo, A. (2013). Full design automation of multi-state RNA devices to program gene expression using energy-based optimization. *PLoS Computational Biology, 9*, e1003172.

Saragliadis, A., Krajewski, S. S., Rehm, C., Narberhaus, F., & Hartig, J. S. (2013). Thermozymes: Synthetic RNA thermometers based on ribozyme activity. *RNA Biology, 10*, 1010–1017.

Schuster, P., Fontana, W., Stadler, P. F., & Hofacker, I. L. (1994). From sequences to shapes and back: A case study in RNA secondary structures. *Proceedings of the Royal Society B, 255*, 279–284.

Shu, W., Liu, M., Chen, H., Bo, X., & Wang, S. (2010). ARDesigner: A web-based system for allosteric RNA design. *Journal of Biotechnology, 150*, 466–473.

Soukup, G. A., & Breaker, R. R. (1999). Relationship between internucleotide linkage geometry and the stability of RNA. *RNA, 5*, 1308–1325.

Suess, B., & Weigand, J. E. (2008). Engineered riboswitches: Overview, problems and trends. *RNA Biology, 5*, 24–29.

Taneda, A. (2011). MODENA: A multi-objective RNA inverse folding. *Advances and Applications in Bioinformatics and Chemistry, 4*, 1–12.

Wachsmuth, M., Findeiß, S., Weissheimer, N., Stadler, P. F., & Mörl, M. (2013). De novo design of a synthetic riboswitch that regulates transcription termination. *Nucleic Acids Research, 41*, 2541–2551.

Waldminghaus, T., Kortmann, J., Gesing, S., & Narberhaus, F. (2008). Generation of synthetic RNA-based thermosensors. *Biological Chemistry, 389*, 1319–1326.

Washio, T., Sasayama, J., & Tomita, M. (1998). Analysis of complete genomes suggests that many prokaryotes do not rely on hairpin formation in transcription termination. *Nucleic Acids Research*, *26*, 5456–5463.

Weigand, J. E., Sanchez, M., Gunnesch, E.-B., Zeiher, S., Schroeder, R., & Suess, B. (2008). Screening for engineered neomycin riboswitches that control translation initiation. *RNA*, *14*, 89–97.

Wickiser, J. K., Winkler, W. C., Breaker, R. R., & Crothers, D. M. (2005). The speed of RNA transcription and metabolite binding kinetics operate an FMN riboswitch. *Molecular Cell*, *18*, 49–60.

Wilkinson, K. A., Merino, E. J., & Weeks, K. M. (2006). Selective 2′-hydroxyl acylation analyzed by primer extension (SHAPE): Quantitative RNA structure analysis at single nucleotide resolution. *Nature Protocols*, *1*, 1610–1616.

Winkler, W., Nahvi, A., & Breaker, R. R. (2002). Thiamine derivatives bind messenger RNAs directly to regulate bacterial gene expression. *Nature*, *419*, 952–956.

Xayaphoummine, A., Bucher, T., & Isambert, H. (2005). Kinefold web server for RNA/DNA folding path and structure prediction including pseudoknots and knots. *Nucleic Acids Research*, *33*, W605–W610.

Zadeh, J. N., Wolfe, B. R., & Pierce, N. A. (2011). Nucleic acid sequence design via efficient ensemble defect optimization. *Journal of Computational Chemistry*, *32*, 439–452.

Zhang, J., & Deutscher, M. P. (1992). A uridine-rich sequence required for translation of prokaryotic mRNA. *Proceedings of the National Academy of Sciences of the United States of America*, *89*, 2605–2609.

CHAPTER TWO

Ligand-Dependent Exponential Amplification of Self-Replicating RNA Enzymes

Charles Olea Jr., Gerald F. Joyce[1]
Department of Chemistry, The Skaggs Institute for Chemical Biology, The Scripps Research Institute, La Jolla, California, USA
[1]Corresponding author: e-mail address: gjoyce@scripps.edu

Contents

Abstract

A general analytical method for the detection of target ligands has been developed, based on a special class of self-replicating aptazymes. These "autocatalytic aptazymes" are generated by linking an aptamer domain to the catalytic domain of a self-replicating RNA enzyme. Ligand-dependent self-replication of RNA proceeds in a self-sustained manner, undergoing exponential amplification at a constant temperature without the assistance of any proteins or other biological materials. The rate of exponential amplification is dependent on the concentration of the ligand, thus enabling quantitative ligand detection. This system has the potential to detect any ligand that can be recognized by an aptamer, including small molecules and proteins. The instability of RNA in biological samples due to the presence of ribonucleases can be overcome by employing the enantiomeric L-RNA form of the self-replicating enzyme. Methods for real-time fluorescence monitoring over the course of exponential amplification are currently being developed.

Methods in Enzymology, Volume 550
ISSN 0076-6879
http://dx.doi.org/10.1016/bs.mie.2014.10.030

1. INTRODUCTION

An analytical method for quantitative ligand detection should be general for various types of ligand molecules, sensitive and specific for the target ligand, robust in the context of biological and environmental samples, inexpensive, and simple to operate. The two most prominent methods in this regard are the enzyme-linked immunosorbent assay (ELISA; Engvall & Perlman, 1971) and the quantitative polymerase chain reaction (qPCR; Wang, Doyle, & Mark, 1989). ELISA has broader generality, being applicable to any ligand for which one can generate a corresponding high-affinity antibody. Quantitative PCR is specific for nucleic acid targets, but has greater sensitivity due to its exponential signal amplification compared to the linear amplification of ELISA. Both methods are inexpensive and simple to operate and are carried out routinely in a wide variety of settings.

The discovery of synthetic and naturally occurring riboswitches (Tang & Breaker, 1997; Winkler, Nahvi, Roth, Collins, & Breaker, 2004) has inspired application of this class of molecules to ligand detection, exploiting the ability of RNA to recognize both small molecule and macromolecule targets, and harnessing the ligand-dependent structural rearrangement of RNA to generate a measurable signal. For example, the ligand-recognition (aptamer) domain of a riboswitch can be linked to an RNA enzyme (ribozyme) such that the enzyme is active only when the ligand is bound (Soukup & Breaker, 1999). The resulting "aptazyme" can be used to generate a signal that reflects the abundance of the corresponding ligand.

This chapter describes a special class of aptazymes for which the catalytic component is an RNA enzyme that catalyzes its own replication. As with qPCR, these "autocatalytic aptazymes" undergo exponential amplification with a growth rate that reflects the concentration of the ligand. Unlike qPCR, however, RNA self-replication proceeds at a constant temperature and does not require the assistance of proteins. Ligand-dependent, self-replicating RNA enzymes have not yet been applied outside the research setting, but are now sufficiently robust that they can be designed and utilized by nonspecialists for many applications.

The self-replicating RNA enzyme was derived from the R3C ligase, a simple RNA motif that catalyzes the templated ligation of two RNA substrates (Rogers & Joyce, 2001). If those substrates, when joined together, form another copy of the enzyme, then self-replication is achieved (Lincoln & Joyce, 2009; Paul & Joyce, 2002). In the aptazyme format, the enzyme is active only in the presence of the target ligand, this being the case for both the parental enzyme and its copies that are generated through replication (Lam & Joyce, 2009). Thus ligand dependency is maintained throughout the course of exponential amplification.

The self-replicating RNA enzyme is linked to the aptamer domain via a short stem region, enabling installation of various aptamers that form a closed stem in the presence of their cognate ligand. Contact with the ligand enhances the structural organization of the aptamer domain, typically resulting in stabilization of the linking stem. Closure of the stem results in activation of the conjoined enzyme, which in turn results in replication. Even if the difference in activity is less than absolute for the ligand-bound compared to ligand-free state, its effect is felt as an exponential growth parameter, resulting in signal generation that is strongly ligand dependent.

This chapter discusses the construction and operation of autocatalytic aptazymes, using the well-studied theophylline aptamer as an example (Jenison, Gill, Pardi, & Polisky, 1994). Procedures are described for conducting quantitative exponential amplification assays. Also discussed is a nuclease-resistant version of the amplification system, based on L-RNA molecules that can replicate in biological samples (Olea, Horning, & Joyce, 2012). Finally, a brief overview is given of efforts to develop a real-time fluorescence assay for ligand detection.

2. EXPONENTIAL AMPLIFICATION OF RNA ENZYMES

The self-replicating RNA enzyme (E) catalyzes the ligation of two RNA substrates (A and B) to form another copy of itself (Fig. 1A). The ligation reaction involves attack of the 3′-hydroxyl of substrate A on the 5′-triphosphate of substrate B, forming a phosphodiester linkage and releasing inorganic pyrophosphate. The resulting E·E complex dissociates in a non-rate-limiting manner at a constant temperature of 40–50 °C, completing the replication cycle (Fig. 1B). Replication is self-sustaining, requiring only the starting RNA enzyme, RNA substrates, $MgCl_2$, and a suitable buffer (Lincoln & Joyce, 2009).

Based on insight from kinetic studies (Ferretti & Joyce, 2013), an improved version of the self-replicating RNA enzyme was recently developed (Robertson & Joyce, 2014). This improved enzyme has an exponential growth rate of 0.14 min^{-1}, corresponding to a doubling time of 5 min. The sequence of the improved enzyme is shown in Fig. 1A, with the six mutations relative to the original version noted in the figure legend. Both

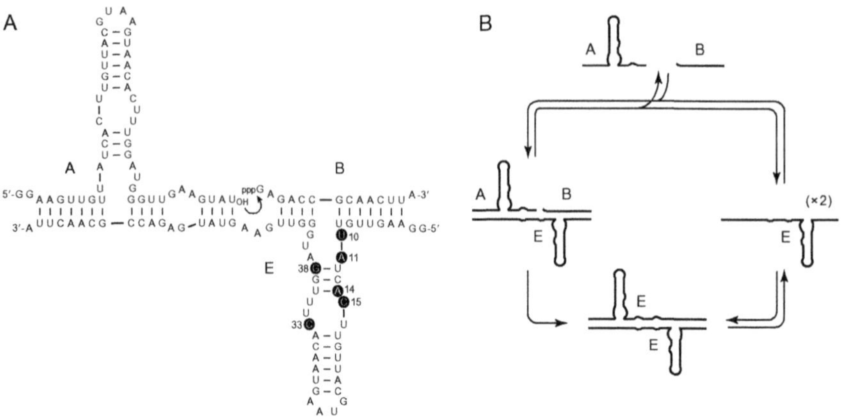

Figure 1 Self-replicating RNA enzyme. (A) Sequence and secondary structure of the enzyme (E), which catalyzes ligation of two substrates (A and B) to form a new copy of E. Curved arrow indicates the site of ligation. Mutations in the improved compared to original form of the enzyme are numbered and highlighted by black circles. These six mutations are: G10→U, C11→A, G14→A, A15→C, G33→C, and A38→G. (B) Self-replication cycle involving binding of A and B to E, ligation of A and B to form of a new copy of E, and dissociation of the E·E complex.

the original and improved versions of the enzyme also can undergo cross-replication, whereby two different RNA enzymes catalyze each other's synthesis from a total of four component substrates (Kim & Joyce, 2004). In cross-replication, the templating regions of the paired enzymes are nonidentical. This makes it possible for many distinct replicating pairs of enzymes to operate in the same reaction mixture, each drawing upon its corresponding substrates, and thus enabling multiplexed ligand-dependent amplification.

2.1. Materials

The RNA enzymes and substrates are prepared in the usual manner by *in vitro* transcription of corresponding synthetic DNA templates. The substrates also can be prepared by chemical synthesis using D-RNA phosphoramidites (Glen Research, Sterling, VA) or L-RNA phosphoramidites (ChemGenes, Wilmington, MA). Note that preparation of the synthetic B substrate necessitates chemical triphosphorylation of the 5′-hydroxyl, which is a technique that requires expertise in synthetic organic chemistry (Ludwig & Eckstein, 1989). Thus, unless purchased from a commercial vendor (e.g., TriLink, San Diego, CA), the B substrate is normally prepared (in the D-RNA form) by *in vitro* transcription. Following transcription or synthesis, the enzymes and substrates are purified by denaturing polyacrylamide gel electrophoresis (PAGE) and desalted.

The 3′-terminal homogeneity of the transcribed A or B substrate can be enhanced by processing these materials using either RNase P or a *cis*-cleaving hammerhead ribozyme, respectively. In the presence of a suitable guide RNA (Forster & Altman, 1990), the RNA subunit of RNase P cleaves at a defined position within an extended-length A substrate to give the desired product, having a reactive 3′-hydroxyl at its terminus (Lincoln & Joyce, 2009). A hammerhead ribozyme located within an extended-length B substrate cleaves at a defined position to remove that extension and yield the desired product, having a 2′,3′-cyclic phosphate at its terminus (Ferré-D'Amaré & Doudna, 1996; Ferretti & Joyce, 2013). The RNA enzyme (E) does not require processing and can be prepared by either run-off transcription or E-catalyzed ligation of A and B.

2.2. Procedure for RNA self-replication

Standard conditions for self-replication employ 0.01–0.25 μ*M* E, 5–10 μ*M* each of A and B, 25 m*M* $MgCl_2$, and 50 m*M* *N*-(2-hydroxyethyl)-piperazine-*N*′-3-propanesulfonic acid (EPPS; pH 8.5) at 42 °C. Different forms of the self-replicating enzyme have somewhat different temperature optima, ranging from 40 to 50 °C (Robertson & Joyce, 2014). The original replicator generally performs best at the lower end of this range, whereas the improved replicator generally performs best at the higher end.

1. Mix water, EPPS buffer, and RNA in a 97.5-μL volume; heat at 70 °C for 2 min, then hold on ice.
2. Just prior to use, incubate this mixture at 42 °C for 3 min.
3. Initiate replication by adding 2.5 μL of a solution of 1 *M* $MgCl_2$ that has been prewarmed at 42 °C; mix quickly and continue incubation at 42 °C.
4. Samples are removed at various times during the course of replication and quenched by the addition of EDTA in excess of the concentration of Mg^{2+}.
5. Sampled materials are separated by PAGE, distinguishing unreacted substrates from ligated products.

Typically, the A substrate is 5′-labeled using either ^{32}P or a fluorescent dye. The course of amplification is followed by monitoring the increasing concentration of E (and corresponding decreasing concentration of A). The rate of exponential amplification is determined by fitting these data to the logistic growth equation:

$$[E]_t = a/(1 + be^{-ct}),$$

where $[E]_t$ is the concentration of enzyme at time t, a is the maximum extent of growth, b is the degree of sigmoidicity, and c is the exponential growth rate.

3. LIGAND-DEPENDENT EXPONENTIAL AMPLIFICATION

The catalytic domain of the self-replicating RNA enzyme can be linked to an aptamer domain so that amplification occurs in a ligand-dependent manner (Lam & Joyce, 2009). The link is achieved through the central stem-loop of the enzyme, which is made contiguous with a stem that helps define the aptamer motif (Fig. 2A). Analogous to naturally occurring riboswitches, binding of the ligand by the aptamer domain results in a conformational change that stabilizes the joining stem, thereby stabilizing the catalytic center of the enzyme.

Most aptamers contain a stem element that provides a suitable point of attachment to the RNA enzyme. However, some fine-tuning is usually necessary to determine the optimal stability of the connecting stem. As a starting point, the predicted $-\Delta G^\circ$ value for the connecting stem should be ~4 kcal/mol. The preferred composition of the stem, both length and sequence, then should be determined empirically. Although the enzyme is a constant, the idiosyncratic nature of various aptamer structures makes it difficult to predict the optimal composition for the connecting stem. In addition, one must choose a balance between minimal activity in the absence of the ligand and maximal activity in the presence of saturating ligand. This can be assessed in a simple ligation assay before testing the preferred composition in a full replication assay.

Ligand-dependent replication proceeds at a constant temperature, with an exponential growth rate that depends on the fractional occupancy of the ligand-binding domain. The maximum extent of amplification is independent of the ligand concentration, being determined by the concentration of the A and B substrates. Thus, the concentration of ligand is assessed based on the time required to reach a defined partial extent of amplification. This is analogous to a C_t value for qPCR, which is the number of thermal cycles required to reach a defined threshold of PCR amplification.

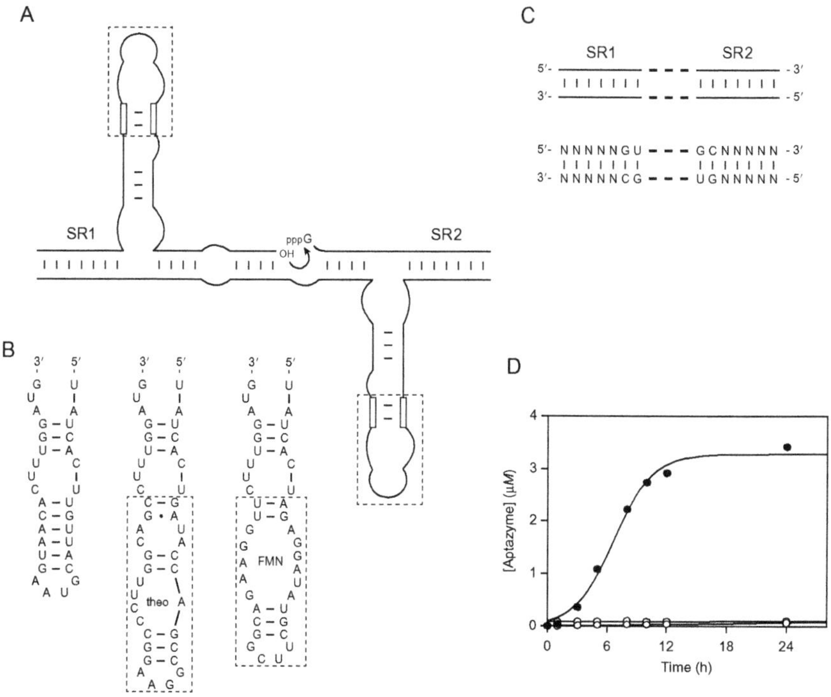

Figure 2 Self-replicating aptazymes. (A) An aptamer domain (dashed box) is linked to the central stem-loop of the self-replicating RNA enzyme via a connecting stem (open lines). (B) Modification of the central stem-loop to include an aptamer that binds either theophylline (theo) or FMN. The connecting stem has been optimized to maximize activity in the presence compared to the absence of the ligand (Lam & Joyce, 2009). (C) The two substrate-recognition domains of the enzyme (SR1 and SR2) form G·C and U·G pairs at the two innermost positions and can vary in sequence at the other positions (N·N), so long as complementarity is maintained. (D) Cross-replicating aptazymes that undergo exponential amplification in the presence of both theophylline and FMN (filled circles), but only linear amplification in the presence of either ligand alone (open circles). Reaction conditions: 0.02 μ*M* enzymes (original version), 5 μ*M* substrates, 2 m*M* theophylline and/or 1 m*M* FMN, 25 m*M* $MgCl_2$, and 50 m*M* EPPS (pH 8.5) at 42 °C (Lam & Joyce, 2009).

3.1. Procedure for quantitative ligand detection

Ligand-dependent exponential amplification is carried out under the same conditions that are employed in RNA self-replication, as described above. The ligand-containing sample is prewarmed and introduced just prior to the addition of $MgCl_2$ which initiates amplification. A standardized plot is generated by carrying out amplification in the presence of various concentrations of ligand. For each concentration of ligand, the time required to reach a defined threshold of amplification is determined, typically chosen as 25% of the maximum extent. Then a semilog plot is constructed of time-to-threshold versus ligand concentration. For a sample containing an unknown concentration of ligand, the time-to-threshold is measured and related to the standardized plot to infer the concentration of ligand in the sample.

As an example, the theophylline aptamer (Jenison et al., 1994) was appended to the central stem of the self-replicating RNA enzyme via a short connecting stem that was contiguous with the closing stem of the aptamer (Fig. 2B). Shorter connecting stems resulted in lower exponential growth rates even in the presence of saturating theophylline, whereas longer stems resulted in detectable amplification even in the absence of theophylline (Lam & Joyce, 2009).

As an alternative to measuring the time-to-threshold for various ligand concentrations, one can instead follow the entire course of exponential amplification for each ligand concentration, then determine the exponential growth rate by fitting the data to the logistic growth equation, as described above. Then, one constructs a saturation plot of exponential growth rate versus ligand concentration. The half-maximal rate corresponds to the apparent K_d of the ligand-binding domain and the maximal rate corresponds to the behavior at saturation. For a sample containing an unknown concentration of ligand, the exponential growth rate is determined and related to the saturation plot to infer the concentration of ligand in the sample. In most cases, however, it is more expedient simply to measure the time required to reach a defined threshold.

3.2. Multiplexed ligand detection

It is possible to detect multiple ligands simultaneously in a sample, using multiple pairs of cross-replicating RNA enzymes, each containing a different aptamer domain (Lam & Joyce, 2009). Orthogonality is achieved by employing distinct substrate-recognition domains in each pair of cross-replicating aptazymes (Fig. 2C). There should be at least two base pairs of discrimination between each combination of paired enzymes. There are five base pairs of variable sequence within each of the two substrate-recognition domains of the enzyme, making it straightforward to design orthogonal combinations. Discrimination is most sensitive for those nucleotides that lie closest to the site of ligation. Highly GC-rich sequences should be avoided because they may impede the rate of dissociation of the E·E complex, which would reduce the maximum rate of exponential amplification.

It is also possible to require the simultaneous detection of two ligands by a cross-replicating pair to achieve exponential amplification (Lam & Joyce, 2009). This requires that each member of the pair contain a different aptamer domain. If only one of the two ligands is present, then only one of the paired enzymes will be active and the system will exhibit linear amplification. If both ligands are present, then the system becomes autocatalytic and exponential amplification is observed.

As an example, the theophylline aptamer was linked to one enzyme and the FMN aptamer (Burgstaller & Famulok, 1994) to the other enzyme of a cross-replicating pair (Lam & Joyce, 2009). The FMN aptamer was linked to the central stem of the RNA enzyme via the same short connecting stem that was used to link the theophylline aptamer (Fig. 2B). Exponential growth was observed only in the presence of both ligands (Fig. 2D). One can construct a plot of either time-to-threshold or exponential growth rate as a function of ligand concentration, doing so for each of the two ligands while maintaining a saturating concentration of the other ligand.

3.3. Coupling ligand recognition to ligand-independent amplification

The potential dynamic range of ligand-dependent exponential amplification, that is, the range of ligand concentrations over which the system exhibits saturation behavior, is determined by the K_d of the ligand-binding domain. Ligand detection is most precise at concentrations that are within 10-fold of the K_d and becomes imprecise when the ligand concentration is more than 100-fold below the K_d (due to slow amplification) or 100-fold above the K_d (due to saturation). A further constraint is that the K_m of the replicating enzyme for the A and B substrates is in the range of 1–10 μM (Ferretti & Joyce, 2013). This means that the substrates must be present at such a concentration to achieve a fast rate of amplification. However, because the aptamer domain lies within the A substrate, the requirement for saturating substrates sets a minimum concentration at which the ligand can be sensed.

Quantitative PCR employs a ligand-recognition molecule (PCR primer) that is present at micromolar concentrations, yet can detect subattomolar concentrations of the target nucleic acid. This is because ligand recognition is required to initiate amplification, but subsequent rounds of amplification proceed in a ligand-independent manner. The same principle can be applied to the self-replicating RNA enzyme (Lam & Joyce, 2011). First, a simple aptazyme is used to carry out ligand-dependent joining of two RNA molecules to form seed copies of the self-replicating enzyme. Then, these seed copies are used to initiate exponential amplification, which proceeds in a ligand-independent manner. The disadvantage of this approach, like PCR, is that the ligand is not sensed throughout the course of amplification, potentially giving rise to false-positive signals. The advantage is that the dynamic range of the assay is no longer restricted by the K_m of the replicating enzyme, instead depending on the number of seed copies of replicating enzyme that are generated in the ligand-dependent ligation reaction.

The same R3C ligase motif used in replication also is used to generate seed copies of the replicating enzyme, but is made to perform ligation at a different nucleotide position to avoid interference with the replication cycle (Fig. 3). Two substrates are provided: one corresponding to the 5′ half of substrate A (designated A_1), and the other corresponding to the 3′-half of substrate A conjoined to substrate B (designated A_2–B). The ligated product, A_1–A_2–B, is equivalent to E, the self-replicating RNA enzyme. The aptamer domain is linked to the central stem of the ligase enzyme in the same manner as for the autocatalytic aptazyme. In principle, ligand-dependent ligation and ligand-independent exponential amplification could be carried out in the same reaction mixture, but it is preferable to carry out these two steps sequentially so that the conditions can be optimized for each reaction. It is not necessary to carry out any purification or processing steps between the two reactions.

3.4. Procedure for coupled amplification

The reaction mixtures for both ligand-dependent ligation and ligand-independent exponential amplification contain 25 m*M* $MgCl_2$ and 50 m*M* EPPS (pH 8.5) at 42 °C. The first reaction employs an excess of the ligase aptazyme and a modest concentration of substrates. The second reaction employs whatever seed enzymes are carried over from the first reaction and a higher concentration of substrates. The RNA substrates for the second reaction can be maintained at dryness until rehydrated by the addition of the completed first reaction mixture.

1. Ligand-dependent generation of seed enzyme:
   - **1.1.** Mix water, EPPS buffer, 500 pmol ligase aptazyme, 50 pmol A_1, and 50 pmol A_2–B in a volume of 97.5-μL (less that of the ligand-containing sample); heat at 70 °C for 2 min, then hold on ice.
   - **1.2.** Just prior to use, incubate this mixture at 42 °C for 3 min.
   - **1.3.** Add ligand-containing sample that has been prewarmed at 42 °C.

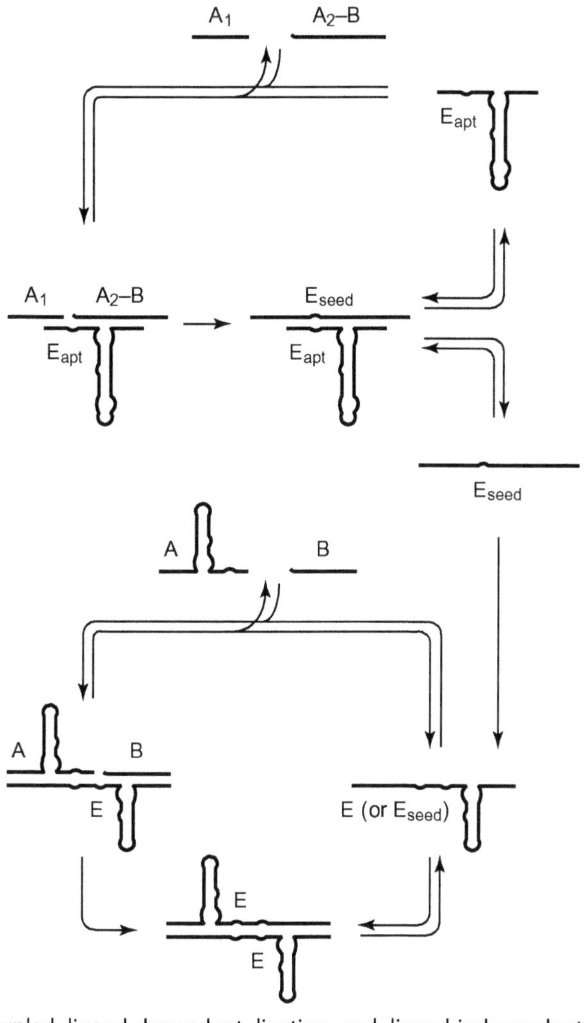

Figure 3 Coupled ligand-dependent ligation and ligand-independent exponential amplification. An aptazyme (E_{apt}) catalyzes ligand-dependent ligation of two substrates (A_1 and A_2–B) to form seed copies of the self-replicating RNA enzyme (E_{seed}), which then initiate exponential amplification utilizing substrates A and B.

1.4. Initiate ligation reaction by adding 2.5 μL of a solution of 1 *M* $MgCl_2$ that has been prewarmed at 42 °C; mix quickly and continue incubation at 42 °C for 20 min.

2. Ligand-independent exponential amplification:

2.1. Prepare a separate aqueous solution containing 200 pmol each of A and B; evaporate to dryness.

2.2. Transfer the products of the completed first reaction mixture directly to the second reaction vessel; mix to dissolve and resume incubation at 42 °C.

2.3. Samples are removed at various times during the course of replication and quenched by the addition of EDTA in excess of the concentration of Mg^{2+}.

2.4. Sampled materials are separated by PAGE, distinguishing unreacted substrates from ligated products.

The extent of the first reaction should not be allowed to exceed more than 20% relative to the starting amount of substrate. Otherwise, the production of seed enzymes would begin to reflect the maximum extent of reaction rather than the concentration of ligand. A saturation plot can be constructed based on the yield of seed enzymes as a function of ligand concentration. The yield is determined either at a fixed time for all samples or at various times (depending on reaction rate) and normalized to a fixed time. These data are fit to the equation:

$$[\mathrm{E}] = \varepsilon + \left\{[\mathrm{E}]_{\mathrm{max}}[\mathrm{ligand}] \,/\, (K_{\mathrm{d}} + [\mathrm{ligand}])\right\},$$

where [E] is the yield of seed enzyme for a given concentration of ligand, ε is the yield in the absence of ligand, and $[\mathrm{E}]_{\mathrm{max}}$ is the yield at saturating ligand concentration.

Even if no seed enzyme is produced in the first reaction mixture, exponential amplification can still occur in the second reaction mixture due to the spontaneous initiation of amplification. The A and B substrates can form an A·B·A·B tetramolecular complex that undergoes ligation at a reduced rate to produce E molecules (Ferretti & Joyce, 2013). This is analogous to the spontaneous initiation of PCR amplification that can occur in the absence of a target nucleic acid, largely as a consequence of primer-dimer formation (Rychlik, 1995). Spontaneous initiation of RNA amplification can be reduced by decreasing the stability of the tetramolecular complex relative to the E·A·B complex. Alternatively, it may be preferable to carry out RNA amplification in the cross-replication format to enable the use of more discriminating substrates that are less prone to spontaneous initiation. As with qPCR, the spontaneous initiation of amplification cannot be eliminated entirely and is instead managed as a control reaction that helps to define the sensitivity of ligand detection.

The ligand-dependent generation of seed enzyme is carried out in the presence of various concentrations of ligand (including no ligand), and each of these reaction mixtures is then used to initiate ligand-independent exponential amplification. For each concentration of ligand, the time required to reach a defined threshold is determined in the subsequent exponential amplification reaction, typically chosen as 25% of the maximum extent. Then a semilog plot is constructed of time-to-threshold versus ligand concentration. To be more precise, one can plot the time-to-threshold as a function of the yield of seed enzyme, normalized to the yield obtained at saturating ligand concentration (Lam & Joyce, 2011). For a sample containing an unknown concentration of ligand, the time-to-threshold is measured and related to the standardized plot to infer the concentration of ligand in the sample.

4. NUCLEASE-RESISTANT AUTOCATALYTIC APTAZYMES

RNA is a fragile molecule in the presence of biological or environmental samples due to its susceptibility to degradation by ribonucleases. For the detection of small-molecule ligands, such as drugs or metabolites, it is possible to deproteinize the samples by phenol extraction prior to analysis. Ligand-dependent exponential amplification of RNA proceeds readily in human serum that has been pretreated in this manner (Lam & Joyce, 2009). For the detection of protein ligands, however, a different approach is required. The tactic that is commonly applied to RNA aptamers is to replace the nucleotides that are most susceptible to ribonuclease degradation by nuclease-resistant nucleotide analogs. Unpaired pyrimidine residues are especially vulnerable in this regard, although almost any RNA sequence is susceptible to cleavage by ribonucleases. The challenge when introducing nuclease-resistant nucleotide analogs is to avoid impairing RNA function, which is especially difficult for a highly optimized molecule such as the self-replicating RNA enzyme discussed in this chapter.

An alternative approach that can be applied to the autocatalytic aptazymes is to construct the enzyme and its substrates entirely from nonnatural L-ribonucleotides (Olea et al., 2012). The mirror-image enzyme behaves identically as its D-RNA counterpart, but is completely resistant to ribonucleases. The L-RNA molecules must be prepared by chemical synthesis, which poses three special challenges. First, the A substrate cannot be [5′-^{32}P]-labeled using T4 polynucleotide kinase and instead is labeled with a fluorophore following oligonucleotide synthesis. Second, the B substrate must be chemically 5′-triphosphorylated, which requires expertise in synthetic organic chemistry (see Section 2.1). Third, the size of the enzyme plus the appended aptamer domain approaches the limit of what can be prepared by chemical synthesis. Thus, it may be more expedient to prepare starting copies of the enzyme using the autocatalytic reaction itself, employing the two L-RNA substrates.

The protocols for self-replication, quantitative ligand detection, and coupled amplification are identical for the D- and L-RNA self-replication systems. The aptamer domain that is linked to the enzyme also must be constructed of L-nucleotides. For achiral ligands, such as theophylline, the same sequence can be used for the D- and L-aptamers. For chiral ligands, including proteins, it is necessary to obtain a corresponding L-RNA aptamer, usually referred to as a "Spiegelmer" (Klussmann, Nolte, Bald, Erdmann, & Furste, 1996). These compounds are obtained by first selecting D-aptamers against the enantiomer of the target ligand, then preparing an L-RNA of the same sequence that binds the desired target. Protein-binding L-aptamers usually are obtained by selecting D-RNAs that bind a D-peptide corresponding to the enantiomer of a structural epitope within the target protein (Leva et al., 2002).

5. REAL-TIME FLUORESCENCE ASSAYS

It is time consuming to use PAGE analysis to follow the course of exponential amplification. As with modern qPCR, it would be preferable to monitor amplification in real-time based on the increasing production of a fluorescent signal. However, unlike PCR which generates new copies of the DNA amplicon from dNTPs, the self-replicating RNA enzyme generates new copies of itself from oligonucleotide building blocks that are already present in the reaction mixture. Thus, the detection method must distinguish the ligated products from these unligated substrates. Several approaches are being investigated that would provide such a detection method, but none is yet ready for general application. Three approaches are discussed briefly here.

The first approach utilizes a molecular beacon that binds across the ligation junction, hybridizing to E, but not appreciably to A or B (Fig. 4A). A molecular beacon is an oligodeoxynucleotide that contains both a fluorescent label and a fluorescence quencher and adopts either a hairpin conformation in the absence of the target or an extended conformation when hybridized to the target (Tyagi & Kramer, 1996). In the hairpin conformation, the label and quencher are in close proximity and fluorescence is quenched, whereas in the extended conformation the label and quencher

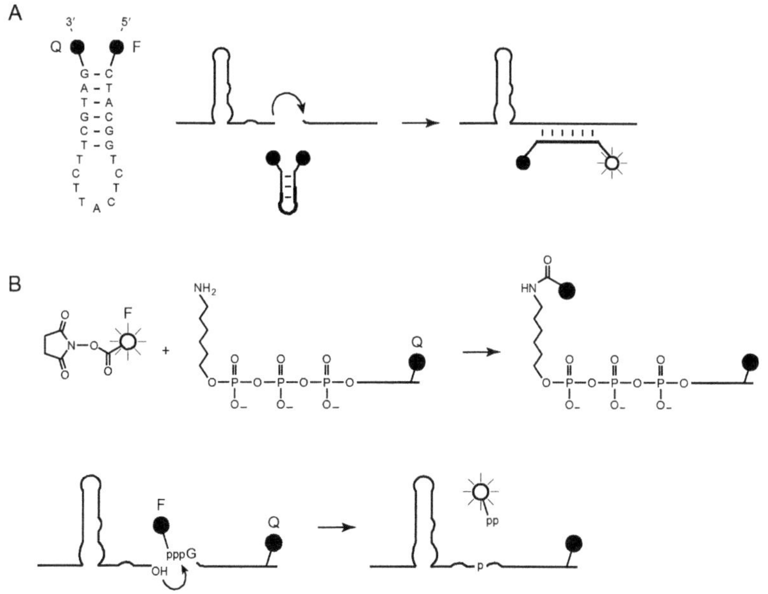

Figure 4 Two approaches for real-time florescence monitoring of exponential amplification. (A) A molecular beacon contains a fluorescent label (F) and fluorescence quencher (Q) and forms either a nonfluorescent hairpin structure in the presence of unligated substrates or a fluorescent extended structure when hybridized to the ligated product. The sequence of the beacon is shown at the left with the region of hybridization in bold. (B) A fluorescent label is linked to the γ-phosphate of substrate B, which also contains a fluorescence quencher. Ligation of the two substrates results in the release of labeled pyrophosphate, now separated from the quencher.

are well separated and a fluorescent signal occurs. The preferred molecular beacon for the self-replicating RNA binds to seven nucleotides on either side of the ligation junction, thereby separating a 5′-fluorescein label from a 3′-dabcyl quencher. Amplification can be monitored using a standard qPCR instrument, but operating at constant temperature. There are two disadvantages to employing molecular beacons: first, they somewhat inhibit the rate of exponential amplification; and second, they cannot distinguish among multiplexed pairs of cross-replicating enzymes because these molecules all have the same sequence surrounding the ligation junction.

A second approach exploits the fact that with each RNA-catalyzed ligation event a molecule of inorganic pyrophosphate is released. The released pyrophosphate can be used to generate a luminescent signal based on an ATP-regenerative luciferase assay (Ronaghi, Karamohamed, Pettersson, Uhlén, & Nyrén, 1996). This approach has been used to monitor the course of ligand-dependent exponential amplification (Lam & Joyce, 2009). However, it was necessary to withdraw samples from the amplification mixture due to differences in the preferred conditions for exponential amplification and luciferase-mediated signaling. Furthermore, this method does not enable multiplexed detection of different ligands.

A third approach also takes advantage of the release of inorganic pyrophosphate, but links a fluorescent dye to the γ-phosphate of the B substrate (Kumar et al., 2005) and incorporates a fluorescence quencher within that substrate. Fluorescence is quenched for the unreacted substrate, but not for the dye-labeled pyrophosphate that is released (Fig. 4B). The labeled B substrate is prepared by *in vitro* transcription in the presence of 5 m*M* γ-aminohexyl-GTP (Jena Bioscience, Jena, Germany) and 2 m*M* GTP, followed by reaction with the *N*-hydroxysuccinimide ester of various fluorophores (e.g., TAMRA, Cy5, Alexa Fluor 488, and Alexa Fluor 610). In principle, different labels could be employed to allow multiplex analysis. At present, however, the method requires further optimization to obtain a rapid rate of amplification with the various dye-labeled substrates.

6. CONCLUSIONS

The autocatalytic aptazymes described in this chapter offer a new tool for the general purpose, quantitative detection of target ligands. The exponential growth rate is dependent on the concentration of ligand, enabling quantitative detection. Moreover, the system can operate in biological samples by employing the L-RNA form of the replicating enzyme and its substrates. Additional effort is needed to develop a real-time fluorescence assay and to apply the system to a broad range of complex targets.

ACKNOWLEDGMENTS

This work was supported by grant no. GM065130 from the National Institutes of Health. C. O. was supported by Ruth L. Kirschstein National Research Service Award no. F32CA165430 from the National Institutes of Health.

REFERENCES

Burgstaller, P., & Famulok, M. (1994). Isolation of RNA aptamers for biological cofactors by *in vitro* selection. *Angewandte Chemie, 33*, 1084–1087.

Engvall, E., & Perlman, P. (1971). Enzyme-linked immunosorbent assay (ELISA). Quantitative assay of immunoglobulin G. *Immunochemistry, 8*, 871–874.

Ferré-D'Amaré, A. R., & Doudna, J. A. (1996). Use of *cis*- and *trans*-ribozymes to remove 5′ and 3′ heterogeneities from milligrams of *in vitro* transcribed RNA. *Nucleic Acids Research, 24*, 977–978.

Ferretti, A. C., & Joyce, G. F. (2013). Kinetic properties of an RNA enzyme that undergoes self-sustained exponential amplification. *Biochemistry, 52*, 1227–1235.

Forster, A. C., & Altman, S. (1990). External guide sequence for an RNA enzyme. *Science, 249*, 783–786.

Jenison, R. D., Gill, S. C., Pardi, A., & Polisky, B. (1994). High-resolution molecular discrimination by RNA. *Science, 263*, 1425–1429.

Kim, D.-E., & Joyce, G. F. (2004). Cross-catalytic replication of an RNA ligase ribozyme. *Chemistry & Biology, 11*, 1505–1512.

Klussmann, S., Nolte, A., Bald, R., Erdmann, V. A., & Furste, J. P. (1996). Mirror-image RNA that binds D-adenosine. *Nature Biotechnology, 14*, 1112–1115.

Kumar, S., Sood, A., Wegener, J., Finn, P. J., Nampalli, S., Nelson, J. R., et al. (2005). Terminal phosphate labeled nucleotides: Synthesis, applications, and linker effect on incorporation by DNA polymerases. *Nucleosides, Nucleotides & Nucleic Acids, 24*, 401–408.

Lam, B. J., & Joyce, G. F. (2009). Autocatalytic aptazymes enable ligand-dependent exponential amplification of RNA. *Nature Biotechnology, 27*, 288–292.

Lam, B. J., & Joyce, G. F. (2011). An isothermal system that couples ligand-dependent catalysis to ligand-independent exponential amplification. *Journal of the American Chemical Society, 133*, 3191–3197.

Leva, S., Lichte, A., Burmeister, J., Muhn, P., Jahnke, B., Fesser, D., et al. (2002). GnRH binding RNA and DNA Spiegelmers. *Chemistry & Biology, 9*, 351–359.

Lincoln, T. A., & Joyce, G. F. (2009). Self-sustained replication of an RNA enzyme. *Science, 323*, 1229–1232.

Ludwig, J., & Eckstein, F. (1989). Rapid and efficient synthesis of nucleoside 5′-*O*-(1-thiotriphosphates), 5′-triphosphates and 2′,3′-cyclophosphorothioates using 2-chloro-4H-1,3,2-benzodioxaphosphorin-4-one. *The Journal of Organic Chemistry, 54*, 631–635.

Olea, C., Jr., Horning, D. P., & Joyce, G. F. (2012). Ligand-dependent exponential amplification of a self-replicating L-RNA enzyme. *Journal of the American Chemical Society, 134*, 8050–8053.

Paul, N., & Joyce, G. F. (2002). A self-replicating ligase ribozyme. *Proceedings of the National Academy of Sciences of the United States of America, 99*, 12733–12740.

Robertson, M. P., & Joyce, G. F. (2014). Highly efficient self-replicating RNA enzymes. *Chemistry & Biology, 21*, 238–245.

Rogers, J., & Joyce, G. F. (2001). The effect of cytidine on the structure and function of an RNA ligase ribozyme. *RNA, 7*, 395–404.

Ronaghi, M., Karamohamed, S., Pettersson, B., Uhlén, M., & Nyrén, P. (1996). Real-time DNA sequencing using detection of pyrophosphate release. *Analytical Biochemistry, 242*, 84–89.

Rychlik, W. (1995). Selection of primers for polymerase chain-reaction. *Molecular Biotechnology, 3*, 129–134.

Soukup, G. A., & Breaker, R. R. (1999). Engineering precision RNA molecular switches. *Proceedings of the National Academy of Sciences of the United States of America, 96*, 3584–3589.

Tang, J., & Breaker, R. R. (1997). Rational design of allosteric ribozymes. *Chemistry & Biology, 4*, 453–459.

Tyagi, S., & Kramer, F. R. (1996). Molecular beacons: Probes that fluoresce upon hybridization. *Nature Biotechnology, 14*, 303–308.

Wang, A. M., Doyle, M. V., & Mark, D. F. (1989). Quantitation of mRNA by the polymerase chain reaction. *Proceedings of the National Academy of Sciences of the United States of America, 86*, 9717–9721.

Winkler, W. C., Nahvi, A., Roth, A., Collins, J. A., & Breaker, R. R. (2004). Control of gene expression by a natural metabolite-responsive ribozyme. *Nature, 428*, 281–286.

CHAPTER THREE

Design of Modular "Plug-and-Play" Expression Platforms Derived from Natural Riboswitches for Engineering Novel Genetically Encodable RNA Regulatory Devices

Jeremiah J. Trausch, Robert T. Batey[1]

Department of Chemistry and Biochemistry, University of Colorado at Boulder, Boulder, Colorado, USA

[1]Corresponding author: e-mail address: robert.batey@colorado.edu

Contents

Abstract

Genetically encodable RNA devices that directly detect small molecules in the cellular environment are of increasing interest for a variety of applications including live cell imaging and synthetic biology. Riboswitches are naturally occurring sensors of intracellular metabolites, primarily found in the bacterial mRNA leaders and regulating their expression. These regulatory elements are generally composed of two domains: an aptamer that binds a specific effector molecule and an expression platform that informs the transcriptional or translational machinery. While it was long established that

Methods in Enzymology, Volume 550
ISSN 0076-6879
http://dx.doi.org/10.1016/bs.mie.2014.10.031

riboswitch aptamers are modular and portable, capable of directing different output domains including ribozymes, switches, and fluorophore-binding modules, the same has not been demonstrated until recently for expression platforms. We have engineered and validated a set of expression platforms that regulate transcription through a secondary structural switch that can host a variety of different aptamers, including those derived through *in vitro* selection methods, to create novel chimeric riboswitches. These synthetic switches are capable of a highly specific regulatory response both *in vitro* and *in vivo*. Here we present the methodology for the design and engineering of chimeric switches using biological expression platforms.

1. INTRODUCTION

Genetically encodable synthetic RNAs are increasingly considered important tools for a diverse set of applications. These include live cell imaging (Paige, Nguyen-Duc, Song, & Jaffrey, 2012; Paige, Wu, & Jaffrey, 2011; Stojanovic & Kolpashchikov, 2004), directed protein evolution (Michener & Smolke, 2012), control of engineered biosynthetic pathways (Carothers, Goler, Juminaga, & Keasling, 2011), elucidation of biochemical pathways (Fowler, Brown, & Li, 2010), selection of organisms that overproduce useful compounds (Yang et al., 2013), and molecular scaffolds for organizing enzymes (Delebecque, Lindner, Silver, & Aldaye, 2011). There are a number of reasons why RNA is a powerful route to engineering a functional device. Primarily, it is relatively straightforward to computationally model or biochemically assess its secondary structure and use this information as the basis for designing functional molecules (reviewed in Chappell et al., 2013). Just as important, robust and powerful methods exist for the selection or evolution of RNAs that bind ligands—including small molecules and proteins—or that catalyze a chemical reaction (reviewed in Stoltenburg, Reinemann, & Strehlitz, 2007). Thus, rational and evolutionary approaches can be combined to create novel RNAs. Finally, over the past decade, it has been realized that RNA is a central player in the regulation of a wide variety of biological processes (reviewed in Caldelari, Chao, Romby, & Vogel, 2013; Clark, Choudhary, Smith, Taft, & Mattick, 2013; Storz, Vogel, & Wassarman, 2011). This reinforces the idea that RNA devices can function as robustly in the cellular environment as their protein counterparts. Among the diverse set of designed RNAs, some of the potentially most useful are those that control gene expression in response to a small-molecule effector.

An example of a genetic regulatory RNA device that has been intensively developed and studied is the allosteric ribozyme (often called an aptazyme) (Fig. 1A) (reviewed in Silverman, 2003). These RNAs generally consist of three components: a *receptor domain* that binds a small molecule, protein, or metal ion; a *readout domain* composed of a self-cleaving ribozyme, most often a hammerhead ribozyme; and a *communication module* that couples the two functional domains (Soukup & Breaker, 1999; Tang & Breaker, 1997). Binding of a signal (ligand) to the aptamer of the receptor domain activates the self-cleaving activity of the ribozyme, which can be easily observed using various methods. Transmission of the occupancy status of

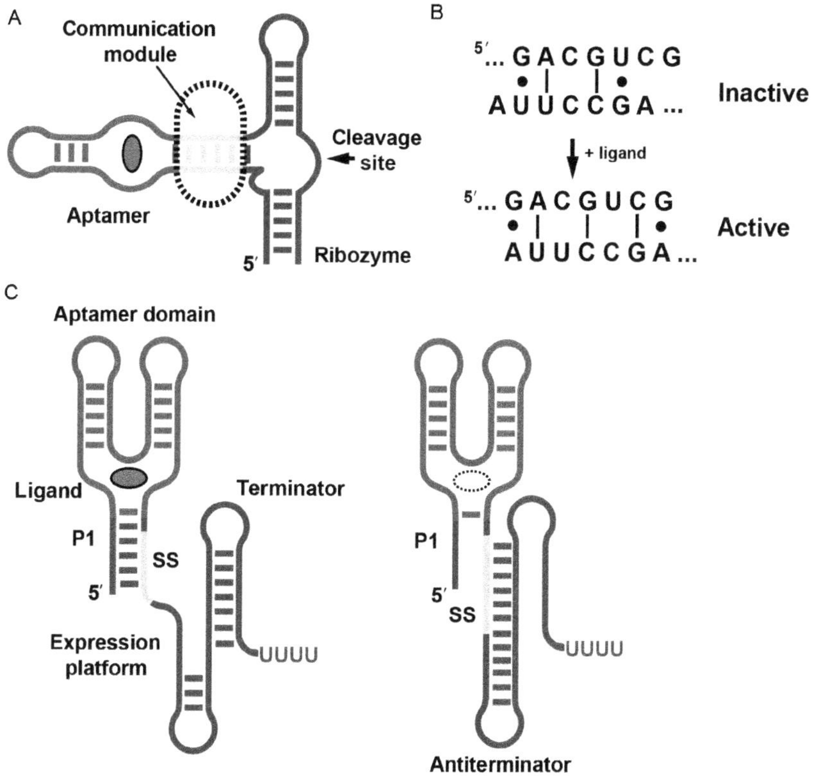

Figure 1 Cartoon representation of two RNA devices. (A) Domain organization of an engineered allosteric ribozyme (aptazyme) highlighting the aptamer (red), ribozyme (blue), and communication module (green). Binding of an effector ligand to the aptamer (magenta) activates the ribozyme via a structural change in the communication module that in turn organizes the active site for strand scission. (B) Sequence of a representative communication module (Soukup & Breaker, 1999) showing the two alternative structures adopted. (C) Domain organization of a natural riboswitch that regulates transcriptional termination. Binding of ligand to the aptamer domain (red) directs the secondary structural switch of the expression platform (blue). A sequence element that is found in one of the two mutually exclusive helices (of which the P1 helix is one) is referred to as the "switching sequence" (green). (See the color plate.)

the aptamer domain to the readout domain is achieved by the communication module—a simple helix designed or evolved to adopt one of two different pairing schemes in response to an effector molecule binding the receptor (Fig. 1B) (Kertsburg & Soukup, 2002; Koizumi, Soukup, Kerr, & Breaker, 1999). While these devices have seen some success, they face a number of challenges in their design and implementation, in particular, with regard to interdomain communication. For example, one significant limitation is that the "slippery" secondary structure of the communication module contributes to leakiness and slow allostery (de Silva & Walter, 2009). It is notable that the one example of a ligand-activated ribozyme found in biology, the *glmS* ribozyme, does not employ alternative secondary structures for activation (Winkler, Nahvi, Roth, Collins, & Breaker, 2004). Instead, the ligand directly participates in the catalytic mechanism as a cofactor (Viladoms & Fedor, 2012), such that the aptamer and ribozyme domains are integrated into a single structural unit (Cochrane, Lipchock, & Strobel, 2007; Klein & Ferre-D'Amare, 2006). This fundamental difference in the natural and engineered ligand-activated ribozyme suggests that it is both instructive and useful to adopt strategies based upon highly evolved RNA devices found in nature.

The biological solution to a ligand-activated regulatory RNA element is the *riboswitch*. These are RNA sequences most often found in the leader sequence of bacterial mRNAs, regulating gene expression through their ability to directly bind small molecules in the cellular environment (reviewed in Breaker, 2012; Garst, Edwards, & Batey, 2011). Most riboswitches contain two structurally and spatially distinct domains: the *aptamer domain* (receptor) and the *expression platform* (readout) (Fig. 1C). The aptamer domain, which is always situated on the 5′-side of the riboswitch, creates a complex structural architecture capable of binding a small-molecule ligand, typically with both high affinity and specificity. The expression platform is responsible for transforming the occupancy status of the aptamer domain into the appropriate regulatory response at either the transcriptional or translational level. In transcriptional regulation, the expression platform contains two mutually exclusive secondary structures: an intrinsic (rho-independent) transcriptional terminator hairpin and an antiterminator hairpin. These two structures dictate the expression fate of the mRNA, and depending upon their relationship with the aptamer, can activate or repress gene expression in response to ligand binding. At the heart of these RNAs is a sequence shared between the two alternative structures, one being the first helix (generally denoted as the P1 helix) of the aptamer domain, called the *switching sequence* ("ss" denoted in green, Fig. 1C). Given that these regulatory devices are both broadly distributed across bacteria and entrusted to regulate essential processes (Barrick & Breaker, 2007), these RNAs have vast potential as the foundation for synthetic devices that robustly function in the cellular context.

One key difference between aptazymes and riboswitches is the nature of the helical element linking the two functional domains. In the case of aptazymes, the communication module is often a "slippery" helical element with non-Watson–Crick or mismatch pairs (Fig. 1B) that are hard to design *de novo* and that can contribute to low fidelity of the resultant device. Conversely, the helical element (called the "P1 helix") connecting the two functional domains in many riboswitches is a Watson–Crick paired helix (Fig. 1C). This simple element is easy to predict computationally and is readily integrated into various design strategies. In addition, we observed that within the P1 helix of many riboswitches the sequence elements essential for ligand binding to the aptamer domain and the secondary structural switch in the expression platform do not overlap (Ceres, Garst, Marcano-Velazquez, & Batey, 2013; Ceres, Trausch, & Batey, 2013). Thus, it should in principle be straightforward to design novel "chimeric" riboswitches rationally using modules composed of aptamer domains and expression platforms with minimal design effort by simply coupling them through a Watson–Crick paired helix of the appropriate length.

To test this concept, we developed a mix-and-match strategy using modular components of natural riboswitches to create novel functional chimeras that either repress (Ceres, Garst, et al., 2013) or activate (Ceres, Trausch, et al., 2013) gene expression in response to ligand. While it was previously appreciated that the aptamer domains of natural riboswitches can be used in various RNA devices (Nomura & Yokobayashi, 2007; Paige et al., 2012; Wieland, Benz, Klauser, & Hartig, 2009), the same was not true for the expression platforms. Importantly, we also found that synthetic aptamers derived from *in vitro* selection approaches can interface with natural expression platforms, greatly extending the potential of these switch modules. Currently, a wide variety of aptamers have been fused to five different expression platforms to create functional switches (Ceres, Garst, et al., 2013; Ceres, Trausch, et al., 2013), and a recent study has demonstrated a sixth (Robinson et al., 2014). The function of these artificial switches can be easily monitored using either a single-turnover *in vitro* transcription assay or an *in vivo* reporter assay. In this chapter, we outline our strategy for the design of modular expression platforms, coupling to small-molecule-binding aptamers and testing of activity using different approaches.

2. DESIGN OF RIBOSWITCH MODULES

2.1. Design strategy

In the development of expression platform modules, the most important property is *composability*, defined as the ability of a given expression platform to be linked with different aptamer domains and yet retain its capacity to direct the expression machinery in a ligand-dependent fashion (Win & Smolke, 2007). Not all biological riboswitches have aptamer domains and expression platforms that fit this requirement. For example, the SAM-II (Gilbert, Rambo, Van Tyne, & Batey, 2008) and SAM-III (Lu et al., 2008) classes of riboswitches invariably have their aptamer domains and expression platforms as single structural units (Fig. 2A). In each case, a ribosome-binding site (RBS) is embedded entirely within the aptamer such that ligand binding inhibits 30*S* subunit binding (green, Fig. 2). Thus, these classes of riboswitches are single composite structures. A second type of riboswitch architecture that is also unsuitable for modular design is that in which the effector makes direct contacts with both domains, as observed in cobalamin-binding riboswitches (Johnson, Reyes, Polaski, & Batey, 2012) (Fig. 2B). In these RNAs, the two functional domains are separable (J.T. Polaski & R.T. Batey, unpublished observations), but the ligand directly mediates the interdomain interaction that represses translation (Johnson et al., 2012). Thus, these expression platforms are highly specific for the effector molecule. A third type of riboswitch architecture unsuitable

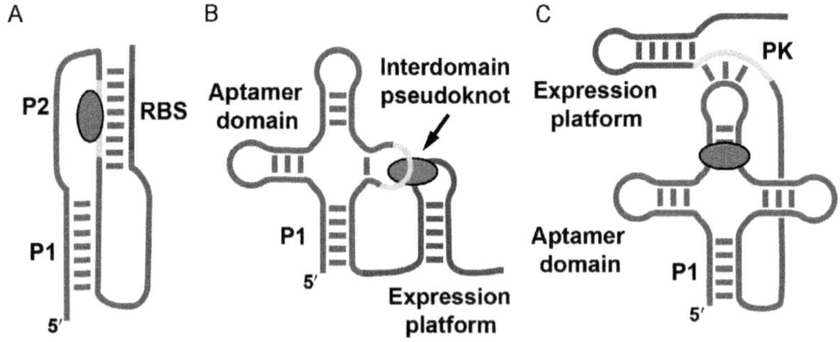

Figure 2 Domain organizations of natural riboswitches unsuitable for development of modular expression platforms. Riboswitch architectures in which the ligand-dependent structural switch is incorporated into a pseudoknot motif are generally unsuitable for modular aptamers or expression platforms without significant redesign. Examples include the (A) SAM-II, (B) cobalamin, and (C) SAM-IV riboswitches. Note that the switching sequence (green) in each case is part of a pseudoknot. (See the color plate.)

for modular design is that utilized by a number of riboswitch classes in which the two domains are linked through a pseudoknot rather than a helix, as observed in the SAM-I/IV (Trausch et al., 2014) and c-di-AMP classes (Nelson et al., 2013) (Fig. 2C). In these RNAs, sequence elements within the aptamer domain that are essential for ligand binding overlap with those involved with the structural switch of the expression platform. The common theme in all of the above riboswitches is that the switching sequence or other aspect of the structural switch is part of a pseudoknot motif. Since this motif is often complicated by tertiary interactions mediated by the crossing strands of the helices (reviewed in Giedroc & Cornish, 2009), they are difficult to engineer and thus we have avoided them in our design approach. It should be noted, however, that the pseudoknotted SAH riboswitch has been successfully redesigned as a modular transcriptional "ON" switch, but required extensive redesign of the expression platform (Ceres, Trausch, et al., 2013).

The most ideal expression platforms for development of composable regulatory switches are found in the purine, SAM-I, and lysine riboswitch classes, although other classes do contain members that could be used (Ceres, Garst, et al., 2013). In these classes, the aptamer domain is linked to the expression platform through the P1 helix (Fig. 1C). However, even within these classes, not all members have separable domains. The boundaries of the aptamer domain can be readily determined using information from structural and biochemical studies, in addition to phylogenetic alignments found in Rfam (Griffiths-Jones, Bateman, Marshall, Khanna, & Eddy, 2003). For example, the 5′-/3′-boundaries of the aptamer domain of all purine riboswitches are defined by two base pairs in the P1 helix adjacent to the three-way junction that harbors the ligand-binding site. These two pairs form base triple interactions with nucleotides in the adjacent three-way junction upon ligand binding (Batey, Gilbert, & Montange, 2004) which is proposed to be central to the mechanism of ligand-dependent regulation (Stoddard, Gilbert, & Batey, 2008). Phylogenetic alignments reveal no strong sequence conservation in any base pair beyond the two pairs proximal to the purine-binding site (Mandal, Boese, Barrick, Winkler, & Breaker, 2003; Mandal & Breaker, 2004). Thus, the aptamer domain boundaries for these riboswitches are firmly established as just the two base pairs of the P1 helix most proximal to the three-way junction of the aptamer domain. For the SAM-I and lysine riboswitches, similar considerations yield boundaries restricted to the 3′-side of the P1 helix (Ceres, Garst, et al., 2013). Other riboswitch classes containing members with a clear

P1-mediated division of domains are tetrahydrofolate (THF) (Trausch, Ceres, Reyes, & Batey, 2011), thiamine pyrophosphate (TPP) (Edwards & Ferre-D'Amare, 2006; Serganov, Polonskaia, Phan, Breaker, & Patel, 2006), flavin mononucleotide (Serganov, Huang, & Patel, 2009), and c-di-GMP class I (Kulshina, Baird, & Ferre-D'Amare, 2009; Smith et al., 2009).

For this application, RNA aptamers derived from *in vitro* selection also require the equivalent of a P1 helix adjacent to the ligand-binding site. Unfortunately, current methods do not incorporate a selection pressure for such an arrangement. For example, the biotin (Nix, Sussman, & Wilson, 2000) and cyanocobalamin (Sussman, Nix, & Wilson, 2000) aptamers both use a pseudoknot as their structural foundation, making them unsuitable for interfacing with the P1 helix of an expression platform module. Conversely, the theophylline (Zimmermann, Jenison, Wick, Simorre, & Pardi, 1997) and tetracycline (Xiao, Edwards, & Ferre-D'Amare, 2008) aptamers have a favorable architecture for modularity and have been used in numerous devices. For both of these aptamers, structural and biochemical data enable precise definition of the sequence requirements of the ligand-binding domain. Furthermore, because both aptamers lack significant tertiary structure, it is possible to circularly permute the aptamer as a means of optimizing the aptamer–expression platform chimera. The sequences of modular aptamers that we have employed for making chimeric riboswitches are given in Table 1.

The design of expression platform modules from natural riboswitches requires several considerations. First, as detailed above, the sequence involved in the secondary structural switch must not overlap with those required for ligand binding by the aptamer. The *Bacillus subtilis metE* *S*-adenosylmethionine (SAM-I)-sensing riboswitch exemplifies this principal requirement (Fig. 3). It is a transcriptional "OFF" switch that forms a stable terminator downstream of the aptamer domain upon ligand binding. In the absence of SAM, the alternative antitermination stem between residues within P1 and the terminator stem forms and enables the polymerase to escape the riboswitch and synthesize the entire message (Epshtein, Mironov, & Nudler, 2003; McDaniel, Grundy, Artsimovitch, & Henkin, 2003; Winkler, Nahvi, Sudarsan, Barrick, & Breaker, 2003). In this case, the switching sequence (green, Fig. 3) shared between the P1 helix (ligand-bound state) and the antiterminator (ligand-free state) does not overlap with sequences required for SAM binding (orange, Fig. 3). It should be noted that not all riboswitches fit this requirement, even though they might have the same arrangement of P1 and antiterminator. For example, simulation of the

Table 1 Sequence of aptamer and expression platform modules

Aptamer domain

Source	Ligand	Sequence
B. subtilis xpt-pbuX	Guanine, hypoxanthine	ATATAATCGCGTGGATATGGCACGCAAGTT TCTACCGGGCACCGTAAATGTCCGACTAT
B. subtilis pbuE	Adenine, 2-aminopurine	ATATAACCTCAATAATATGGTTTGAGGGTG TCTACCAGGAACCGTAAAATCCTGATTAT
B. subtilis metE	*S*-adenosylmethionine	ATCGAGAGTTGGGCGAGGGATTGGCCTTTT GACCCCAACAGCAACCGACCGTAATACCAT TGTGAAATGGGGCGCACTGCTTTTCGCGCC GAGACTGATGTCTCATAAGGCACGGTGCTA ATTCCATCAGATTGTGTCTGAGAGAT
B. subtilis yitJ	*S*-adenosylmethionine	TCAAGAGAAGCAGAGGGACTGGCCCGACGA AGCTTCAGCAACCGGTGTAATGGCGATCAG CCATGACCAAGGTGCTAAATCCAGCAAGCT CGAACAGCTTGGAAGA
B. subtilis lysC	Lysine	GAGGTGCGAACTTCAAGAGTATGCCTTTGG AGAAAGATGGATTCTGTGAAAAAGGCTGAA AGGGGAGCGTCGCCGAAGCAAATAAAACCC CATCGGTATTATTTGCTGGCCGTGCATTGA ATAAATGTAAGGCTGTCAAGAAATCATTTT CTTGGAGGGC
Streptococcus mutans folT	Tetrahydrofolate	AGTAGATGATTCGCGTTAAGTGTGTGTGAA GGGATGTCGTCACACAACGAAGCGAGAGCG CGGTGAATCATAGCATCCG
B. subtilis ribD	Flavin mononucleotide	TCGGGGCAGGGTGGAAATCCCGACCGGCGG TAGTAAAGCACATTTGCTTTAGAGCCCGTG ACCCGTGTGCATAAGCACGCGGTGGATTCA GTTTAAGCTGAAGCCGACAGTGAAAGTCTG GATGGGAGA
Theophylline[a]	Theophylline	ATACCAGCTTCGAAAGAAGCCCTTGGCAG
Tetracycline[a]	Tetracycline	GGGAGAGGTGAAGAATACGACCACCTAGGT AGAAATACCTAAAACATACCC

Expression platform

Source	Type of switch	Sequence[b]
B. subtilis metE	Transcriptional "OFF"	CAAAAAATTAATAACATTTTCTCTT(aptamer) GAGAGAGGCAGTGTTTTACGTAGAAAAGCC TCTTTCTCTCATGGGAAAGAGGCTTTTTGT TGTGAGAAAACCTCTTAGCAGCCTGTATCC GCGGGTGAAAGAGAGTGTTTTACATATAAA GGAGGAGAAACA**ATG**

Continued

Table 1 Sequence of aptamer and expression platform modules—cont'd
Expression platform

Source	Type of switch	Sequence
B. subtilis yitJ	Transcriptional "OFF"	CTTCCTGACACGAAAATTTCATATCCGTTC TTA(aptamer) TAAGAAGAGACAAAATCACTGACAAAGTCT TCTTCTTAAGAGGACTTTTTTTATTTCTCT TTTTTCCTTGCTGATGTGAATAAAGGAGGC AGACA**ATG**
B. subtilis lysC	Transcriptional "OFF"	AATTTCATAGTTAGATCGTGTTATATGGTG AAGATA(aptamer) TATCTCGTTGTTCATAATCATTTATGATGA TTAATTGATAAGCAATGAGAGTATTCCTCT CATTGCTTTTTTTATTGTGGACAAAGCGCT CTTTCTCCTCACCCGCACGAACCAAAATGT AAAGGGTGGTAATAC**ATG**
B. subtilis pbuE[c]	Transcriptional "ON"	AATTAAATAGCTATTATCACGATTTT (aptamer) AAAATCCTGATTACAAAATTTGTTTATGAC ATTTTTTGTAATCAGGATTTTTTATTTATC AAAACATTTAAGTAAAGGAGTTTGTT**ATG**
Ralstonia solanacearum metH	Transcriptional "ON"	CTCAGGCGGCCCTTCGGGTCGCTCCCGCCT TCCGTTCGAGCTCCGCTG(aptamer) CGGCGCTCGCAAACCCGCGTTTTCCTTGCC CCGGTTTGCGGCGCCGTTTTTTTTTGCGGC GTCCCCGGTGCAATTTCGCTTTGCCGAGGT GATCC**ATG**

[a]Derived from *in vitro* selection.
[b]The ATG initiating codon of the gene controlled by the riboswitch is denoted in bold.
[c]The sequence shown is for the "6T" variant of the polyuridine tract of the intrinsic terminator.

secondary structure of the *B. subtilis yxjG* and *metI* SAM-I riboswitches using Sfold (http://sfold.wadsworth.org/cgi-bin/index.pl) suggests that their switching sequences overlap with the two critical A–U pairs for SAM binding in the P1 helix.

Another useful property of a potential expression platform module is that it be able to direct the ligand-dependent switch using both *in vitro* and *in vivo* assays. The rationale for using *in vitro* methods is that this enables rapid analysis of a set of variants for those that display robust ligand-dependent switching along with detailed characterization of the best variants. For riboswitches regulating transcription, ligand-dependent regulation can be

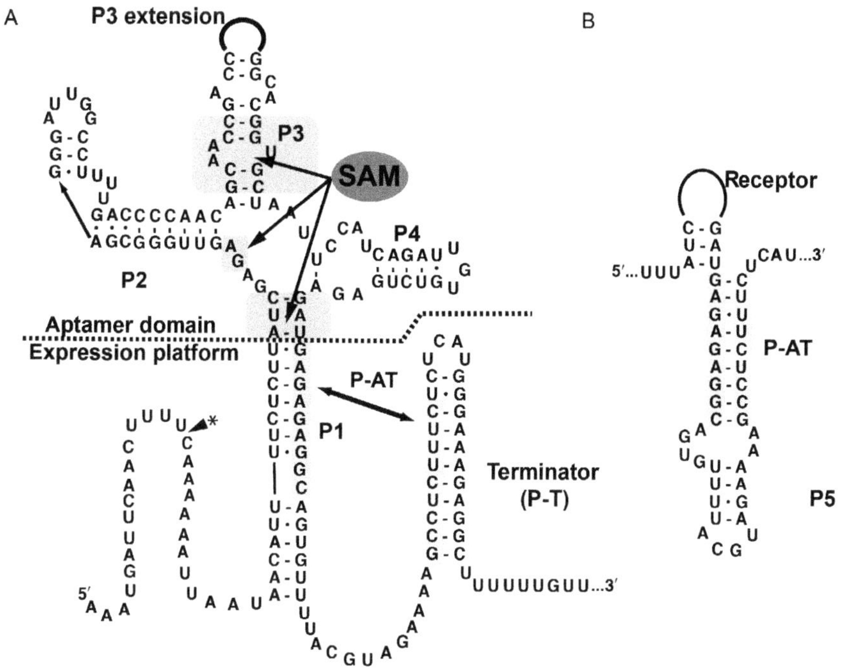

Figure 3 Example riboswitch with separable domains. (A) The secondary structure of the *S*-adenosylmethionine bound "OFF" state of the *B. subtilis metE* SAM-I riboswitch. The regions of the RNA that are in direct contact with SAM are highlighted in orange; of particular note are two essential A–U pairs proximal to the four-way junction. The switching sequence of the expression platform is highlighted in green and its complementary sequence in blue. (B) Secondary structure of the alternative antiterminator helix (P-AT). (See the color plate.)

easily monitored using a single-turnover transcription assay with *Escherichia coli* RNA polymerase (Artsimovitch & Henkin, 2009; Dann et al., 2007). For riboswitches that regulate at the level of translation, there are coupled transcription–translation assays that monitor the production of a reporter protein (Lemay et al., 2011; Mishler & Gallivan, 2014). Importantly, both the intrinsic terminator of transcriptional expression platforms and the Shine–Dalgarno sequence (RBS) of translation regulatory elements are well conserved in bacteria. Thus, a chimera could in principle be developed and optimized using an *E. coli*-based system *in vitro* and imported into the bacterium of choice for further refinement of function in the cellular context. We have developed five modular expression platforms, both repressive (Ceres, Garst, et al., 2013) and activating (Ceres, Trausch, et al., 2013), whose sequences are given in Table 1. These sequences represent only a

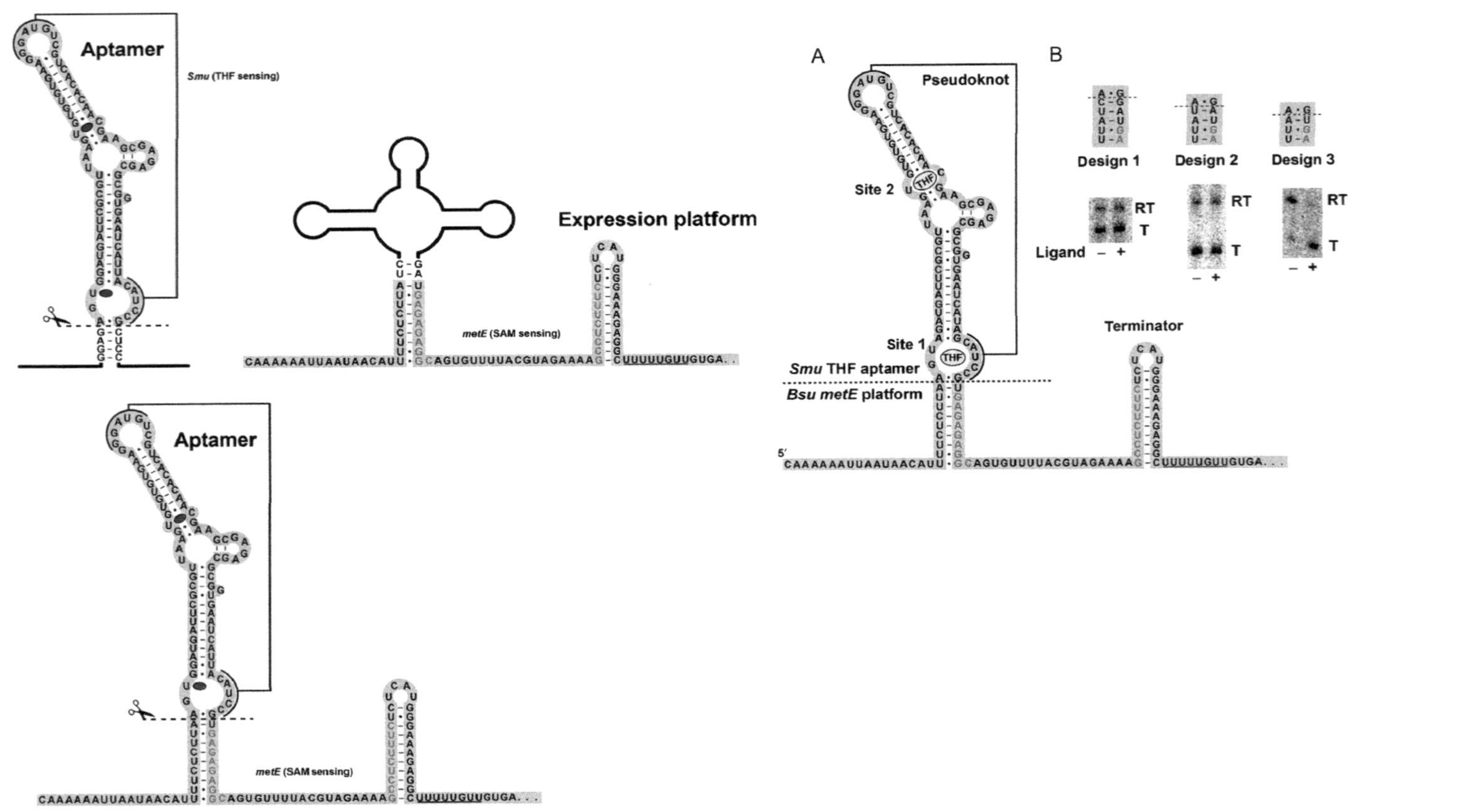

Figure 4 (See legend on opposite page.)

few possible modules, and certainly there are other natural sequences that may be significantly superior regulatory elements.

The coupling of aptamer and expression platform modules into a chimeric riboswitch is illustrated by the construction of a novel riboswitch used to probe the importance of the two ligand-binding sites in the THF aptamer. The *Streptococcus mutans* (*Smu*) *folT* THF aptamer, whose crystal structure had been determined, was fused to the *B. subtilis metE* expression platform (Fig. 4A and B) (Trausch et al., 2011). The boundary of the aptamer domain was defined by inspection of the structure along with consideration of phylogenetic conservation patterns to identify residues within P1 that are unnecessary for ligand binding. Therefore, the 5′- and 3′-boundaries of the aptamer were determined to occur at a conserved A·G mismatch adjacent to the 3′-end of the P1 helix. It should be noted that to bind THF efficiently there likely must be a few Watson–Crick base pairs preceding the A·G pair to stabilize the aptamer domain alone (analogous to what was observed for the purine riboswitch aptamer domain). As an initial starting point for the creation of a functional chimera, the entire P1 of the *metE* expression platform was retained, despite the fact that the critical switching sequence stops three residues from the 3′-side of P1 (Fig. 4). Initial testing of this sequence referred to as the *SmufolT/Bsu metE* chimera was done using an *in vitro* transcription assay (see Section 3). Note that the nomenclature we use for chimeric riboswitches is the name of the organism and regulated gene from which each part is derived in the order *aptamer domain/expression platform*. This revealed only a slight amount of read-through (RT) product in the absence of ligand and a negligible increase in terminated (T) product in the presence of ligand ("design 1," Fig. 4B). Using this poor riboswitch as a starting point, we have developed simple designs for sequence optimization to maximize the dynamic range and to achieve full repression in the presence of effector.

Figure 4 Structure and activity of a designed chimeric riboswitch. (A) The chimeric riboswitch formed by the fusion of the *Streptococcus mutants folT* THF aptamer domain (orange) and the *Bacillus subtilis metE* SAM-I class riboswitch expression platform (cyan). The secondary structure is of "design 3" variant in the ligand-bound "OFF" state. Note that the aptamer domain is not completely wild type; P4 has been shortened and capped with a tetraloop, corresponding to the sequence used for structural analysis (PDB ID 4LVV). (B) Secondary structure of the three design variants in the P1 stem around the aptamer/expression platform fusion site (dashed line). Below each is the products of an *in vitro* transcription assay in the absence and presence of ligand; "RT" denotes the read-through product ("ON" state) and "T" denotes terminated form ("OFF" state). While "design 1" and "design 2" show little or no ligand-dependent switching, "design 3" (also shown in panel A) is a robust regulatory element. (See the color plate.)

2.2. Design optimization

It is currently hypothesized that the relative thermodynamic stability of the mutually exclusive helices of the structural switch (in the *metE* module, P1 and P-AT) is critical for efficient riboswitch function (Ceres, Garst, et al., 2013). Since many riboswitches, including all that regulate at the transcriptional level, obligatorily function in the context of transcription, we have proposed that the folding decision is in large part dictated by the principle of "encoded cotranscriptional folding" (Xayaphoummine, Viasnoff, Harlepp, & Isambert, 2007). This model hypothesizes that RNA folding is dictated by the 5′-to-3′ polarity of the helices along with the relative stability of the competing helices. Further, *in vivo* studies have suggested that alternative structures exchange rapidly and with a low energy barrier between them (Mahen, Watson, Cottrell, & Fedor, 2010). Thus, an important design feature is modulation of the length of one of the helices to alter their relative stability.

The above considerations of the nature of the P1 helix reflect two different perspectives on the mechanism of secondary structural switching. The first is thermodynamic in that the relative stabilities of the two competing helices are the key property, reflecting $\Delta\Delta G$. Since it is simple to predict the thermodynamic stabilities of secondary structural elements, this is generally the property this is considered in computationally designed riboswitches. However, there is growing evidence that for many riboswitches secondary structure exchange is a kinetic process, governed by $\Delta G^{\ddagger}$. For example, in the *B. subtilis* adenine-responsive *pbuE* riboswitch the competing helix to P1 is significantly more stable, even when ligand is bound. In this case, ligand binding presents a kinetic roadblock for secondary structure exchange giving RNA polymerase sufficient time to escape the terminator element (Frieda & Block, 2012). For the modular expression platforms that we have developed, only $\Delta\Delta G$ is considered in the optimization of their performance when fused to a specific aptamer.

The simplest way to alter systematically the relative stabilities of the mutually exclusive pair of helices without changing the ligand-binding properties of the aptamer domain is to make stabilizing or destabilizing mutations to the P1 helix. In the initial *Smu folT/Bsu metE* chimera, transcriptional termination was dominant under both low and high concentrations of ligand ("design 1," Fig. 4B) using the single-turnover *in vitro* transcription assay (Section 3). This indicated that P1 was too strong relative to the antiterminator stem, even in the absence of effector. To weaken P1, a series of constructs were constructed and tested that systematically weaken P1 by sequentially deleting base pairs ("design 2" and "design 3"). This approach yielded a chimera displaying robust ligand-dependent regulation (Fig. 4B).

As an alternative approach, we have also mutagenized nucleotides at the 5′-end of the 5′-strand of the P1 helix. The advantage to this approach is that the aptamer–expression platform fusion site is not perturbed. This strategy was illustrated in the fusion of a 2-aminopurine binding mutant of the *B. subtilis xpt-pbuX* purine aptamer to an engineered variant of the *Ralstonia solanacearum metH* SAH expression platform (*Bsu* (C74U)*xpt/Rso metH*) (Ceres, Trausch, et al., 2013). This chimeric "ON" switch initially displayed ligand-dependent switching but strongly favored the RT state. In the absence of guanine, the switch displayed approximately 30% RT transcription, an unfavorable property for an "ON" switch that should be completely repressed in the absence of the effector. This observation indicated that the P1 helix is too stable and must be weakened to allow the terminator to robustly form without ligand. In this case, 3 nt at the 5′ side of the P1 helix were mutated to remove base pairs. Weakening the P1 helix resulted in a significant drop in RT without ligand.

A second approach to optimizing performance is to engineer accessory elements in the expression platform. For example, when the optimized *Bsu* (C74U)*xpt/Rso metH* chimera was placed upstream of a GFP reporter for use in *E. coli*, a very low dynamic range was observed (Ceres, Trausch, et al., 2013). Again, we observed significant leaky RT expression of the riboswitch, indicating that the terminator helix was not efficiently forming. To fix this issue, a stabilizing tetraloop motif was added to the distal end of the terminator to drive efficient nucleation of the helix. This strategy was inspired by the observation that in the *B. subtilis lysC* lysine riboswitch, the antiterminator helix contains a "nucleator" stem-loop at its distal tip that facilitates its efficient formation, as monitored by an *in vitro* transcription assay (Blouin, Chinnappan, & Lafontaine, 2011). It should be noted that this sequence lies outside the region directly involved in the structural switch and therefore can be engineered without affecting other properties of the RNA. Another site that can be modified in the expression platform of transcriptional regulators is the polyuridine tract of the intrinsic terminator. In this case, the number of uridines in this tract can be altered to affect termination efficiency (Peters, Vangeloff, & Landick, 2011). In the case of engineered "ON" switches based upon the *B. subtilis pbuE* adenine riboswitch, we found that decreasing the number of uridines in this tract from eight to five or six greatly improved the dynamic range of the riboswitch while not significantly compromising the termination efficiency in the absence of ligand (Ceres, Trausch, et al., 2013).

In summary, designing modular riboswitches is a two-part process. First, the appropriate aptamer domain and expression platform are linked together in a fashion that maintains the length of the P1 helix at approximately that of the wild-type riboswitch. This should be tested using one of the assays discussed below. While in many cases the initial chimera displays some ligand-conditional regulatory activity, depending upon the observed behavior, a set of rational mutations maybe need to be made to optimize its performance. These can be introduced in three key regions: the P1 helix, the polyuridine tract of the intrinsic terminator, and the capping stem-loop of the helical element that directly competes with the P1 helix. Lastly, design variants should then be again tested for activity and the most favorable riboswitch retained. It should be noted that alterations to multiple sites in the expression platform can be made in a chimera to fine-tune the regulatory properties. In the next sections, we describe our current methodology for assessing the regulatory activity of designed riboswitches using *in vitro* and cell-based approaches.

3. ANALYSIS OF RIBOSWITCH ACTIVITY USING AN *IN VITRO* SINGLE-TURNOVER TRANSCRIPTION ASSAY

The *in vitro* single-turnover transcription assay is a robust method for rapid analysis of riboswitches that regulate transcription (Artsimovitch & Henkin, 2009; Dann et al., 2007). From this experiment, the T_{50} (concentration of ligand required to achieve a half-maximal regulatory response) and dynamic range (difference between percentage regulatory response in the presence and absence of ligand) can be determined. For transcriptional "OFF" switches, the most important property for effective function is complete repression in the presence of ligand, while for transcriptional "ON" switches, it is complete repression in the absence of ligand. Put another way, it is important to fully suppress expression in the "OFF" state, but partial activation, rather than full, can still lead to a clearly observable regulatory response. It should be noted that for riboswitches that regulate translation, the equivalent assay is a coupled transcription/translation assay, which has been used by several groups to investigate regulatory activity (Lemay et al., 2011; Mishler & Gallivan, 2014).

3.1. Template construction

Success of the *in vitro* transcription reaction is critically dependent upon the quality of the template. Generally, we use a strong T7A1 promoter (5′-GGTTATCAAAAAGAGTA**TTGACT**TAAAGTCTAACCTATAG**GATACT**TACAGCC; -35 and -10 sites are in bold) that initiates transcription using *E. coli* RNA polymerase (Artsimovitch, Svetlov, Anthony, Burgess, & Landick, 2000). The next nucleotide following this sequence is the transcription start site of the RNA. Finally, the template should extend 50–100 residues past the polyuridine tract of the transcriptional terminator to ensure sufficient separation between the RT and T products on a denaturing polyacrylamide gel. An annotated sequence of the *Smu folT/Bsu metE* chimera DNA transcription template that spans the promoter and riboswitch is given in Fig. 5A.

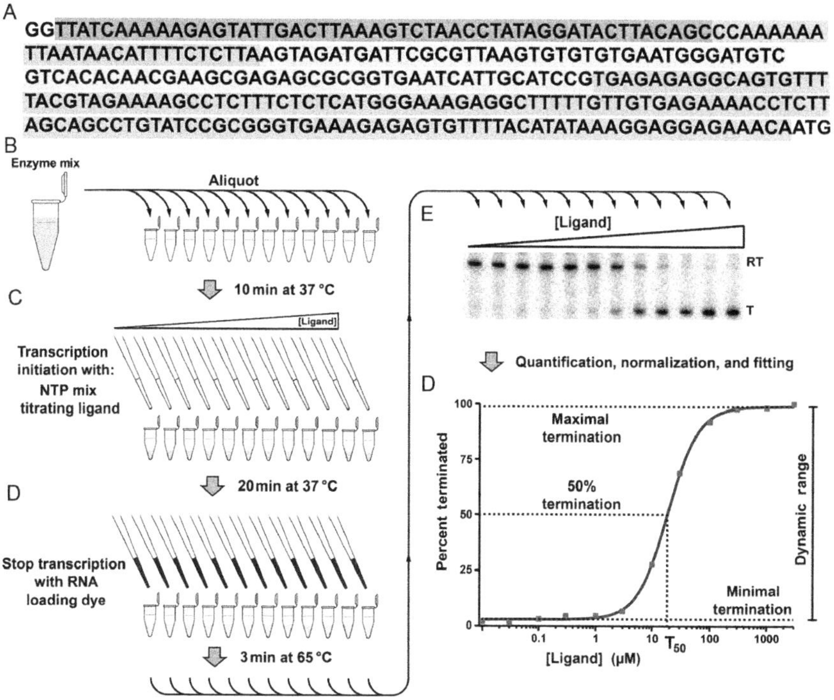

Figure 5 Workflow for single-turnover transcription assay. (A) Annotated sequence of the DNA template for the *Smu folT/Bsu metE* "design 3" riboswitch. The T7A1 promoter is highlighted in red, the *Smu folT* aptamer in orange, the *Bsu metE* expression platform in cyan, and the initiating ATG codon in yellow. (B) The enzyme mix is aliquoted into thin-walled PCR tubes and incubated in a thermocycler. (C) The reaction is initiated by the addition of NTP mix. Because this solution contains ligand, a separate NTP mix must be made for every concentration tested. (D) The reaction is terminated by the addition of loading dye and a raise in temperature to 65 °C prior to separation of the products by 6% (29:1) denaturing polyacrylamide gel electrophoresis. (E) Example of a gel using the chimeric *Smu folT/Bsu metE* riboswitch showing the read-through (RT) from the terminated (T) products as a function of ligand concentration. (F) Fit of data to a two-state-binding isotherm to calculate the maximum and minimum percent terminated as well as the dynamic range and T_{50}. (See the color plate.)

3.1.1 Overlapping extension PCR

Transcription templates of chimeric riboswitches are most easily generated using a set of overlapping synthetic DNA oligonucleotides (Integrated DNA Technologies) assembled using recursive PCR (Prodromou & Pearl, 1992). The use of overlapping oligonucleotides has the advantage of not requiring a genomic or other source, and therefore, natural or unnatural sequences can be created very easily *de novo*. The simplest means of rapidly generating a set of oligonucleotides for a DNA sequence is computationally; our laboratory most often uses the GeneDesign Web site (http://54.235.254.95/gd/), which automatically generates a set of DNA oligonucleotides of the specified length and overlap (Table 2 gives the set of oligonucleotide sequences used to create the *folT/metE* chimera). In this example, the first oligonucleotide (oligo A) specifies the first 60 nt of the sense strand. The second (oligo B) encodes the antisense strand; its sequence overlaps 20 nt with oligo A and continues for an additional 40 nt. The third (oligo C) encodes sense strand

Table 2 Oligonucleotides used to construct the transcription template for the *Smu folT/Bsu metE* (design 3) transcription template

Oligonucleotide	Sequence
5′ outer primer	GGTTATCAAAAAGAGTATTGACT
3′ outer primer	CATTGTTTCTCCTCCTTTATATG
Oligo A	GGTTATCAAAAAGAGTATTGACTTAAAGTC TAACCTATAGGATACTTACAGCCCAAAAAA
Oligo B	TTAACGCGAATCATCTACTTAAGAGAAAAT GTTATTAATTTTTTGGGCTGTAAGTATC
Oligo C	AGTAGATGATTCGCGTTAAGTGTGTGTGAA TGGGATGTCGTCACACAACGAAGCGAGAG
Oligo D	AAACACTGCCTCTCTCACGGATGCAATGAT TCACCGCGCTCTCGCTTCGTTGTGTGAC
Oligo E	GTGAGAGAGGCAGTGTTTTACGTAGAAAAG CCTCTTTCTCTCATGGGAAAGAGGCTTT
Oligo F	ACCCGCGGATACAGGCTGCTAAGAGGTTTT CTCACAACAAAAAGCCTCTTTCCCATGAGA
Oligo G	AGCAGCCTGTATCCGCGGGTGAAAGAGAGT GTTTTACATATAAAGGAGGAGAAACAATG

sequence, overlapping with oligo B by 20 nt and continuing for an additional 40 nt. This strategy is continued until the desired sequence is fully encompassed. To amplify the assembled product, we also use two shorter 5′- and 3′-primers of 20–30 nt in length (5′ outer and 3′ outer, Table 2).

Recursive PCR is similar to a standard PCR. Typically, the DNA template is generated using a 100 μL reaction (Table 3) with 1 μ*M* of the 5′ and 3′ outer primers and the 10 n*M* of each overlapping oligonucleotides (oligo A, B, C, etc.). Amplification is achieved using a cycle of 30 s at 95 °C, 30 s at 52 °C, and 1 min at 72 °C for 30 rounds. The product is assessed on a 1.5% agarose gel with the appropriate size markers as a control (Sambrook & Russel, 2001).

It is critical that a single DNA product be observed in the PCR to minimize the uninterpretable products resulting from unpredicted transcription. If the reaction yields bands not of the correct length (295 base pairs for *folT/metE*), the reaction is optimized using gradient PCR in which the annealing temperature is varied between 45 and 65 °C. Another approach that can assist in synthesizing difficult templates is to divide the PCR into two separate reactions, each amplifying only half of the full-length product. The resulting two halves can then be amplified together to create a single product in the second stage of PCR.

3.1.2 Purification of the template

It is critical to purify the DNA transcription template before use in the *in vitro* transcription reaction to ensure unincorporated primers and dNTPs do not alter initiation or termination efficiency. Once the PCR has been optimized to yield a single product, the resultant transcription template is purified using

Table 3 Recursive PCR

Stock	**Volume**
5 × Phusion buffer[a]	20 μL
10 m*M* each dNTP[a]	2 μL
5′ outer primer, 100 μ*M*	1 μL
3′ outer primer, 100 μ*M*	1 μL
Inner oligonucleotides, 1 μ*M*	1 μL each (total 7 μL)
Phusion DNA polymerase[a]	1 μL
ddH_2O	68 μL

[a]Source: New England Biolabs.

a PCR purification kit (Qiagen, #28104). The template is eluted in 50 μL of 10 m*M* Tris–Cl, pH 8.5, and should be at a working stock concentration of between 50 and 100 ng/μL (as measured by absorbance at 260 nm and calculated using one absorbance unit equals 50 μg/mL DNA) (Sambrook & Russel, 2001).

3.2. Single-turnover *in vitro* transcription assay

1. *Preparation of the transcription solution.* Prepare a 17.5 μL reaction containing 50–100 ng of DNA template, 2× transcription buffer (140 m*M* Tris–Cl, pH 8.0, 140 m*M* NaCl, 0.2 m*M* EDTA, 28 m*M* β-mercaptoethanol, and 70 μg/mL BSA), $MgCl_2$, α-[^{32}P]-ATP, and *E. coli* RNA polymerase σ^{70} holoenzyme (Epicentre, #S90050) (Table 4). This mixture is prepared for each individual transcription reaction. A minimum of two reactions should be assembled (plus and minus ligand), but multiple reactions can be set up for a ligand titration. For this type of experiment, the transcription solution can be scaled up and subsequently aliquoted into individual PCR tubes. The final magnesium chloride concentration is variable, depending upon the riboswitch and must be optimized, but RNA polymerase requires at least 2.5 m*M* (at least 1.5 m*M* in excess of the total NTP concentration). It is important to realize that altering the magnesium concentration will potentially affect the rates of RNA polymerization and folding of the resulting RNA structure.

Table 4 Formulation of the enzyme mix

Stock	Volume (1 reaction) (μL)[a]	Final concentration
2× transcription buffer	12.5	~1.4×
10× $MgCl_2$[b]	2.5	~1.4×
50–100 ng/μL DNA template	1	50–100 ng/reaction
25 μCi α-[32P]-ATP	0.5	~0.7 μCi
1 U/μL *E. coli* RNA polymerase	0.25	0.25 U/reaction
H_2O	0.75	
Total[c]	17.5	

[a]The volumes should be scaled as needed, usually making extra to account for imperfect pipetting.
[b]The concentration of $MgCl_2$ can be varied based on several considerations reviewed in the text.
[c]The final volume of the enzyme mix is 17.5 μL, but note that the final volume for the reaction once the NTP mix is added is 25 μL.

2. *Prereaction incubation.* The transcription solution is incubated at 37 °C for 10 min to allow RNA polymerase to bind to the promoter.
3. *Transcription initiation.* Transcription is initiated by adding a mixture of NTP, heparin, and ligand (Table 5). The resultant reactions are incubated at 37 °C for 20 min. The concentration of NTPs can be varied to affect the rate of polymerization; note that higher levels of NTPs will lower the incorporation of labeled ATP in each RNA strand and reduce observed signal. In an alternative method, transcription can be initiated with a dinucleotide (Grundy, Winkler, & Henkin, 2002; Landick, Wang, & Chan, 1996). Adding only three of the four nucleotides results in the formation of a stable stalled complex (~10–20 nt). The transition from initiation to stable elongation is then triggered by the addition of the fourth nucleotide.
4. *Stopping the reaction.* Transcription is quenched with the addition of 25 μL RNA loading dye (95% (v/v) deionized formamide, 10 m*M* EDTA, pH 8.0, 0.25% (w/v) bromophenol blue, and 0.25% (w/v) xylene cyanol) followed by immediately raising the temperature to 65 °C for 3 min. At this stage, the reactions can be stored at −20 °C or directly loaded onto a denaturing polyacrylamide gel.
5. *Separating T and RT products.* A 0.8-mm thick, 6% 29:1 acrylamide: bisacrylamide denaturing (8 *M* urea) gel is used to separate RT and T products. Following loading 6 μL of each sample into the wells, the gel is run at 12 W for about 1.5 h in 1× TBE buffer (0.1 *M* Tris, 0.045 m*M* boric acid, and 1 m*M* EDTA).
6. *Drying the gel.* The gel is transferred to filter paper and dried for 1 h at 80 °C under negative pressure.
7. *Exposing, imaging, and quantification.* The dried gel is exposed overnight to a phosphor screen and imaged using a Typhoon 9400 (or equivalent phosphorimager). Individual bands can be quantified using ImageQuant

Table 5 Formulation of NTP mix

Stock	**Volume (μL)**	**Concentration in NTP mix**
2.5 m*M* each NTP	10	250 μ*M*
2 mg/mL heparin	10	0.2 mg/mL
5× ligand[a]	66.6	3.33×
H_2O	13.4[b]	

[a]The ligand concentration should be varied when doing a titration.
[b]This volume is in large excess to improve pipetting accuracy (only 7.5 μL are needed per reaction).

5.2 (Molecular Dynamics) or ImageJ (National Institutes of Health; http://imagej.nih.gov/ij/). An example of a typical gel is shown in Fig. 5E.

8. *Data processing.* To quantify the fractional amount (Θ) of RNA in the RT and T bands, first the intensity of each band is background corrected. Second, it is important to account for differential amount of labeling of each band by normalizing intensity values to the number of adenines within the sequence. Finally, the percent T or RT transcription (depending on whether it is an "OFF" or "ON" switch) is calculated by dividing either T or RT by the total RNA in both bands. Error values and error bars are reported as the standard deviation of three independent experiments.

To calculate T_{50}, the average percent T or RT for each concentration of ligand is fit to a two-state model using the equation

$$\theta = a + b\left(\frac{L^n}{L^n + T_{50}^n}\right) \tag{1}$$

where L is the ligand concentration, n is the Hill coefficient, a is the minimum percent T or RT, and b is the dynamic range (such that $a+b$ is the maximum percent T or RT). Fitting can be performed with KaleidaGraph (Synergy Software), IGOR Pro (WaveMetrics), or equivalent program that performs nonlinear least squares analysis. Error in T_{50} is reported as the error in the fitting of the average values at each ligand concentration. Data fit and calculated values for T_{50}, T_{min} (percentage termination in the absence of ligand), and T_{max} (percentage termination at high ligand concentration) are shown in Fig. 5F.

4. CELL-BASED GFP REPORTER ASSAY

While the above assay is useful for rapidly assaying the regulatory activity of designed chimeric riboswitches, most practical applications will require function in the cellular environment. Switches that perform well *in vitro* have been observed to be highly effective in regulating gene expression in a cellular context when cloned upstream of a reporter, but this is not universally true. Therefore, candidate switches that function *in vitro* still need to be screened and potentially optimized for performance in the bacterium of choice. The following describes the approach that we use to assay the function of chimeric riboswitches in *E. coli*.

4.1. Reporter design

Riboswitch activity can be monitored in *E. coli* using a variety of different reporter genes, including widely adopted fluorescent proteins and reporter enzymes. We have developed a set of low-copy vectors derived from pBR327 that contain a reporter gene (encoding GFPuv, mCherry, or β-galactosidase) downstream of a multiple cloning site for insertion of a desired riboswitch (Fig. 6). The riboswitch is cloned between unique *Nsi*I and *Hin*dIII sites (note that the -10 site of the promoter and the first few amino acids of the reporter protein are within this region and need to be part of the insert). The switch should be positioned immediately downstream from a promoter. A critical consideration is the strength of the promoter, which needs to be optimized for each specific riboswitch. While we often use a strong T7A1 or *tac* promoter *in vivo*, in some cases we observe toxicity effects of overexpressing the reporter. For this reason, a weaker promoter is also tested (-35,5′-TTTACG; -10,5′-TATAAT) (Davis, Rubin, & Sauer, 2011).

For cell-based analysis, the preferred strain of *E. coli* is BW25113, the parental strain of the Keio knockout collection (Baba et al., 2006). In our experience, this strain grows robustly and rapidly on chemically defined media and generates consistent results. Importantly, one of the few genetic modifications is a deletion of *lacZ*, enabling use of the Miller assay. Furthermore, a large set of single-gene knockouts is commercially available within this strain (GE Dharmacon, #OEC4987) that can assist with riboswitch analysis. For example, a strain containing a deletion of the MetJ repressor protein, which regulates sulfur metabolism in *E. coli*, was used to elevate intracellular SAM concentrations to facilitate experiments investigating the regulatory behavior of the SAM-I/IV variant riboswitch (Trausch et al., 2014). Another consideration at this stage is whether the concentration of small molecule of interest can be readily varied in the cell to enable comparison of riboswitch activity in its absence and presence. This is not practical for some natural riboswitches, as the effector molecule is essential for cell growth (see Section 4.3).

4.2. Protocol

1. *Overnight growth.* Single colonies are grown to saturation in a rotating drum overnight at 37 °C in 5 mL of a rich defined growth medium: 37 m*M* sodium phosphate, 66 m*M* dipotassium phosphate, 15 m*M* ammonium sulfate, 0.6% glucose (w/v), 0.1% casamino acids (w/v),

A

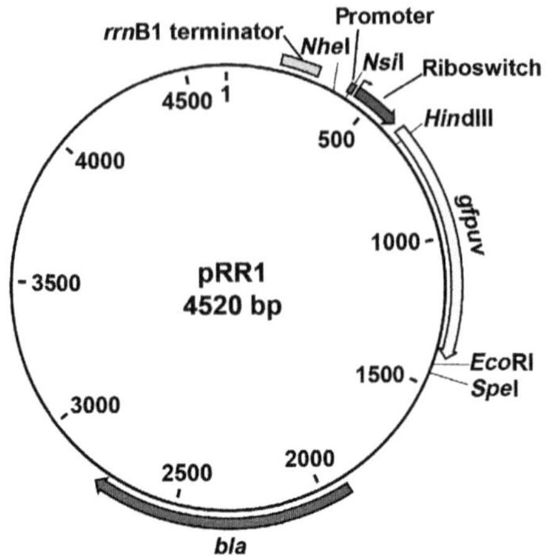

B

GCTAGCCACAGCTAACACCACGTCGTCCCTATCTGCTGC
NheI

CCTAGGTCTATGAGTGGTTGCTGGATAACTTGACAGGCA
-35

*
TGCATAAGGCTCGTATAATATATTCAGGGAGACCACAAC
NsiI -10 TxS

GGTTTCCCAAGAAGGAGATATACC ATG ACC ATG ATT ACG
RBS Met

CCA AGC TTG CAT GCC TGC AGG TCG ACT CTA GAG
HindIII SphI

GAT CCC CGG GTA CCG GTA GAA AAA ATG AGT
KpnI

Figure 6 Plasmid vector for cell-based riboswitch activity assays. (A) Map of the parental vector used for cell-based riboswitch activity assays in *E. coli*. Unique restriction sites used for modular cloning of different promoters, riboswitches, and reporter genes are denoted. (B) Sequence of the promoter/riboswitch region of the pRR1 vector highlighting the promoter (yellow (light gray in the print version)), unique restriction sites (underlined), transcription (asterisk), and translation start sites.

2 mg/mL disodium citrate, 0.1 mg/mL magnesium sulfate, 0.01 mg/mL thiamine, 30 μg/mL iron(III) chloride $6H_2O$, 6 μg/mL zinc(II) chloride $4H_2O$, 6 μg/mL cobalt(II) chloride $6H_2O$, 6 μg/mL sodium molybdate $2H_2O$, 3 μg/mL calcium chloride $2H_2O$, 3 μg/mL copper(II) chloride, 3 μg/mL manganese(II) chloride, and 1.5 μg/mL boric acid. This medium is supplemented with 100 μg/mL ampicillin to ensure positive selection for plasmid.

2. *Inoculation and growth.* Five microliters of saturated overnight culture is used to inoculate 5 mL of fresh rich defined media containing 100 μg/mL ampicillin. For a titration or plus/minus assay, the media can be supplemented with a small-molecule metabolite. Cells should grow to mid-log in about 6 h as monitored by the optical density at 600 nm (OD_{600}).

3a. *Quantification (fluorescence protein reporter).* During mid-log phase, 300 μL of media are added to a Greiner 96-well polystyrene plate (Cat. #655079), and OD_{600} is measured using a Tecan Infinite M200 Pro plate reader. When monitoring GFPuv, fluorescence is measured using the same plate and plate reader exciting at 395 nm and collecting emissions at 510 nm.

3b. *Quantification (β-galactosidase reporter).* When monitoring β-galactosidase production (Miller, 1992), cells are placed on ice for 60 min and the OD_{600} is measured as in Step 3a. In a 2-mL tube, 20 μL of culture is mixed with 80 μL of permeabilization solution (200 m*M* Na_2HPO_4, 20 m*M* KCl, 2 m*M* $MgSO_4$, 0.8 mg/mL hexadecyltrimethylammonium bromide (CTAB), 0.4 mg/mL sodium deoxycholate, and 5.4 μL/mL β-mercaptoethanol). With all solutions equilibrated to 30 °C, 600 μL of substrate solution (60 m*M* Na_2HPO_4, 20 m*M* NaH_2PO_4, 10 m*M* KCl, 20 μg/mL CTAB, 10 μg/mL sodium deoxycholate, 2.7 μL/mL β-mercaptoethanol, and 1 mg/mL *ortho*-nitrophenyl-β-galactoside) is added and mixed thoroughly. The reaction is allowed to commence for 20–60 min and is stopped by the addition of 700 μL stop solution (1 *M* Na_2CO_3). Cell debris is spun down and 300 μL of the solution is transferred to the plate described above. The absorbance at 420 nm (ABS_{420}) is measured and should be less than 1.

4a. *Data processing (fluorescence protein reporter).* For GFPuv, fluorescence data are normalized to OD_{600}. Empty pBR327-normalized fluorescence is used to subtract background. Data are fit to a two-state-binding model to determine the EC_{50} and maximal fold induction. If possible, a binding knockout riboswitch should be used as a negative control. This construct should show no significant changes in fluorescence when the concentration of ligand is altered.

4b. *Data processing (β-galactosidase reporter).* For β-galactosidase, the ABS_{420} and OD_{600} from Step 3b are used to calculate the Miller units using the equation.

$$1 \text{ Miller unit} = \frac{Abs_{420}}{OD_{600} \times \text{min} \times \text{mL}} \times 1000 \quad (2)$$

where min is incubation time with the substrate solution in minutes and mL is the volume of cells added to the permeabilization solution in milliliters. These data can be processed similar to normalized fluorescence with background subtraction and fit to the appropriate binding equation.

4.3. Other considerations using *in vivo* reporters

In some instances, titration of a small molecule may not be practical. This issue could arise with a ligand that cannot cross the membrane or a ligand that is so essential for cell viability that it cannot be significantly varied. In these cases, we introduce a single-point mutation into the ligand-binding site of the aptamer domain that does not perturb the RNA structure (for example, U74A for the adenine riboswitch). By comparing the activity of the wild-type riboswitch to that of the binding knockout, the fold repression or induction can be inferred. It is important to note that the degree that the wild-type switch is turned "on" or "off" is unknown. It could be that under saturating ligand concentrations the fold change would be more dramatic. In the case of the *folT*/*metE* chimeric riboswitch, this approach was used to test the importance of each individual binding site for its ability to repress gene expression *in vivo*. While ligand binding at the three-way junction site (U7C) is unable to repress expression of the downstream LacZ reporter, binding at the pseudoknot site (U25C) significantly terminates transcription (Fig. 7). Knocking out binding at each individual site established that THF binding at the pseudoknot-binding site is critical for regulation while binding to the three-way junction site is not essential for regulatory activity. This result mirrors observations obtained using the single-turnover *in vitro* transcription assay.

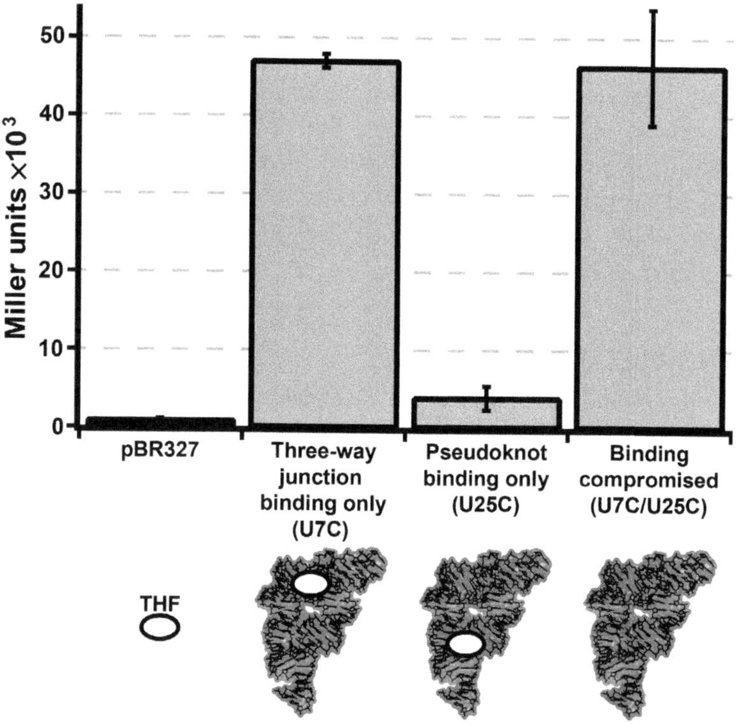

Figure 7 Assessment of regulatory activity of the *folT/metE* chimeric riboswitch in the cellular context. Bar graph of β-galactosidase activity in Miller units produced from cells expressing the riboswitch cloned upstream of LacZ. The switch containing only a THF-binding site at the three-way junction (U7C) fails to repress transcription and displays similar reporter activity as a complete knockout (U7C/U25C). Alternatively, a switch containing a single THF-binding site near the pseudoknot (U25C) is capable of repressing expression of LacZ. Note that this repression is due to endogenous concentrations of THF in the cell.

5. CONCLUDING REMARKS

We have shown that regulatory domains of natural riboswitches that act at the level of transcription can be used as modular elements for the creation of novel genetically encodable RNA elements. Importantly, these modules can be fused with aptamers generated from *in vitro* selection, opening the possibility of the facile creation of diverse array of chimeric riboswitches for applications in synthetic biology. The expression platforms that we have explored represent only a very small subset of those that have the potential for being reengineered. For example, it will be useful to develop a set of "ON" and "OFF" switches that control gene expression at the translational level for bacteria whose RNA polymerase may not be strongly regulated by a consensus standard intrinsic terminator. For the development of regulatory elements functional in Eukarya, it might be possible to reengineer TPP riboswitches that regulate alternative splicing or 3′-end processing in fungi and plants (Cheah, Wachter, Sudarsan, & Breaker, 2007; Wachter et al., 2007), which would greatly broaden the applicability of these RNA devices.

ACKNOWLEDGMENT

This work was supported by a grant from the National Science Foundation to R. T. B.

REFERENCES

Artsimovitch, I., & Henkin, T. M. (2009). In vitro approaches to analysis of transcription termination. *Methods, 47*, 37–43.

Artsimovitch, I., Svetlov, V., Anthony, L., Burgess, R. R., & Landick, R. (2000). RNA polymerases from Bacillus subtilis and Escherichia coli differ in recognition of regulatory signals in vitro. *Journal of Bacteriology, 182*, 6027–6035.

Baba, T., Ara, T., Hasegawa, M., Takai, Y., Okumura, Y., Baba, M., et al. (2006). Construction of Escherichia coli K-12 in-frame, single-gene knockout mutants: The Keio collection. *Molecular Systems Biology, 2*, 2006.0008.

Barrick, J. E., & Breaker, R. R. (2007). The distributions, mechanisms, and structures of metabolite-binding riboswitches. *Genome Biology, 8*, R239.

Batey, R. T., Gilbert, S. D., & Montange, R. K. (2004). Structure of a natural guanine-responsive riboswitch complexed with the metabolite hypoxanthine. *Nature, 432*, 411–415.

Blouin, S., Chinnappan, R., & Lafontaine, D. A. (2011). Folding of the lysine riboswitch: Importance of peripheral elements for transcriptional regulation. *Nucleic Acids Research, 39*, 3373–3387.

Breaker, R. R. (2012). Riboswitches and the RNA world. *Cold Spring Harbor Perspectives in Biology, 4*, a003566.

Caldelari, I., Chao, Y., Romby, P., & Vogel, J. (2013). RNA-mediated regulation in pathogenic bacteria. *Cold Spring Harbor Perspectives in Medicine, 3*, a010298.

Carothers, J. M., Goler, J. A., Juminaga, D., & Keasling, J. D. (2011). Model-driven engineering of RNA devices to quantitatively program gene expression. *Science, 334*, 1716–1719.

Ceres, P., Garst, A. D., Marcano-Velazquez, J. G., & Batey, R. T. (2013). Modularity of select riboswitch expression platforms enables facile engineering of novel genetic regulatory devices. *ACS Synthetic Biology, 2*, 463–472.

Ceres, P., Trausch, J. J., & Batey, R. T. (2013). Engineering modular 'ON' RNA switches using biological components. *Nucleic Acids Research, 41*, 10449–10461.

Chappell, J., Takahashi, M. K., Meyer, S., Loughrey, D., Watters, K. E., & Lucks, J. (2013). The centrality of RNA for engineering gene expression. *Biotechnology Journal, 8*, 1379–1395.

Cheah, M. T., Wachter, A., Sudarsan, N., & Breaker, R. R. (2007). Control of alternative RNA splicing and gene expression by eukaryotic riboswitches. *Nature, 447*, 497–500.

Clark, M. B., Choudhary, A., Smith, M. A., Taft, R. J., & Mattick, J. S. (2013). The dark matter rises: The expanding world of regulatory RNAs. *Essays in Biochemistry*, *54*, 1–16.

Cochrane, J. C., Lipchock, S. V., & Strobel, S. A. (2007). Structural investigation of the GlmS ribozyme bound to Its catalytic cofactor. *Chemistry & Biology*, *14*, 97–105.

Dann, C. E., 3rd., Wakeman, C. A., Sieling, C. L., Baker, S. C., Irnov, I., & Winkler, W. C. (2007). Structure and mechanism of a metal-sensing regulatory RNA. *Cell*, *130*, 878–892.

Davis, J. H., Rubin, A. J., & Sauer, R. T. (2011). Design, construction and characterization of a set of insulated bacterial promoters. *Nucleic Acids Research*, *39*, 1131–1141.

Delebecque, C. J., Lindner, A. B., Silver, P. A., & Aldaye, F. A. (2011). Organization of intracellular reactions with rationally designed RNA assemblies. *Science*, *333*, 470–474.

de Silva, C., & Walter, N. G. (2009). Leakage and slow allostery limit performance of single drug-sensing aptazyme molecules based on the hammerhead ribozyme. *RNA*, *15*, 76–84.

Edwards, T. E., & Ferre-D'Amare, A. R. (2006). Crystal structures of the thi-box riboswitch bound to thiamine pyrophosphate analogs reveal adaptive RNA-small molecule recognition. *Structure*, *14*, 1459–1468.

Epshtein, V., Mironov, A. S., & Nudler, E. (2003). The riboswitch-mediated control of sulfur metabolism in bacteria. *Proceedings of the National Academy of Sciences of the United States of America*, *100*, 5052–5056.

Fowler, C. C., Brown, E. D., & Li, Y. (2010). Using a riboswitch sensor to examine coenzyme B(12) metabolism and transport in E. coli. *Chemistry & Biology*, *17*, 756–765.

Frieda, K. L., & Block, S. M. (2012). Direct observation of cotranscriptional folding in an adenine riboswitch. *Science*, *338*, 397–400.

Garst, A. D., Edwards, A. L., & Batey, R. T. (2011). Riboswitches: Structures and mechanisms. *Cold Spring Harbor Perspectives in Biology*, *3*, a003533.

Giedroc, D. P., & Cornish, P. V. (2009). Frameshifting RNA pseudoknots: Structure and mechanism. *Virus Research*, *139*, 193–208.

Gilbert, S. D., Rambo, R. P., Van Tyne, D., & Batey, R. T. (2008). Structure of the SAM-II riboswitch bound to S-adenosylmethionine. *Nature Structural & Molecular Biology*, *15*, 177–182.

Griffiths-Jones, S., Bateman, A., Marshall, M., Khanna, A., & Eddy, S. R. (2003). Rfam: An RNA family database. *Nucleic Acids Research*, *31*, 439–441.

Grundy, F. J., Winkler, W. C., & Henkin, T. M. (2002). tRNA-mediated transcription antitermination in vitro: Codon-anticodon pairing independent of the ribosome. *Proceedings of the National Academy of Sciences of the United States of America*, *99*, 11121–11126.

Johnson, J. E., Jr., Reyes, F. E., Polaski, J. T., & Batey, R. T. (2012). B12 cofactors directly stabilize an mRNA regulatory switch. *Nature*, *492*, 133–137.

Kertsburg, A., & Soukup, G. A. (2002). A versatile communication module for controlling RNA folding and catalysis. *Nucleic Acids Research*, *30*, 4599–4606.

Klein, D. J., & Ferre-D'Amare, A. R. (2006). Structural basis of glmS ribozyme activation by glucosamine-6-phosphate. *Science*, *313*, 1752–1756.

Koizumi, M., Soukup, G. A., Kerr, J. N., & Breaker, R. R. (1999). Allosteric selection of ribozymes that respond to the second messengers cGMP and cAMP. *Nature Structural Biology*, *6*, 1062–1071.

Kulshina, N., Baird, N. J., & Ferre-D'Amare, A. R. (2009). Recognition of the bacterial second messenger cyclic diguanylate by its cognate riboswitch. *Nature Structural & Molecular Biology*, *16*, 1212–1217.

Landick, R., Wang, D., & Chan, C. L. (1996). Quantitative analysis of transcriptional pausing by Escherichia coli RNA polymerase: His leader pause site as paradigm. *RNA Polymerase and Associated Factors, Part B*, *274*, 334–353.

Lemay, J. F., Desnoyers, G., Blouin, S., Heppell, B., Bastet, L., St-Pierre, P., et al. (2011). Comparative study between transcriptionally- and translationally-acting adenine riboswitches reveals key differences in riboswitch regulatory mechanisms. *PLoS Genetics*, *7*, e1001278.

Lu, C., Smith, A. M., Fuchs, R. T., Ding, F., Rajashankar, K., Henkin, T. M., et al. (2008). Crystal structures of the SAM-III/S(MK) riboswitch reveal the SAM-dependent translation inhibition mechanism. *Nature Structural & Molecular Biology*, *15*, 1076–1083.

Mahen, E. M., Watson, P. Y., Cottrell, J. W., & Fedor, M. J. (2010). mRNA secondary structures fold sequentially but exchange rapidly in vivo. *PLoS Biology*, *8*, e1000307.

Mandal, M., Boese, B., Barrick, J. E., Winkler, W. C., & Breaker, R. R. (2003). Riboswitches control fundamental biochemical pathways in Bacillus subtilis and other bacteria. *Cell*, *113*, 577–586.

Mandal, M., & Breaker, R. R. (2004). Adenine riboswitches and gene activation by disruption of a transcription terminator. *Nature Structural & Molecular Biology*, *11*, 29–35.

McDaniel, B. A., Grundy, F. J., Artsimovitch, I., & Henkin, T. M. (2003). Transcription termination control of the S box system: Direct measurement of S-adenosylmethionine by the leader RNA. *Proceedings of the National Academy of Sciences of the United States of America*, *100*, 3083–3088.

Michener, J. K., & Smolke, C. D. (2012). High-throughput enzyme evolution in Saccharomyces cerevisiae using a synthetic RNA switch. *Metabolic Engineering*, *14*, 306–316.

Miller, J. H. (1992). *A short course in bacterial genetics: A laboratory manual and handbook for Escherichia coli and related bacteria*. Plainview, NY: Cold Spring Harbor Laboratory Press.

Mishler, D. M., & Gallivan, J. P. (2014). A family of synthetic riboswitches adopts a kinetic trapping mechanism. *Nucleic Acids Research*, *42*, 6753–6761.

Nelson, J. W., Sudarsan, N., Furukawa, K., Weinberg, Z., Wang, J. X., & Breaker, R. R. (2013). Riboswitches in eubacteria sense the second messenger c-di-AMP. *Nature Chemical Biology*, *9*, 834–839.

Nix, J., Sussman, D., & Wilson, C. (2000). The 1.3 A crystal structure of a biotin-binding pseudoknot and the basis for RNA molecular recognition. *Journal of Molecular Biology*, *296*, 1235–1244.

Nomura, Y., & Yokobayashi, Y. (2007). Reengineering a natural riboswitch by dual genetic selection. *Journal of the American Chemical Society*, *129*, 13814–13815.

Paige, J. S., Nguyen-Duc, T., Song, W., & Jaffrey, S. R. (2012). Fluorescence imaging of cellular metabolites with RNA. *Science*, *335*, 1194.

Paige, J. S., Wu, K. Y., & Jaffrey, S. R. (2011). RNA mimics of green fluorescent protein. *Science*, *333*, 642–646.

Peters, J. M., Vangeloff, A. D., & Landick, R. (2011). Bacterial transcription terminators: The RNA 3'-end chronicles. *Journal of Molecular Biology*, *412*, 793–813.

Prodromou, C., & Pearl, L. H. (1992). Recursive PCR: A novel technique for total gene synthesis. *Protein Engineering*, *5*, 827–829.

Robinson, C. J., Vincent, H. A., Wu, M. C., Lowe, P. T., Dunstan, M. S., Leys, D., et al. (2014). Modular riboswitch toolsets for synthetic genetic control in diverse bacterial species. *Journal of the American Chemical Society*, *136*(30), 10615–10624.

Sambrook, J., & Russel, D. W. (2001). In *Molecular cloning: A laboratory manual*. (3rd ed.). Cold Spring Harbor, NY: Cold Spring Harbor Laboratory Press.

Serganov, A., Huang, L., & Patel, D. J. (2009). Coenzyme recognition and gene regulation by a flavin mononucleotide riboswitch. *Nature*, *458*, 233–237.

Serganov, A., Polonskaia, A., Phan, A. T., Breaker, R. R., & Patel, D. J. (2006). Structural basis for gene regulation by a thiamine pyrophosphate-sensing riboswitch. *Nature*, *441*, 1167–1171.

Silverman, S. K. (2003). Rube Goldberg goes (ribo)nuclear? Molecular switches and sensors made from RNA. *RNA*, *9*, 377–383.

Smith, K. D., Lipchock, S. V., Ames, T. D., Wang, J., Breaker, R. R., & Strobel, S. A. (2009). Structural basis of ligand binding by a c-di-GMP riboswitch. *Nature Structural & Molecular Biology, 16*, 1218–1223.

Soukup, G. A., & Breaker, R. R. (1999). Engineering precision RNA molecular switches. *Proceedings of the National Academy of Sciences of the United States of America, 96*, 3584–3589.

Stoddard, C. D., Gilbert, S. D., & Batey, R. T. (2008). Ligand-dependent folding of the three-way junction in the purine riboswitch. *RNA-A Publication of the RNA Society, 14*, 675–684.

Stojanovic, M. N., & Kolpashchikov, D. M. (2004). Modular aptameric sensors. *Journal of the American Chemical Society, 126*, 9266–9270.

Stoltenburg, R., Reinemann, C., & Strehlitz, B. (2007). SELEX-A (r)evolutionary method to generate high-affinity nucleic acid ligands. *Biomolecular Engineering, 24*, 381–403.

Storz, G., Vogel, J., & Wassarman, K. M. (2011). Regulation by small RNAs in bacteria: Expanding frontiers. *Molecular Cell, 43*, 880–891.

Sussman, D., Nix, J. C., & Wilson, C. (2000). The structural basis for molecular recognition by the vitamin B 12 RNA aptamer. *Nature Structural Biology, 7*, 53–57.

Tang, J., & Breaker, R. R. (1997). Rational design of allosteric ribozymes. *Chemistry & Biology, 4*, 453–459.

Trausch, J. J., Ceres, P., Reyes, F. E., & Batey, R. T. (2011). The structure of a tetrahydrofolate-sensing riboswitch reveals two ligand binding sites in a single aptamer. *Structure, 19*, 1413–1423.

Trausch, J. J., Xu, Z., Edwards, A. L., Reyes, F. E., Ross, P. E., Knight, R., et al. (2014). Structural basis for diversity in the SAM clan of riboswitches. *Proceedings of the National Academy of Sciences of the United States of America, 111*, 6624–6629.

Viladoms, J., & Fedor, M. J. (2012). The glmS ribozyme cofactor is a general acid-base catalyst. *Journal of the American Chemical Society, 134*, 19043–19049.

Wachter, A., Tunc-Ozdemir, M., Grove, B. C., Green, P. J., Shintani, D. K., & Breaker, R. R. (2007). Riboswitch control of gene expression in plants by splicing and alternative 3' end processing of mRNAs. *The Plant Cell, 19*, 3437–3450.

Wieland, M., Benz, A., Klauser, B., & Hartig, J. S. (2009). Artificial ribozyme switches containing natural riboswitch aptamer domains. *Angewandte Chemie, 48*, 2715–2718.

Win, M. N., & Smolke, C. D. (2007). A modular and extensible RNA-based gene-regulatory platform for engineering cellular function. *Proceedings of the National Academy of Sciences of the United States of America, 104*, 14283–14288.

Winkler, W. C., Nahvi, A., Roth, A., Collins, J. A., & Breaker, R. R. (2004). Control of gene expression by a natural metabolite-responsive ribozyme. *Nature, 428*, 281–286.

Winkler, W. C., Nahvi, A., Sudarsan, N., Barrick, J. E., & Breaker, R. R. (2003). An mRNA structure that controls gene expression by binding S-adenosylmethionine. *Nature Structural Biology, 10*, 701–707.

Xayaphoummine, A., Viasnoff, V., Harlepp, S., & Isambert, H. (2007). Encoding folding paths of RNA switches. *Nucleic Acids Research, 35*, 614–622.

Xiao, H., Edwards, T. E., & Ferre-D'Amare, A. R. (2008). Structural basis for specific, high-affinity tetracycline binding by an in vitro evolved aptamer and artificial riboswitch. *Chemistry & Biology, 15*, 1125–1137.

Yang, J., Seo, S. W., Jang, S., Shin, S. I., Lim, C. H., Roh, T. Y., et al. (2013). Synthetic RNA devices to expedite the evolution of metabolite-producing microbes. *Nature Communications, 4*, 1413.

Zimmermann, G. R., Jenison, R. D., Wick, C. L., Simorre, J. P., & Pardi, A. (1997). Interlocking structural motifs mediate molecular discrimination by a theophylline-binding RNA. *Nature Structural Biology, 4*, 644–649.

CHAPTER FOUR

Integrating and Amplifying Signal from Riboswitch Biosensors

Michael S. Goodson[1], Svetlana V. Harbaugh, Yaroslav G. Chushak, Nancy Kelley-Loughnane

711th Human Performance Wing, Air Force Research Laboratory, Wright-Patterson Air Force Base, Dayton, Ohio, USA

[1]Corresponding author: e-mail address: michael.goodson.1.ctr.gb@us.af.mil

Contents

Abstract

Biosensors offer a built-in energy supply and inherent sensing machinery that when exploited correctly may surpass traditional sensors. However, biosensor systems have been hindered by a narrow range of ligand detection capabilities, a relatively low signal output, and their inability to integrate multiple signals. Integration of signals could increase the specificity of the sensor and enable detection of a combination of ligands that may indicate environmental or developmental processes when detected together. Amplifying biosensor signal output will increase detector sensitivity and detection range. Riboswitches offer the potential to widen the diversity of ligands that may be detected, and advances in synthetic biology are illuminating myriad possibilities in signal processing using an orthogonal parts-based engineering approach. In this chapter, we describe the design, building, and testing of a riboswitch-based Boolean logic AND

Methods in Enzymology, Volume 550
ISSN 0076-6879
http://dx.doi.org/10.1016/bs.mie.2014.10.032

gate in bacteria, where an output requires the activation of two riboswitches, and the biological circuitry required to amplify the output of the AND gate using natural extracellular bacterial communication signals to "wire" cells together.

1. INTRODUCTION

Nature has developed myriad solutions to detecting and responding to external environmental ligands. By exploiting a cell's natural sensing and signal processing capabilities, modular, sensitive, and selective engineered sensors can be produced. The discovery and development of riboswitches has opened the door to their usage in ligand-dependent reportable devices, or biosensors. The ideal biosensor would be self-sustaining and would generate a signal that could be detected from afar without the need for energy-hungry electronics, i.e., a passive sensor. Indeed, the major advantage of biosensors is that they can overcome the limitations of traditional sensors since they supply their own energy and utilize sensing machinery already present in nature (Bhatia, 2006). The advantage of using riboswitches as sensors is that the output of the sensor can also be modular, i.e., it can be the expression of any gene. This modularity provides a wealth of options regarding the functionalization of the sensor and its integration into a wider sensing array. The simplest detectable biological sensor output is the expression of a fluorescent protein. This is advantageous since the fluorescent signal is additive, meaning that signal intensity is dependent upon the amount of protein produced; it generally requires illumination by a specific wavelength to emit a signal, meaning that knowledge of the correct excitation wavelength can be controlled; and the signal is often outside the visible spectrum, making it covert. However, a limitation of biological sensors that produce a fluorescent output is that the amount of fluorescence produced is usually sufficient only for detection by sensitive lab-based equipment. In addition, unlike electronic-based sensors, the output of most of the currently available biological sensors is generally not conducive to signal integration. Thus, improved signal integration and signal amplification in biological sensors is required for the development of the next generation of fieldable, sensor platforms.

1.1. Biological circuits

Synthetic biology has taken the lead from electrical engineering to construct circuits and logic gates that have been used in pharmaceutical and biotechnological applications (Anderson, Voigt, & Arkin, 2007; Ellis, Wang, & Collins, 2009; Guet, Elowitz, Hsing, & Leibler, 2002; Mayo, Setty, Shavit, Zaslaver, & Alon, 2006; Rinaudo et al., 2007; Tamsir, Tabor, & Voigt, 2011; Weber et al., 2008). Indeed, multiple logic gates have been combined to perform more complex "programs" within cells (Friedland et al., 2009; Lou et al., 2010; Moon, Lou, Tamsir, Stanton, & Voigt, 2012; Tabor et al., 2009), and even between cells using bacterial cell–cell communication pathways as "wires" (Tamsir et al., 2011). These recent advances have opened the way to incorporate biological circuitry into biosensors to address integration of information and signal amplification. To attain biological sensor integration and signal amplification, we consulted biology. In nature, multiple inputs are integrated into one signal via a signaling cascade. For example, the phosphorylation cascade network incorporating many inputs to activate a common transcriptional activator such as in the NF-kappaB pathway (Kawai & Akira, 2007). In addition, signals are amplified by a positive feedback mechanism whereby activation of an output signal causes more signal to be produced (Afroz & Beisel, 2013). This often leads to run-away production of the output which is generally detrimental to natural systems (e.g., cancerous growth), but is beneficial to signal amplification. Using inference from natural systems, we will describe methods to utilize the modularity of our sensor output and synthetic biology to integrate the output of riboswitches and amplify their output signal. In our method, each compartment of the biological "program" is a group of bacterial cells that have the same genetic identity. This has the benefit of producing a more reliable computation by population averaging the response. We will use plasmids to act as the computational units of each compartment (Tamsir et al., 2011). The plasmids are constructed so that they contain all of the required information and are compatible in the same cell. However, between cells, components can be reused because plasmid function is "insulated" within the cell. This allows for reuse of transcription factors, for example, in multiple cells such that a program to be built from a smaller number of orthogonal parts (Tamsir et al., 2011).

There are a number of considerations when designing the circuit and the plasmids that will convey the information. These include not only the genes to be activated, but also genetic controllers such as strength of promoters and riboswitch binding sites, degradation of the product, how tightly a riboswitch is regulated, and the level of output required. Many of these genetic circuit components are gathered in publicly accessible repositories, such as the "Registry of Standard Biological Parts" (http://parts.igem.org).

Computational modeling is achieving huge strides in being able to predict which genetic components will form a reliable circuit, and many of the "industrial" scale synthetic biology laboratories successfully utilize programs such as Clotho (Xia et al., 2011) to design complex circuitry. However, for many researchers that want to apply synthetic biology principles to enhance their research, there is currently no better way than to follow the engineering mantra of "design-build-test." We apply this three-step paradigm first to riboswitch signal integration, and then to signal amplification.

2. RIBOSWITCH SIGNAL INTEGRATION

2.1. Design

The specificity of biological sensors would be increased, and potentially dangerous chemical interactions could be detected if two (or more) signals from signature sensing elements could be integrated into a single report. For example, the likelihood of false positives could be reduced if two sensing elements that detected different parts of the same molecular signature were integrated such that a reporter was expressed only if both were activated. Similarly, such a circuit would also be advantageous in the detection of two chemicals that together produce hazardous consequences although they may be independently innocuous. Traditionally, biological sensors have been unable to be integrated. However, using recent advances in synthetic biology, we plan to achieve sensor integration by developing a genetic program that utilizes an orthogonal genetic Boolean logic gate in individual cells. In current genetic logic systems, inputs are promoters that activate expression of a protein (the output) that is either detected or becomes the input for another logic gate. In our system, we plan to use riboswitches to activate expression of transcription factors that serve as the input to the logic gate.

2.1.1 AND gate

Using data mining and directed evolution synthetic biologists have built transcriptional AND gates (Moon et al., 2012; Wang, Barahona, & Buck, 2013; Wang, Kitney, Joly, & Buck, 2011). The inputs to these gates induce expression of two proteins that are both required to activate a promoter. Activation of this promoter results in expression of the output of the gate. Thus both inputs must be present to produce an output. Orthogonal AND gates have been connected to form programs, the largest of which is a 4-input AND gate that consists of three circuits that integrate four inducible systems, requiring 11 regulatory proteins, all within the same cell (Moon et al., 2012). We will describe integrating two riboswitches in *E. coli* using a hetero-regulation module from the hrp (hypersensitive response and pathogenicity) system for Type III secretion in the bacterium *Pseudomonas syringae* (Hutcheson, Bretz, Sussan, Jin, & Pak, 2001; Wang et al., 2011). This device comprises riboswitch control of two coactivating genes *hrpR* and *hrpS*, and a sigma^{54}-dependent *hrpL* promoter driving the output, a fluorescent reporter gene (Figs. 1 and 2). The *hrpL* promoter is activated only when both *hrpR* and *hrpS* are expressed. Riboswitch integration using AND gate logic opens the path to increased layering and complexity of genetic programming in biological sensors. In addition, because of the modularity of biological circuits, the integration circuits described here and the amplification circuits described below may be linked simply by swapping the output of the integration circuit from a fluorescent protein to a bacterial communication molecule.

2.1.2 Selection of riboswitches

The riboswitches that will be used for the signal integration AND gate need to have a tight regulation as well as a well-defined activation ratio (the difference in gene expression when the riboswitch is "ON" compared to when the riboswitch is "OFF"). For the purposes of this study, we will use the theophylline synthetic ribsowitch developed by Justin Gallivan's lab (Desai & Gallivan, 2004; Lynch & Gallivan, 2009), and the 2-aminopurine (2AP) riboswitch described by Dixon et al. (2010). The theophylline riboswitch 12.1 has a large activation ratio of 96 and a half-maximal expression concentration of ~250 μ*M*. The 2AP riboswitch is normally found in the 5′-untranslated region of adenine deaminase in *Vibrio vulnificus* (Mandal & Breaker, 2004) and has an activation ratio of ~12 and a

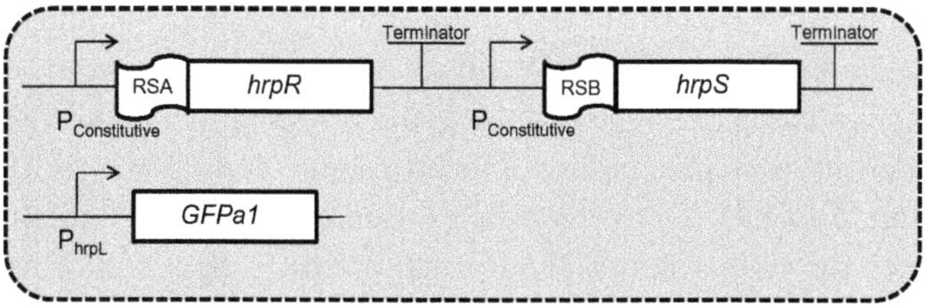

Input A	Input B	Output GFPa1
0	0	0
1	0	0
0	1	0
1	1	1

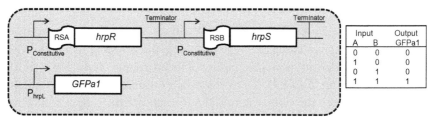

Figure 1 The genetic integration circuit, including the truth table for the circuit. Both riboswitches must be activated for the expression of *hrpR* and *hrpS*. The *hrpL* promoter is only activated in the presence of both hrpR and hrpS. RSA: Riboswitch to ligand A; RSB: riboswitch to ligand B.

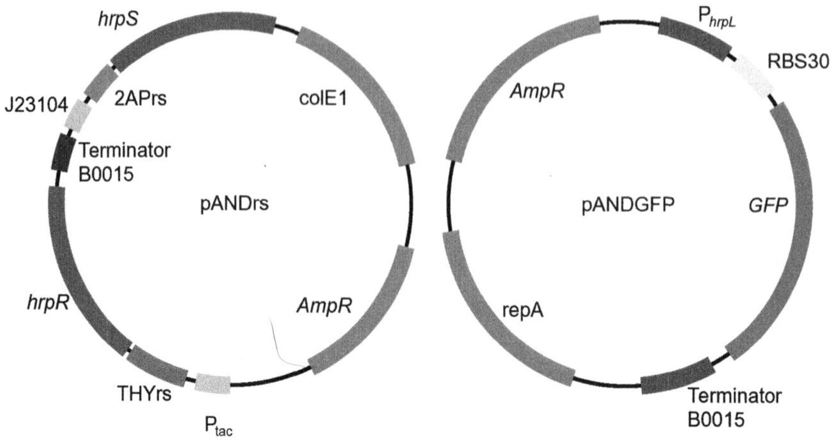

Figure 2 Plasmid maps of genetic integration circuit. *AmpR:* ampicillin resistance; P_{tac}: tac promoter; THYrs: theophylline riboswitch; J23104: strong constitutive promoter; 2APrs: 2-aminopurine riboswitch; colE1: origin of replication; P_{hrpL}: *hrpL* promoter; RBS30; strong ribosome-binding site; repA: origin of replication. (See the color plate.)

half-maximal expression concentration of ~20 μ*M*. As described above, we will be using these riboswitches to induce expression of *hrpR* and *hrpS* genes to form an AND gate. Since these genes have not, to our knowledge, been incorporated into the THY and 2AP riboswitches before, it is essential that some basic modeling using mfold is performed to make sure that the riboswitch will perform as expected. How the gene sequence interferes with the riboswitch can be predicted by inputting a sequence containing the riboswitch and the first 100–150 nucleotides of the downstream gene into mfold (Zuker, 2003) and forcing the aptamer region into configurations corresponding to before and after ligand binding. The energy required to switch the riboswitch on can then be compared to that imparted by the binding of the ligand to the aptamer. The latter is experimentally determined using techniques such as isothermal titration calorimetry (Gilbert, Stoddard, Wise, & Batey, 2006). When the THY riboswitch is placed upstream of *hrpR*, the riboswitch basepairs well with the gene, resulting in a conformation that is unlikely to result in exposure of the ribosome binding sites (RBS) on ligand binding (Table 1). To overcome this issue, we inserted a "linker" sequence between the riboswitch and *hrpR* to ensure the proper conformational change in the presence of the ligand. We used the first 61 amino acids of IS10 transposase, a gene that has been shown to produce a functional switch when placed downstream of THY riboswitch (Desai & Gallivan,

Table 1 Results of modeling riboswitch activation energy

Riboswitch construct	eOFF (kcal/mol)	eON (kcal/mol)	ΔG (kcal/mol)	ΔLigand (kcal/mol)
THYrs-*hrpR*	−37.0	−27.2	−9.8	−9.2[a]
THYrs-linker-*hrpR*	−47.0	−40.7	−6.3	−9.2[a]
2APrs-*hrpS*	−33.0	−32.3	−0.7	−9.4[b]
2APrs-linker-*hrpS*	−43.6	−36.1	−7.5	−9.4[b]

[a]Jenison, Gill, Pardi, and Polisky (1994).
[b]Gilbert et al. (2006).
eOFF: energy required to linearize ribsowitch in OFF state.
eON: energy required to linearize riboswitch in ON state.
Δ*G*: energy difference between eOFF and eON, i.e., energy required to turn on riboswitch.
ΔLigand: energy provided by the ligand binding to the aptamers.

2004). In contrast, when the 2AP riboswitch is placed upstream of *hrpS*, a very weak binding occurs that is likely to result in the riboswitch spontaneously turning "ON" without the presence of the ligand. To rectify this, the first 35 amino acids of eGFP were inserted between the 2AP riboswitch and *hrpS* to ensure proper functioning of the riboswitch (Dixon et al., 2012). The THY riboswitch will be transcribed from the strong *tac* constitutive promoter as described in Desai and Gallivan (2004), while the 2AP riboswitch will be transcribed from a strong constitutive promoter described in the Registry of Standard Biological Parts, J23104 (parts.igem.org/Part:BB_J23104).

2.1.3 Using plasmid backbones available from the Registry of Standard Biological Parts

The components of the AND gate will be carried on plasmids within *E. coli* cells. There are many plasmid backbones described on the Registry of Standard Biological Parts website. The plasmids described herein were generous gifts from synthetic biologists. The plasmid backbone provides a vehicle to get the genetic components into the cell and, as such, is not critical apart from ensuring that the plasmids are compatible if cotransforming multiple different plasmids into a cell.

2.2. Build

2.2.1 Materials for construction of a riboswitch-based AND logic gate

- Plasmids containing the theophylline riboswitch upstream of *LacZ* (pSAL [Lynch & Gallivan, 2009], a generous gift from Dr. Justin Gallivan,

Emory University, Atlanta, GA), an aadA riboswitch with *eGFP* linker (Dixon et al., 2010) preceded by a strong constitutive promoter J23104, and proceeded with *hrpS*-NotI site (generated by DNA2.0, Menlo Park, CA), and plasmids conferring AND gate behavior (one plasmid containing *hrpR* and *hrpS*, and a plasmid containing *GFP* promoted by the *hrpL* promoter [Wang et al., 2013]; generous gifts from Dr. Baojun Wang, Imperial College London, UK).

- Proof-reading high-fidelity DNA polymerase (e.g., Phusion DNA polymerase, New England Biolabs, Ipswich, MA).
- Gibson assembly cloning kit (New England Biolabs, Ipswich, MA).
- Primers designed to amplify genes of interest and including an 18–20 bp overlapping sequence required for Gibson assembly. There are many options available for the synthesis of primers. Ours were purchased from Integrated DNA Technologies (Coralville, IA) and designed using their free oligo design software.
- UltraPure™ DNase/RNase-Free Distilled Water (Invitrogen, Carlsbad, CA).
- Ampicillin (Sigma, St. Louis, MO): 50 mg/mL stock solution in water, final concentration 100 μg/mL of media for all procedures.
- Cloramphenicol (Sigma, St. Louis, MO): 34 mg/mL stock solution in 100% ethanol, final concentration 25 μg/mL of media for all procedures.
- LB Agar: 35 g/L Difco LB Agar (Becton, Dickinson and Company, Franklin Lakes, NJ).
- Sterile Petri dishes, 100 mm (Fisher Scientific, Pittsburg, PA).
- Agarose (Ambion, Austin, TX).
- 1 × Tris–Acetate–EDTA (TAE) buffer prepared from TAE 50 × Solution (Fisher Scientific, Pittsburg, PA).
- 2-Log DNA Ladder (0.1–10.0 kb) and Gel Loading Dye, Blue (6 ×) (New England Biolabs, Ipswich, MA).
- SYBR® Safe DNA Gel Stain (Invitrogen, Carlsbad, CA).
- Nucelospin II gel extraction kit (Clontech, Mountain View, CA).
- NanoDrop 1000 Spectrophotometer (NanoDrop, Wilmington DE).
- Restriction endonucleases *Kpn*I and *Blp*I (New England Biolabs, Ipswich, MA).
- QIAprep plasmid miniprep kit (QIAgen, Valencia, CA).
- Costar 96-well assay plate 3631 (Corning, NY).
- Spectrophotometer (SpectroMax M5; Molecular Devices, Sunnyvale, CA).

Prepare all media and buffer solutions using ultrapure water (prepared by purifying deionized water to obtain a resistance of 18 MΩ cm at 25 °C). For antibiotic solutions and enzymatic reactions, use nuclease-free water. Autoclave media solutions, and filter-sterilize antibiotic solutions. Store media and buffer solutions at room temperature. Store antibiotic solutions at −20 °C.

2.2.2 Methods to construct pANDrs

(1) Using the appropriate plasmid as template, perform PCR to produce three amplicons, ensuring 15–20 bp overlap of each fragment: theophylline riboswitch with *IS10* linker; *hrpR*-terminator B0015-constitutive promoter J23104-aadA riboswitch with *eGFP* linker; *hrpS*.
(2) Digest pSAL with *Kpn*I and *Blp*I.
(3) Run PCR products and pSAL digestion on a 1% agarose–TAE gel and gel extract appropriate bands.
(4) Quantify extracted bands spectrophotometrically.
(5) Assemble gel extracted products using Gibson Assembly method (Gibson et al., 2009). Briefly, this method can join multiple overlapping DNA fragments by using a 5′-exonuclease to create single-stranded 3′-overhangs that anneal and are acted upon by a DNA polymerase to extend the 3′-ends. A DNA ligase then seals the nicks.
(6) Transform into NEB 5-alpha competent *E. coli*.
(7) Plate on LB-ampicillin plates.
(8) Purify the plasmid from overnight cultures of individual colonies.
(9) Confirm correct sequence.
(10) Cotransform pANDrs and plasmid containing P*hrpL* promoted *GFP* (pANDGFP) in *E. coli* BL21 competent cells. These cells were chosen since they are good for protein expression cells because they are deficient in *lon* and *ompT* proteases.

2.3. Test

(1) Inoculate 3.5 mL LB-ampicillin with a single colony of either: (a) cells cotransformed with both pANDrs and pANDGFP; (b) cells transformed pANDrs; and (c) cells transformed with the reporter plasmid, pANDGFP.
(2) Incubate at 37 °C with 220 rpm shaking overnight.

(3) Inoculate 100 μL of overnight cultures into 3 mL aliquots of LB-ampicillin with and without appropriate concentrations of theophylline ([final] = 2.5 m*M*) and 2AP ([final] = 500 μ*M*).
(4) Incubate at 37 °C with 220 rpm shaking overnight.
(5) Record optical density at 600 nm and fluorescence using an excitation wavelength of 480 nm and an emission wavelength of 510 nm (bandpass filter of 495 nm). Cotransformed cultures exposed to both theophylline *and* 2AP will fluoresce at a higher intensity than cotransformed cultures exposed to either theophylline *or* 2AP. No fluorescence is expected from single plasmid transformed bacteria.

3. RIBOSWITCH SIGNAL AMPLIFICATION USING BIOLOGICAL CIRCUITRY

3.1. Design

In biological circuits, the output of an upstream component serves as the input to the next component. Since each computational unit in our system is contained within a different cell, each cell needs to be connected for the circuit to perform. Bacterial cell–cell communication is abundant in nature in a process known as "quorum sensing." Quorum sensing allows appropriate genes to be activated in a high cell density environment. This is achieved by each cell producing a quorum-sensing molecule which is exported out of the cell. When the concentration of this molecule reaches a threshold level, the molecule activates transcription of genes that have a quorum-sensing promoter. This biological wiring allows a signal to be transmitted to surrounding cells, thus propagating any input signal across multiple cell types. Separations of up to 1 cm can be bridged between individual cell types containing different biological circuit components using natural cell–cell communication involved in quorum sensing (Tamsir et al., 2011).

In our design, the input of the amplification circuit will be the output of a sensing device. To form a positive feedback amplification circuit, receipt of the output of the riboswitch must activate transcription of the reporter protein and activate a cascade that provides an intra- and extracellular signal to produce more of the reporter protein. Thus a bicistronic arrangement will be constructed. Quorum-sensing molecules are produced by acyl homoserine lactone (AHL) synthetases LasI and RhlI. Both of these are originally from the bacterium *Pseudomonas aeruginosa* PA01 and have no homologues in the *E. coli* chassis we are using. LasI synthesizes the AHL *N*-3-oxo-dodecanoyl-homoserine lactone (3OC12-HSL), while RhlI synthesizes the AHL *N*-butyryl-homoserine lactone (C4-HSL) (Pesci, Pearson, Seed, & Iglewski, 1997). These AHLs bind to their cognate transcription factors (LasR and RhlR) and activate them. The output of the promoter that is turned on by the transcription factor is used as the input to the next biological circuit component. Thus in our design, activation of the sensor will result in production of an AHL. This signal is received by cells transformed with a plasmid that results in expression of a fluorescent reporter and an AHL (Fig. 3). This signal will propagate throughout the sensor resulting in increased reporter protein production. It will also result in an increased sensitivity of the sensor since, in theory, only one sensor cell is required to detect the molecular signature of interest for all of the cells in the biological sensor to react. The reporter protein we have chosen is GFPa1, a green fluorescent protein from the cephalochordate *Branchistoma floridae*. This protein has been shown to exhibit extremely strong fluorescence (Bomati, Manning, & Deheyn, 2009), and thus makes an ideal reporter.

In designing this circuit we need to tune the production of each protein such that the fluorescent protein is strongly expressed on receipt of the input signal, but HSL production is kept below its threshold until a positive upstream signal is received by the cell. Since the promoters used in this circuit are restricted to those responsive to HSL, the remaining synthetic

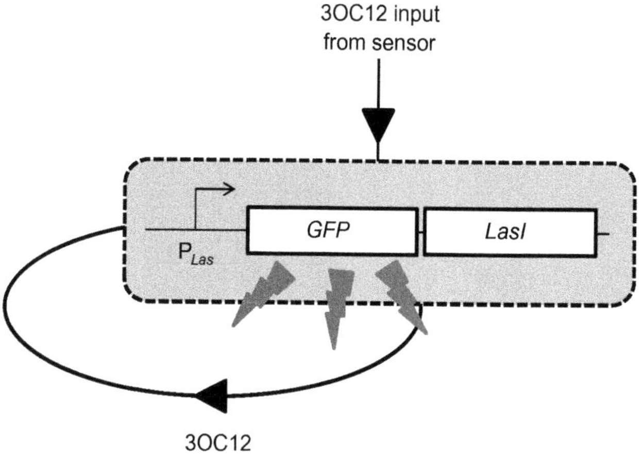

Input	Output	
3OC12	3OC12	GFP
0	0	0
1	1	1

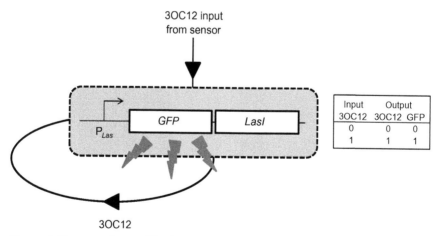

Figure 3 The genetic amplification circuit, including the truth table for the circuit. Detection of a molecular signature by a sensor such as a riboswitch results in expression of *LasI* and production of 3OC12-HSL. 3OC12-HSL activates the *Las* promoter resulting in expression of GFP and LasI. 3OC12-HSL activates the *Las* promoter of the same cell and also diffuses out of the cell resulting in activation of the *Las* promoter in surrounding reporter cells. Thus a progression of GFP and LasI expression is propagated.

biological parts that can tune expression are RBSs and protein degradation tags. Degradation tags are short peptide sequences placed at the C-terminus of proteins that mark them for degradation by intracellular proteases (Keiler, Waller, & Sauer, 1996). We referenced RBS listed in the Registry of Standard Biological Parts (parts.igem.org/Ribosome_Binding_Sites/Prokaryotic/Constitutive/Community_Collection) and also quantified by Wang et al. (2011), and used degradation tags also listed in the registry (parts.igem.org/Protein_domains/Degradation) (Fig. 4).

3.2. Build

3.2.1 Materials for the construction of an amplification circuit

- Plasmids containing *LasR*, P*Las*, *RhlR*, P*Rhl* (pOR30 and pNOR40 [Tamsir et al., 2011], generous gifts from Dr. Christopher Voigt, MIT, MA), and *GFPa1* (Bomati et al., 2009; a generous gift from Dr. Dimitri Deheyn, University of California, San Diego).
- Plasmid backbones pSB3K3 and pSB4A3 (Registry of Standard Biological Parts; a generous gifts from Dr. Christopher Voigt, MIT, MA).
- Primers to amplify *LasI* and *RhlI* from *Pseudomonas aeruginosa* PA01 (ATCC 47085).
- Restriction endonucleases *Kpn*I, *Bsr*GI, *Bst*BI, and *Bmr*I (New England Biolabs, Ipswich, MA).
- PCR and cloning and fluorescent detection materials listed above.
- Synthetic 3OC12-HSL and C4-HSL (Cayman Chemical Co., Ann Arbor, MI).
- DarkReader transilluminator (Clare Chemical Research, Dolores, CO).

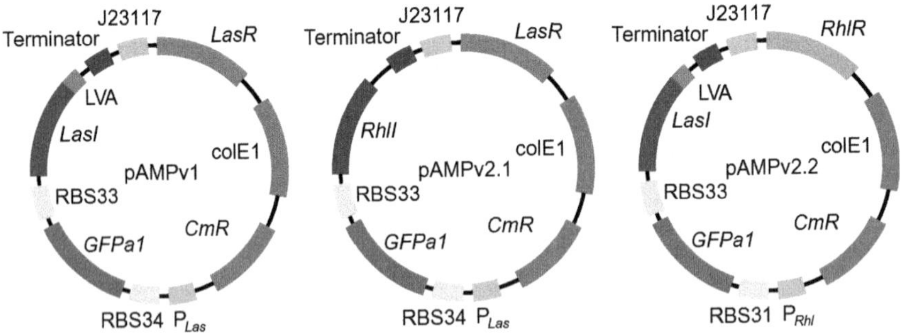

Figure 4 Plasmid maps of amplification circuits. *CmR*: chloramphenicol resistance; P_{Las}: *Las* promoter; RBS34: strong ribosome-binding site; RBS33: weak ribosome-binding site; LVA: protein degradation tag; J23117: weak constitutive promoter; P_{Rhl}: *Rhl* promoter. (See the Color plate.)

3.2.2 Methods to construct an amplification circuit, pAMPv1

3.2.2.1 PLas_RBS34(strong)_GFPa1_Terminator_Weak RBS_LasI +/−degradation tag

(1) Digest pOR30 with *Kpn*I and *Bsr*GI.

(2) Amplify *GFPa1* and *LasI* using primers that have an 18–20 bp overlapping sequence required for Gibson assembly. In this case, our *GFPa1* forward primer included a *Kpn*I site immediately 5′ of the *GFPa1* sequence, and our *LasI* forward primers contained a *GFPa1* overlap-*Bsr*GI site-RBS at its 5′-end and the *LasI*+LVA reverse primers contained a LVA degradation tag (http://parts.igem.org). Two different RBSs were tested: B0032 (weak) and B0033 (very weak).

(3) Separate PCR products and plasmid digestion on a 1% agarose–TAE gel.

(4) Gel extract appropriate amplicon and the large band corresponding to plasmid backbone.

(5) Quantify extracted bands spectrophotometrically.

(6) Assemble gel extracted products using Gibson assembly method using 0.5 pmol of each.

(7) Transform into NEB 5-alpha competent *E. coli*.

(8) Plate on LB-chloramphenicol plates.

(9) Purify the plasmid from overnight cultures of individual colonies.

(10) Confirm correct sequence by sequencing.

3.3. Test

(1) Grow sequence-verified clones in LB-chloramphenicol at 37 °C with 220 rpm shaking overnight.

(2) Record absorbance at 600 nm and fluorescence using an excitation wavelength of 480 nm and an emission wavelength of 510 nm (bandpass filter of 495 nm).

From this experiment, we saw that even when using a weak RBS before *LasI* with or without the addition of a degradation tag, fluorescence was recorded without the addition of an input signal (Fig. 5). In other words, the positive feedback loop was too good and within a cell, any LasI expression resulted in a 3OC12-HSL concentration above threshold level, thus activating the pathway.

3.4. Redesign and build

To overcome amplification circuit self-activation, we made use of the flexibility afforded by utilizing biological wiring to devise a different

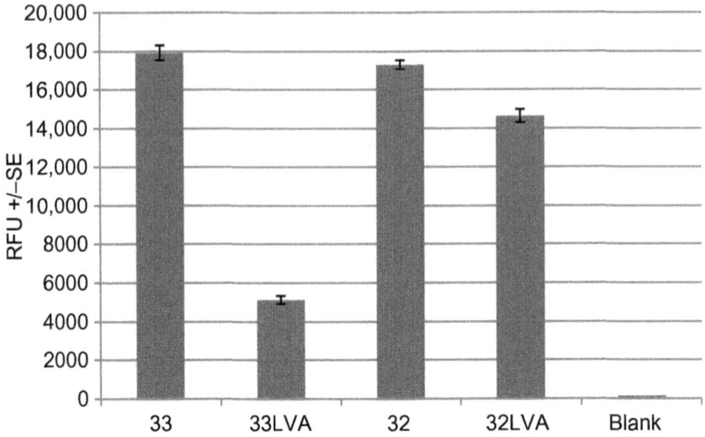

Figure 5 Auto-induction of fluorescence when the amplification circuit is encoded on a single plasmid. The amplification cassette included a weak ribosome-binding site upstream of a *LasI* gene with or without the LVA degradation tag. Mean values of two replicates. 33: P_{Las}_RBS34_*GFPa1*_RBS33_*LasI*; 33LVA: P_{Las}_RBS34_*GFPa1*_RBS33_*LasI*LVA; 32: P_{Las}_RBS34_*GFPa1*_RBS32_*LasI*; 32LVA: P_{Las}_RBS34_*GFPa1*_RBS32_*LasI*LVA; Blank: empty *E. coli* NEB 5-alpha.

amplification circuit (Fig. 6). In this circuit, quorum-sensing molecule production genes and the promoters responsible for quorum-sensing molecule detection were insulated from each other by separating them into different cells. Thus, 3OC12-HSL activates P*Las_GFPa1_RhlI* in a cell to produce fluorescence and C4-HSL. C4-HSL is exported from the cell and activates P*Rhl_GFPa1_LasI* in a different cell type. This results in the production of fluorescence and 3OC12-HSL, and the cycle repeats. This design should reduce the likelihood of self-activation (Fig. 4) since 3OC12-HSL does not bind to RhlR and C4-HSL does not bind to LasR (Pesci et al., 1997), and thus there is no positive feedback loop within a cell. A similar circuit was successfully used to create a microbial consensus consortium that expressed reporter proteins at >100-fold higher level than the responses of the individual cell types in isolation (Brenner, Karig, Weiss, & Arnold, 2007). The *LasI* promoter is more sensitive to 3OC12-HSL than the *RhlI* promoter is to C4-HSL (Fig. 7). Thus, to tune the circuit so that induction results in a comparable fluorescence in each cell type, a degradation tag was added to LasI to dampen the signal production in the P*Rhl_GFPa1_LasI* cassette.

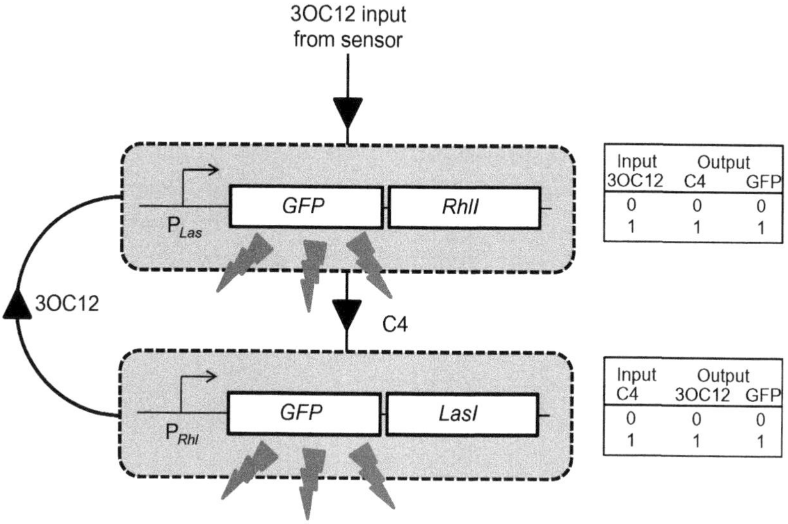

Figure 6 The genetic amplification circuit, including the truth table for the circuit. Detection of a molecular signature by a sensor such as a riboswitch results in expression of LasI and production of 3OC12-HSL. 3OC12-HSL activates the *Las* promoter resulting in expression of GFP and RhlI. RhlI produces C4-HSL. C4-HSL diffuses out of the cell and activates the *Rhl* promoter of a different cell type resulting in expression of GFP and LasI. This initiates a signaling cascade that will spread to all reporter cell types in the sensing device.

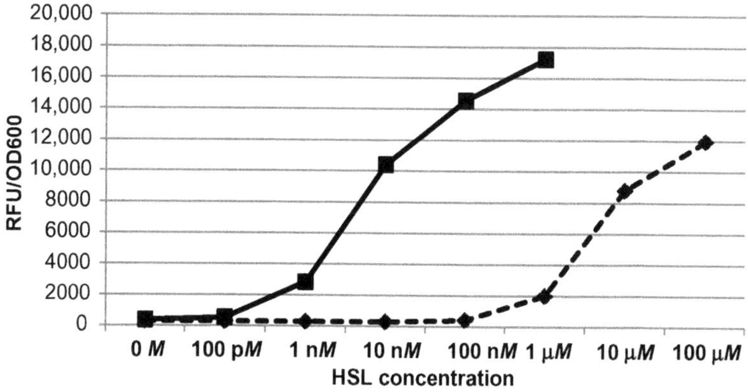

Figure 7 The individual plasmids of amplification circuit 2.0 produce a dose-dependent fluorescence. Solid line, pAMPv2.1 exposed to 3OC12-HSL; dotted line, pAMPv2.2 exposed to C4-HSL.

3.4.1 Methods to construct amplification circuit version 2

3.4.1.1 pAMPv2.1: P_{Las}_RBS 34 (strong)_GFPa1_Terminator_RBS 33 (very weak) _RhlILVA

(1) Digest the plasmid constructed above with *Bsr*GI.
(2) Amplify *RhlI* using primers that have a 18–20 bp overlapping sequence required for Gibson assembly. In this case our forward primer included a *GFPa1* overlap-*Bsr*GI site-RBS 33 at its 5′-end and the *RhlI*+LVA reverse primer contained an LVA degradation tag (Andersen et al., 1998).
(3) Follow protocol Steps (3)–(10) described above.

3.4.1.2 pAMPv2.2: P_{Rhl}_RBS 34 (strong)_GFPa1_Terminator_RBS 33 (very weak) _LasILVA

(1) Digest pNOR40 with *Bst*BI and *Bmr*I.
(2) Amplify *GFPa1* and *LasI* using primers that have a 18–20 bp overlapping sequence required for Gibson assembly. In this case our *GFPa1* forward primer included a *Bst*BI site 5′ of the *GFPa1* gene, and our *LasI* forward primers contained a *GFPa1* overlap-*Bsr*GI site-RBS 33 (very weak) at its 5′-end and the *LasI*+LVA-Terminator-*Bmr*I reverse primers contained a LVA degradation tag.
(3) Follow protocol Steps (3)–(10) described above.

3.5. Test

3.5.1 Fluorescence activation

(1) Grow sequence-verified clones in LB-chloramphenicol with varying concentrations of synthetic HSL at 37 °C with 220 rpm shaking overnight.
(2) Record absorbance at 600 nm and fluorescence using an excitation wavelength of 480 nm and an emission wavelength of 510 nm (bandpass filter of 495 nm).

Unlike the previous iteration, the two plasmids used in this amplification circuit produce a dose-dependent fluorescence (Fig. 7). Therefore, these plasmids can be used in an experiment to show signal progression.

3.5.2 Signal progression

(1) Use a single colony of each cell type to inoculate LB-chloramphenicol.
(2) Incubate at 37 °C with 220 rpm shaking until optical density at 600 nm reaches 0.5.

(3) On an LB-chloramphenicol plate, alternately spot 1 µL of each culture in a straight line with a spacing of 1 cm between spots.
(4) Add 1 µL of 10 µ*M* synthetic 3OC12-HSL 1 cm from first P*Las_GFPa1_RhlI* spot.
(5) For a diffusion-of-signal control, also add 1 µL of 10 µ*M* synthetic 3OC12-HSL 4 cm from first P*Las_GFPa1_RhlI* spot (the limit of diffusion of this concentration of 3OC12-HSL in standard agar plates is ~2 cm; Tamsir et al., 2011).
(6) Incubate at 37 °C for 24 h.
(7) Visualize fluorescence. Signal will progress "along" the line of spots by a cell type receiving a quorum-sensing signal that activates expression of GFPa1 and production of the other quorum-sensing signal, which then activates the next cell type, and so on. Controls that rely on passive diffusion of signal will not exhibit fluorescence.

Thus, using relatively straight-forward molecular biological techniques and "off-the-shelf" biological parts, synthetic biology provides an engineering environment in which to integrate riboswitches into biological circuitry.

ACKNOWLEDGMENTS

We thank Dr. Justin Gallivan at Emory University, Dr. Christopher Voigt at the Massachusetts Institute of Technology, and Dr. Baojun Wang at Imperial College London for their generous gifts of plasmids. We also thank Dr. Dimitri Deheyn at the University of California, San Diego for his generous gift of GFPa1. This work was funded by the Air Force Office of Scientific Research and by the Bio-X STT, Air Force Research Laboratory.

REFERENCES

Afroz, T., & Beisel, C. L. (2013). Understanding and exploiting feedback in synthetic biology. *Chemical Engineering Science*, *103*, 79–90.

Andersen, J. B., Sternberg, C., Poulsen, L. K., Bjorn, S. P., Givskov, M., & Molin, S. (1998). New unstable variants of green fluorescent protein for studies of transient gene expression in bacteria. *Applied and Environmental Microbiology*, *64*, 2240–2246.

Anderson, J. C., Voigt, C. A., & Arkin, A. P. (2007). Environmental signal integration by a modular AND gate. *Molecular Systems Biology*, *3*, 133.

Bhatia, S. N. (2006). Cell and tissue-based sensors. In J. Schultz, M. Mrksich, S. N. Bhatia, D. J. Brady, A. J. Ricco, D. R. Walt, & C. L. Wilkins (Eds.), *Biosensing: International research and development* (pp. 55–65).Dordrecht, Netherlands: Springer, 387 p.

Bomati, E. K., Manning, G., & Deheyn, D. D. (2009). Amphioxus encodes the largest known family of green fluorescent proteins, which have diversified into distinct functional classes. *BMC Evolutionary Biology*, *9*, 77.

Brenner, K., Karig, D. K., Weiss, R., & Arnold, F. H. (2007). Engineered bidirectional communication mediates a consensus in a microbial biofilm consortium. *Proceedings of the National Academy of Science*, *104*, 17300–17304.

Desai, S. K., & Gallivan, J. P. (2004). Genetic screens and selections for small molecules based on a synthetic riboswitch that activates protein translation. *Journal of the American Chemical Society, 126*, 13247–13254.
Dixon, N., Duncan, J. N., Geerlings, T., Dunstan, M. S., McCarthy, J. E. G., Leys, D., et al. (2010). Reengineering orthogonally selective riboswitches. *Proceedings of the National Academy of Science, 107*, 2830–2835.
Dixon, N., Robinson, C. J., Geerlings, T., Duncan, J. N., Drummond, S. P., & Micklefield, J. (2012). Orthogonal riboswitches for tuneable coexpression in bacteria. *Angewandte Chemie International Edition, 51*, 3620–3624.
Ellis, T., Wang, X., & Collins, J. J. (2009). Diversity-based, model-guided construction of synthetic gene networks with predicted functions. *Nature Biotechnology, 27*, 465–471.
Friedland, A. E., Lu, T. K., Wang, X., Shi, D., Church, G., & Collins, J. J. (2009). Synthetic gene networks that count. *Science, 324*, 1199–1202.
Gibson, D. G., Young, L., Chuang, R.-Y., Venter, J. G., Hutchison, C. A., III, & Smith, H. O. (2009). Enzymatic assembly of DNA molecules up to several hundred kilobases. *Nature Methods, 6*, 343–345.
Gilbert, S. D., Stoddard, C. D., Wise, S. J., & Batey, R. T. (2006). Thermodynamic and kinetic characterization of ligand binding to the purine riboswitch aptamers domain. *Journal of Molecular Biology, 359*, 754–768.
Guet, C. C., Elowitz, M. B., Hsing, W., & Leibler, S. (2002). Combinatorial synthesis of genetic networks. *Science, 296*, 1466–1470.
Hutcheson, S. W., Bretz, J., Sussan, T., Jin, S., & Pak, K. (2001). Enhancer-binding proteins HRPR and HRPS interact to regulate hrp-encoded type III protein secretion in Pseudomonas syringae strains. *Journal of Bacteriology, 183*, 5589–5598.
Jenison, R. D., Gill, S. C., Pardi, A., & Polisky, B. (1994). High-resolution molecular discrimination by RNA. *Science, 263*, 1425–1429.
Kawai, T., & Akira, S. (2007). Signaling to NF-kappaB by Toll-like receptors. *Trends in Molecular Medicine, 13*, 460–469.
Keiler, K. C., Waller, P. R., & Sauer, R. T. (1996). Role of a peptide tagging system in degradation of proteins synthesized from damaged messenger RNA. *Science, 271*, 990–993.
Lou, C., Lio, X., Ni, M., Huang, Y., Huang, Q., Huang, L., et al. (2010). Synthesizing a novel genetic sequential logic circuit: A push-on push-off switch. *Molecular Systems Biology, 6*, 350.
Lynch, S. A., & Gallivan, J. P. (2009). A flow cytometry-based screen for synthetic riboswitches. *Nucleic Acids Research, 37*, 184–192.
Mandal, M., & Breaker, R. R. (2004). Adenine riboswitches and gene activation by disruption of a transcription terminator. *Nature Structural & Molecular Biology, 11*, 29–35.
Mayo, A. E., Setty, Y., Shavit, S., Zaslaver, A., & Alon, U. (2006). Plasticity of the cis-regulatory input function of a gene. *PLoS Biology, 4*, e45.
Moon, T. S., Lou, C., Tamsir, A., Stanton, B. C., & Voigt, C. A. (2012). Genetic programs constructed from layered logic gates in single cells. *Nature, 491*, 249–253.
Pesci, E. C., Pearson, J. P., Seed, P. C., & Iglewski, B. H. (1997). regulation of las and rhl quorum sensing in *Pseudomonas aeruginosa*. *Journal of Bacteriology, 179*, 3127–3132.
Rinaudo, K., Bleris, L., Maddamsetti, R., Subramanian, S., Weiss, R., & Benenson, Y. (2007). A universal RNAi-based logic evaluator that operates in mammalian cells. *Nature Biotechnology, 25*, 795–801.
Tabor, J. J., Salis, H. M., Simpson, Z. B., Chevalier, A. A., Levskaya, A., Marcotte, E. M., et al. (2009). A synthetic genetic edge detection program. *Cell, 137*, 1272–1281.
Tamsir, A., Tabor, J. J., & Voigt, C. A. (2011). Robust multicellular computing using genetically encoded NOR gates and chemical 'wires'. *Nature, 469*, 212–215.

Wang, B., Barahona, M., & Buck, M. (2013). A modular cell-based biosensor using engineered genetic logic circuits to detect and integrate multiple environmental signal. *Biosensors and Bioelectronics*, *40*, 368–376.

Wang, B., Kitney, R. I., Joly, N., & Buck, M. (2011). Engineering modular and orthogonal genetic logic gates for robust digital-like synthetic biology. *Nature Communications*, *2*, 508.

Weber, W., Schoenmakers, R., Keller, G. B., Gitzinger, M., Grau, T., Daoud-El Baba, M., et al. (2008). A synthetic mammalian gene circuit reveals antituberculosis compounds. *Proceedings of the National Academy of Sciences of the United States of America*, *105*, 9994–9998.

Xia, B., Bhatia, S., Bubenheim, B., Dadgar, M., Densmore, D., & Anderson, J. C. (2011). Developer's and user's guide to Clotho v2.0: A software platform for the creation of synthetic biological systems. *Methods in Enzymology*, *498*(B), 95–135.

Zuker, M. (2003). Mfold web server for nucleic acid folding and hybridization prediction. *Nucleic Acids Research*, *31*(13), 3406–3415.

CHAPTER FIVE

Simple Identification of Two Causes of Noise in an Aptazyme System by Monitoring Cell-Free Transcription

Norikazu Ichihashi[*,‡,1], **Shungo Kobori**[*], **Tetsuya Yomo**[*,†,‡,¶]

[*]Graduate School of Information Science and Technology, Osaka University, Suita, Osaka, Japan
[†]Graduate School of Frontier Biosciences, Osaka University, Suita, Osaka, Japan
[‡]Exploratory Research for Advanced Technology, Japan Science and Technology Agency, Suita, Osaka, Japan
[¶]Earth-Life Science Institute, Tokyo Institute of Technology, Meguro, Tokyo, Japan
[1]Corresponding author: e-mail address: ichihashi@ist.osaka-u.ac.jp

Contents

Methods in Enzymology, Volume 550
ISSN 0076-6879
http://dx.doi.org/10.1016/bs.mie.2014.10.043

Abstract

Aptazymes are artificially synthesized ribozymes that catalyze reactions in response to ligand binding. Certain types of aptazymes can be utilized as RNA-based regulators of gene expression. These aptazymes contain a sequestered ribosome-binding site (rbs) and release the rbs through self-cleavage in response to ligand binding, inducing the expression of the downstream gene. One of the most important properties of aptazymes as gene expression regulators is their signal-to-noise ratio (S/N ratio), the ratio of target expression in the presence of ligand to that in the absence of ligand. One strategy to improve the S/N ratio is to decrease the noise (expression in the absence of ligand) due to leaky translation without rbs release or ligand-independent rbs release. In this chapter, we describe an easy method to identify the main cause of noise using a cell-free reconstituted transcription–translation system, an ideal platform for the quantitative understanding of biochemical reactions because researchers can strictly control the experimental conditions and the concentrations of all components. This knowledge would be useful for designing aptazymes with high S/N ratios.

1. THEORY

Aptazymes are artificial ribozymes that catalyze reactions upon binding to target ligands. Some aptazymes have been adopted as gene expression switches (Piganeau, Jenne, Thuillier, & Famulok, 2000; Robertson & Ellington, 1999; Soukup & Breaker, 1999, 2000; Thompson, Syrett, Knudsen, & Ellington, 2002; Wieland, Benz, Klauser, & Hartig, 2009; Wieland & Hartig, 2008; Yen et al., 2004). These aptazymes include a ribosome-binding site (rbs) that is inaccessible in the absence of the ligand, inhibiting expression of the downstream gene; however, the aptazymes self-cleave in the presence of the ligand, releasing the rbs and inducing gene expression (Fig. 1A). One of the most important properties of gene-regulating aptazymes is their signal-to-noise ratio (S/N ratio), the ratio of gene expression in the presence of ligand to that in the absence of ligand. One possible strategy to increase the S/N ratio is to decrease noise. In the case of aptazymes, noise has at least two causes (Kobori, Ichihashi, Kazuta, Matsuura, & Yomo, 2012): leaky translation from the noncleaved full-length aptazyme and ligand-independent self-cleavage of the aptazyme (Fig. 1B). Determining which of these causes of noise predominates is critically important to rationally improve the aptazyme S/N ratio. If leaky translation from the noncleaved aptazyme is the main cause of noise, the rbs must be sequestered more completely. However, if ligand-independent self-cleavage is the main cause of noise, the self-cleavage sensitivity of the aptazyme must be manipulated.

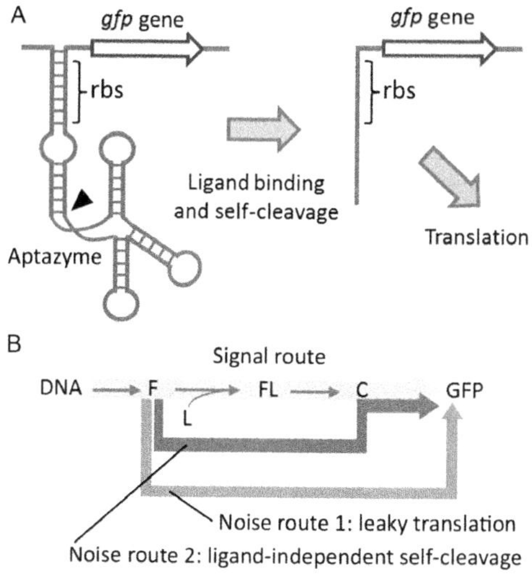

Figure 1 Schematic drawing of an aptazyme that regulates downstream gene expression. (A) An example of an aptazyme containing a ribosome-binding site (rbs) of the downstream *gfp* gene. This aptazyme self-cleaves at the position indicated by the arrowhead upon ligand binding and induces expression of the downstream gene by releasing the rbs. (B) Two possible routes that induce gene expression in the absence of ligand, also known as noise. The noncleaved full-length RNA (F) is transcribed from DNA and binds to the ligand (L) to form the ligand–RNA complex (FL). The complex self-cleaves to produce the self-cleaved RNA (C), which has the exposed rbs and induces the expression of GFP. In the presence of ligand, the induced protein expression represents the signal. In the absence of ligand, expression of the downstream gene represents noise and can occur through two distinct routes: ligand-independent self-cleavage of the aptazyme (noise route 1) or leaky translation from noncleaved full-length RNA (noise route 2).

In a previous study, we showed that the main cause of noise is easily determined by examining the dynamics of the S/N ratio in the steady state (Kobori et al., 2012). If the main cause of noise is ligand-independent self-cleavage, the S/N ratio in the steady state should not change over time because the time dependence of gene expression in the presence (signal) and absence (noise) of ligand are the same. In contrast, if the noise is generated by leaky translation from the noncleaved aptazyme, the S/N ratio in the steady state should increase over time because the kinetics of gene expression in the presence (signal) and absence (noise) of ligand are different (see the previous study for a detailed mathematical description; Kobori et al., 2012).

The S/N ratio kinetics may be easily measured using a reconstituted cell-free transcription–translation system, which involves combining the cell-free system with DNA encoding an aptazyme and GFP in the presence or absence of ligand and measuring the kinetics of GFP fluorescence.

2. EQUIPMENT

Micropipettes
Micropipette tips
0.2 ml microcentrifuge tubes with opaque walls for fluorescence detection
0.2 ml microcentrifuge flat caps for fluorescence detection
QIAquick PCR purification kit
Microcentrifuge
Fluorometer or real-time PCR system
UV spectrophotometer
Polyacrylamide gel electrophoresis equipment
Gel dryer
Autoradiography film
Imager, such as Typhoon FLA 9500 (GE Healthcare)
ImageQuant software (GE Healthcare)

3. MATERIALS

DNA fragment containing an aptazyme and a downstream *gfp* gene
Sol. I, Sol. II, and Sol. III of PUREfrex (Gene Flex)
Urea
Formamide
30% Acrylamide/bis-acrylamide (29:1)
Ammonium persulfate (APS)
Tris base
Hydrochloric acid (HCl)
Boric acid
Ethylenediaminetetraacetic acid (EDTA)
Bromophenol blue
[^{32}P]-UTP
Acetic acid
Ligand for the aptazyme (e.g., thiamine pyrophosphate (TPP))

4. SOLUTIONS AND BUFFERS

Cell-free transcription–translation solution

Component	Final concentration	Stock	Amount
Sol. I of PUREfrex			10 μl
Sol. II of PUREfrex			1 μl
Sol. III of PUREfrex			1 μl
DNA fragment encoding an aptazyme and *gfp* gene	3 n*M*	>30 n*M*	1–2 μl
Water			To a total volume of 20 μl

5 × Tris–borate–EDTA (TBE) buffer

Component	Final concentration	Stock	Amount
Tris	0.45 *M*		54 g
Boric acid	0.45 *M*		27.6 g
EDTA, pH 8.0	10 m*M*	0.5 *M*	20 ml
Water			To a total volume of 1 l

Autoclave the buffer at 121 °C for 20 min.

Loading buffer

Component	Final concentration	Stock	Amount
Formamide	80%		8 ml
EDTA (pH 8.0)	50 m*M*	0.5 *M*	1 ml
Bromophenol blue	0.025%		2.5 mg
Water			To a total volume of 10 ml

4% Denaturing polyacrylamide gel (PAGE) solution

Component	Final concentration	Stock	Amount
Acrylamide/bis-acrylamide (29:1)	4%	30%	1.3 ml
Urea	7 *M*		4.2 g
TBE buffer	1 ×	5 ×	2 ml
APS	0.7 mg/ml	40 mg/ml	175 μl
Tetramethylethylenediamine (TEMED)	1 mg/ml		14 μl
Water			To a total volume of 10 ml

PAGE electrode solution

Component	Final concentration	Stock	Amount
TBE buffer	1 ×	5 ×	100 ml
Water			To a total volume of 500 ml

5. PROTOCOL

5.1. Duration

Preparation	Approximately 2 days
Cell-free transcription and translation	Approximately 3 h
Kinetic analysis	Approximately 2 h
Measurement of the intermediate RNAs	Approximately 5 h

5.2. Preparation

A DNA fragment is prepared using standard PCR techniques and contains a T7 promoter, the target aptazyme gene, and the *gfp* gene arranged $5' \rightarrow 3'$. The fragment must be purified using a conventional column,

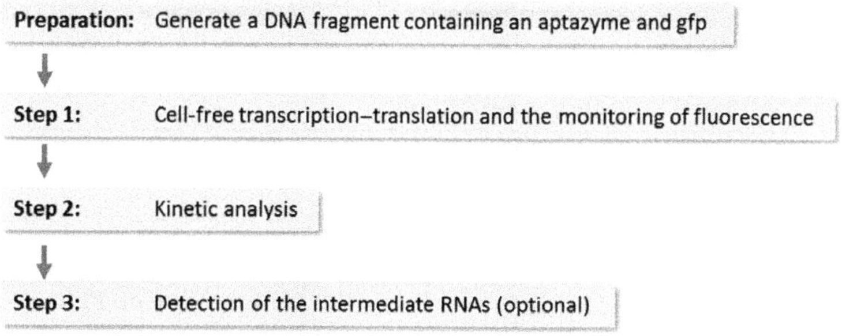

Figure 2 Flowchart of the complete protocol.

such as the QIAquick column (QIAGEN). Approximately 10–50 pmol of PCR product is needed for each transcription–translation reaction. The DNA concentration is measured by UV absorbance. Other genes that encode fluorescent proteins, such as rfp, cfp, yfp, may be used in place of *gfp*.

5.3. Caution

The purity of the DNA fragment is important for efficient translation. Plasmids may also be used instead of PCR fragments, but the translation efficiency of plasmids is usually lower than that of PCR fragments (Fig. 2).

6. STEP 1: CELL-FREE TRANSCRIPTION–TRANSLATION AND FLUORESCENCE MONITORING

6.1. Overview

The DNA fragment, which includes the aptazyme and *gfp* genes, is mixed with the cell-free transcription–translation system and incubated at 37 °C for 1–3 h. During the incubation, the fluorescence of the expressed GFP protein is monitored in real time.

6.2. Duration

3 h

2.1 To measure gene expression in the presence of the ligand, mix the following materials in a 0.2-ml microcentrifuge tube with opaque walls.

DNA fragment	10 pmol
Sol. I of PUREfrex	10 μl
Sol. II of PUREfrex	1 μl
Sol. III of PUREfrex	1 μl
Ligand for the aptazyme	Sufficient amount for the maximum activity (1 m*M* in the case of TPP aptazyme; Kobori et al., 2012)
Water	To a total volume of 20 μl

To measure background fluorescence, the same solution is prepared, except that the DNA fragment is omitted.

2.2 To measure gene expression in the absence of ligand, mix following materials in a 0.2-ml microcentrifuge tube with opaque walls.

DNA fragment	10 pmol
Sol. I of PUREfrex	10 μl
Sol. II of PUREfrex	1 μl
Sol. III of PUREfrex	1 μl
Water	To a total volume of 20 μl

To measure the background fluorescence, the same solution is prepared, but the DNA fragment is omitted.

2.3 Load the tubes into a real-time PCR system and measure the fluorescence every 1–5 min for 1–3 h while incubating at 37 °C.

2.4 When using a fluorometer, incubate the tube at 37 °C for 1–3 h in a heat block and measure the fluorescence manually at 5–15 min intervals.

6.3. Tip

To subtract the background fluorescence, control experiments that omit the DNA fragment are recommended.

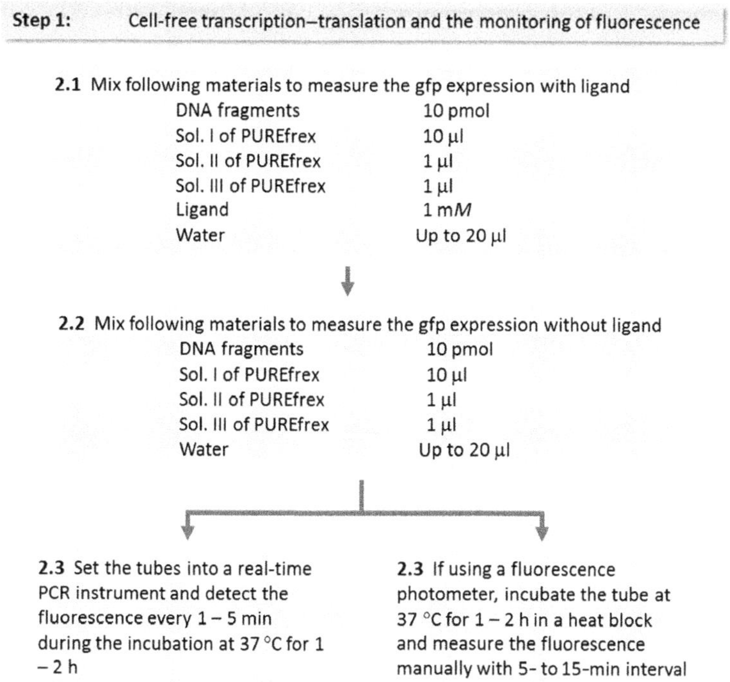

Figure 3 Flowchart of Step 1.

6.4. Tip

To measure kinetics under other buffer conditions, such as reduced magnesium concentration, the composition of Solution I of the PUREfrex kit, which contains all the low-molecular components, can be modified according to a previously reported method (Shimizu et al., 2001; Fig. 3).

7. STEP 2: DATA ANALYSIS

7.1. Overview

To identify the main source of noise, the kinetics of the S/N ratio is analyzed. If the S/N ratio increases, then the most likely main cause of noise is leaky translation; however, if the S/N is constant over time, then ligand-independent self-cleavage is the likely predominant source of noise. To characterize the S/N ratio kinetics, it is necessary to determine the reliable range of the fluorescence data obtained in Step 1. First, the fluorescence range over which the fluorescence level linearly correlates with GFP expression is determined. Second, the background fluorescence level is subtracted from the data. Third, a certain incubation period (usually 40–120 min), over which the mathematical model used in this analysis has been validated, is chosen. Finally, fluorescence in the presence of ligand is divided by fluorescence in the absence of the ligand to obtain the S/N ratio kinetics.

7.2. Duration

2 h

2.1 Determine a fluorescence level under which the fluorescence correlates linearly with GFP concentration

To perform this analysis, it is first necessary to measure the range of GFP concentration over which the fluorescence is linearly correlated with the GFP concentration. The range can be determined by preparing a standard curve in advance that relates concentrations of the fluorescent molecule with the responses of the equipment used in Step 1.

2.2 Subtract background

There is a certain measurable level of fluorescence even in the absence of any GFP expression. To subtract this background fluorescence, the fluorescence of the sample in the absence of the DNA fragment must be subtracted from samples that include DNA fragments, regardless of whether they contain ligand.

2.3. Determine the incubation time available for data analysis

Analysis of the kinetics of the S/N ratio depends on a mathematical model (reported in Kobori et al., 2012) that assumes the RNA species (noncleaved and self-cleaved RNAs) exist in a steady state and the GFP translation rate is constant. These assumptions are only valid over a certain time range during incubation, which generally lasts 40–120 min, because approximately 40 min of incubation is required for RNA transcripts to reach a steady state, and the inactivation of translated proteins may start to become significant after approximately 120 min of incubation. Therefore, only the data from 40 to 120 min of incubation time should be used for further data analysis. If necessary, the minimum incubation time can be determined by measuring the RNA concentrations as described in the next section; the minimum incubation time is the time at which both the noncleaved and self-cleaved RNA concentration become constant. The maximum incubation time can be determined by measuring the kinetics of GFP expression with a DNA fragment that does not include the aptazyme; the maximum incubation time is the time at which GFP fluorescence stops increasing linearly with time.

2.4 Divide the signal by the noise to obtain the S/N ratio

The selected data obtained in the presence of ligand (signal) are divided by the data obtained in the absence of ligand (noise) to calculate the S/N ratio.

2.5 Evaluate the main cause of noise

Based on the kinetics of the S/N ratio, the main cause of noise can be determined. If the S/N ratio remains constant over time, the main cause of noise is most likely ligand-independent self-cleavage. If the S/N ratio increases over time, the noise is most likely caused by leaky translation from noncleaved RNA.

7.3. Tip

If the fluorescence quickly becomes too high and immediately exceeds the linear range of the instrument, we recommend changing the excitation and emission wavelengths to dampen the fluorescence signal.

7.4. Tip

Note that if the analysis is performed using data obtained before the RNA species reach steady state, the S/N ratio will appear to increase over time regardless of the source of the noise. Therefore, using data from the appropriate incubation time is crucial (Fig. 4).

Step 2: Data analysis

2.1 Determine a fluorescence level under which the fluorescence correlates linearly with GFP concentration

↓

2.2 Subtract background

↓

2.3 Determine the incubation time available for data analysis

↓

2.4 Divide signal by noise to obtain the S/N ratio

↓

2.5 Evaluate the main noise route

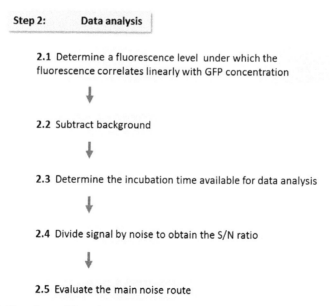

Figure 4 Flowchart of Step 2.

8. STEP 3: (OPTIONAL) QUANTIFICATION OF THE INTERMEDIATE RNAs

8.1. Overview

The data analysis of Step 2 reveals the main cause of gene induction in the absence of ligand. This result can be confirmed experimentally to detect the intermediate RNA species in the absence of ligand. If the self-cleaved RNA fragment is detected in the absence of ligand, the most likely cause of noise is ligand-independent self-cleavage. If the self-cleaved RNA fragment is instead not detected, leaky translation is the most likely cause of noise. It is important to note that the result obtained in this step provides only supporting evidence rather than a final conclusion because undetectable quantities of self-cleaved fragments might be responsible for the observed noise. Measurements of the intermediate RNA species are also important in determining the incubation time over which the RNA species are in the steady state.

8.2. Duration

5 h

2.1 To detect the intermediate RNA species in the absence of ligand, mix the following materials in a 0.2-l microcentrifuge tube.

DNA fragment	10 pmol
Sol. I	10 μl
Sol. II	1 μl
Sol. III	1 μl
α[^{32}P]-isotope-labeled UTP	0.2 μl (0.1 MBq)
Water	To a total volume of 20 μl

2.2 Incubate the solution at 37 °C for 1–2 h.

2.3 Prepare a 4% denaturing polyacrylamide gel.

A solution is prepared as described below and added to an electrophoresis gel prepared with a comb to polymerize for over 30 min.

Component	Final concentration	Stock	Amount
Acrylamide/bis-acrylamide (29:1)	4%	30%	1.3 ml
Urea	7 *M*		4.2 g
TBE buffer	1 ×	5 ×	2 ml
APS	0.7 mg/ml	40 mg/ml	175 μl
Tetramethylethylenediamine (TEMED)	1 mg/ml		14 μl
Water			To a total volume of 10 ml

2.4 Subject an aliquot to 4% denatured polyacrylamide gel electrophoresis.

After incubation, mix an aliquot of the solution with 1 volume of loading buffer and loaded into a well of the gel. Electrophoresis is performed with the electrode buffer described above for approximately 2 h at 100 V. Washing the well before loading is important to obtain a sharp band.

2.5 Fix the gel with 7% acetic acid for 10 min with gentle shaking.

2.6 Dry the gel in a gel dryer at 80 °C for approximately 30–60 min.

Complete drying is crucial for achieving high-sensitivity detection.

2.7 Expose the gel using an imaging plate.

The exposure time should be determined according to the freshness of the isotope used in the experiment. If fresh isotope is used, 30 min of exposure is sufficient.

2.8 Image analysis.

The exposed image is analyzed using a phosphorimager, such as a Typhoon FLA 9500. Self-cleaved RNA appears in a band slightly below that of the intact noncleaved RNA (see the example data in Fig. 5).

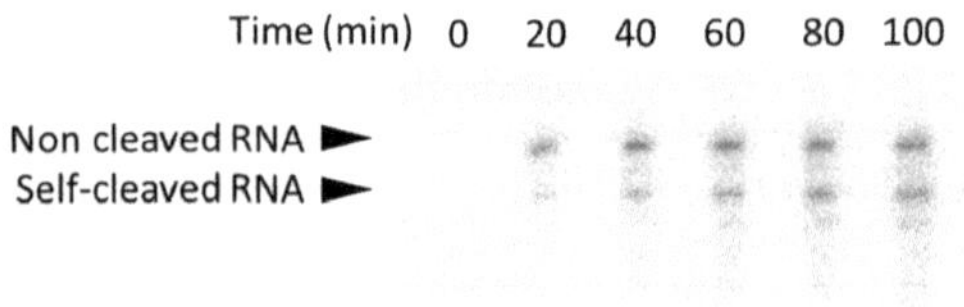

Figure 5 An example of a denaturing PAGE gel. In this experiment, an aptazyme of TPP from our previous study was incubated with 0.03 m*M* TPP for the indicated time and subjected to denatured PAGE as described in the text. Noncleaved (full-length) RNA and self-cleaved RNA were detected as separate bands.

Step 3: (Optional) Quantification of the intermediate RNAs

2.1 For the detection of the intermediate RNA species in the absence of ligand, mix the following materials in 0.2 microcentrifuge tube

DNA fragments	10 pmol
Sol. I of PUREfrex	10 µl
Sol. II of PUREfrex	1 µl
Sol. III of PUREfrex	1 µl
[^{32}P]-UTP	0.1 MBq
Water	Up to 20 µl

↓

2.2 Incubate the solution at 37 °C for 1 – 2 h

2.3 Prepare polyacrylamide gel

↓

2.4 An aliquot is subjected to 4% denatured polyacrylamide gel electrophoresis

↓

2.5 Fix the gel with 7% acetic acid for 10 min with gentle shaking

↓

2.6 Dry the gel with a gel dryer at 80 °C for about 30 – 60 min

↓

2.7 Exposure of the gel to an imaging plate

↓

2.8 Image analysis

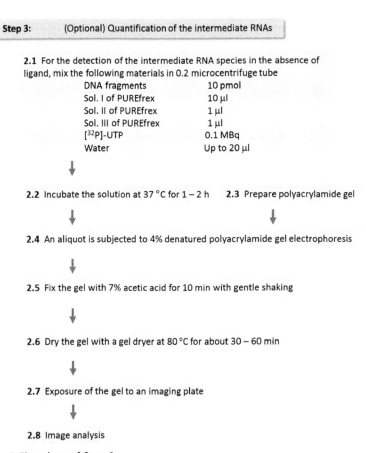

Figure 6 Flowchart of Step 3.

8.3. Tip

Control experiments both in the presence and the absence of ligand are recommended to identify the bands corresponding to intact and self-cleaved RNA.

8.4. Tip

To achieve better separation of the self-cleaved RNA from the noncleaved RNA, a longer electrophoresis might be required (Fig. 6).

REFERENCES

Kobori, S., Ichihashi, N., Kazuta, Y., Matsuura, T., & Yomo, T. (2012). Kinetic analysis of aptazyme-regulated gene expression in a cell-free translation system: Modeling of ligand-dependent and -independent expression. *RNA*, *18*, 1458–1465.

Piganeau, N., Jenne, A., Thuillier, V., & Famulok, M. (2000). An allosteric ribozyme regulated by doxycycline. *Angewandte Chemie International Edition*, *39*, 4369.

Robertson, M. P., & Ellington, A. D. (1999). In vitro selection of an allosteric ribozyme that transduces analytes to amplicons. *Nature Biotechnology*, *17*, 62–66.

Shimizu, Y., Inoue, A., Tomari, Y., Suzuki, T., Yokogawa, T., Nishikawa, K., et al. (2001). Cell-free translation reconstituted with purified components. *Nature Biotechnology*, *19*, 751–755.

Soukup, G. A., & Breaker, R. R. (1999). Nucleic acid molecular switches. *Trends in Biotechnology*, *17*, 469–476.

Soukup, G. A., & Breaker, R. R. (2000). Allosteric nucleic acid catalysts. *Current Opinion in Structural Biology*, *10*, 318–325.

Thompson, K. M., Syrett, H. A., Knudsen, S. M., & Ellington, A. D. (2002). Group I aptazymes as genetic regulatory switches. *BMC Biotechnology*, *2*, 21.

Wieland, M., Benz, A., Klauser, B., & Hartig, J. S. (2009). Artificial ribozyme switches containing natural riboswitch aptamer domains. *Angewandte Chemie International Edition*, *48*, 2715–2718.

Wieland, M., & Hartig, J. S. (2008). Improved aptazyme design and in vivo screening enable riboswitching in bacteria. *Angewandte Chemie International Edition*, *47*, 2604–2607.

Yen, L., Svendsen, J., Lee, J. S., Gray, J. T., Magnier, M., Baba, T., et al. (2004). Exogenous control of mammalian gene expression through modulation of RNA self-cleavage. *Nature*, *431*, 471–476.

CHAPTER SIX

Engineering of Ribosomal Shunt-Modulating Eukaryotic ON Riboswitches by Using a Cell-Free Translation System

Atsushi Ogawa[1]
Proteo-Science Center, Ehime University, Matsuyama, Japan
[1]Corresponding author: e-mail address: ogawa.atsushi.mf@ehime-u.ac.jp

Contents

Abstract

A number of natural and artificial bacterial riboswitches have been reported thus far. However, they generally function only in bacteria, not in eukaryotes. This is because of the differences of expression mechanisms (transcription, translation, and so on)

Methods in Enzymology, Volume 550
ISSN 0076-6879
http://dx.doi.org/10.1016/bs.mie.2014.10.033

between these two main types of organisms. For example, the mechanism of translation initiation is quite different between bacteria and eukaryotes, especially in ribosome loading on mRNA. While the bacterial ribosome binds to a well-conserved, internal sequence some bases before the start codon to initiate translation, the eukaryotic one is loaded on the 5′ terminus with the help of certain eukaryotic initiation factors. This means not only that bacterial riboswitches regulating translation initiation are not available in eukaryotic translation systems, but also that it is physically difficult to construct eukaryotic ON riboswitches that regulate the eukaryotic canonical translation initiation, because an aptamer cannot be inserted upstream of the ribosome loading site. However, the mechanism of noncanonical translation initiation via "ribosomal shunt" enables us to design translation initiation-modulating (specifically, ribosomal shunt-modulating) eukaryotic ON riboswitches. This chapter describes a facile method for engineering these ribosomal shunt-modulating eukaryotic ON riboswitches by using a cell-free translation system. Because these riboswitches do not require hybridization switching thanks to a unique shunting mechanism, they have the major advantages of a low energy requirement for upregulation and relatively straightforward design over common hybridization switch-based ON riboswitches.

1. INTRODUCTION

A riboswitch is a *cis*-regulatory RNA element that regulates expression of a downstream or upstream gene in response to a specific ligand molecule (Roth & Breaker, 2009). It comprises two parts, an aptamer domain and an expression platform, though they are not completely partible. The ligand specifically binds to the former part to induce conformational changes of the latter, which turns on or off the gene expression (specifically, transcription termination, translation initiation, splicing, or mRNA degradation). Many types of natural riboswitches have been discovered thus far, mainly in bacteria (Breaker, 2011), and almost all of them are OFF riboswitches, in which a ligand dose-dependently downregulates the gene expression.

Although natural aptamers in natural riboswitches recognize only endogenous metabolite ligands, an artificial RNA aptamer for an arbitrary molecule can be obtained from a randomized RNA library via an *in vitro* selection method (Ellington & Szostak, 1990; Tuerk & Gold, 1990). Therefore, an *in vitro*-selected aptamer could potentially be used to contribute ligand-dependency to the engineering of a synthetic riboswitch. In fact, several artificial riboswitches with *in vitro*-selected aptamers have been reported (Chang, Wolf, & Smolke, 2012; Topp & Gallivan, 2010; Wieland & Hartig, 2008; Wittmann & Suess, 2012). The most commonly used are bacterial riboswitches regulating translation initiation, which is relatively easy to control. They have generally been obtained through a screening from a library constructed by inserting an *in vitro*-selected aptamer with some randomized nucleotides into the region upstream of the ribosome binding site (RBS) (Davidson, Harbaugh, Chushak, Stone, & Kelley-Loughnane, 2013; Lynch & Gallivan, 2009; Nomura & Yokobayashi, 2007). Not only OFF riboswitches but also ON riboswitches, which ligand dose-dependently upregulate the gene expression, can be identified under an appropriate screening condition. In these translation initiation-regulating bacterial riboswitches, the accessibility of the 30S ribosome to the RBS is a key of regulation. For example, in the ON riboswitches, the RBS is sequestered by forming a duplex with a part of the aptamer domain in the absence of the ligand, while the RBS becomes accessible to the ribosome in a ligand dose-dependent manner, because it is opened by aptamer–ligand complex formation (Ogawa, 2014).

On the other hand, in a eukaryotic canonical translation system, the 40S ribosome is loaded onto the 5′ terminus (i.e., the 5′ cap) with the help of several eukaryotic initiation factors, so that it is impossible to insert an aptamer upstream of the ribosome loading site. This limitation is a major disadvantage for the design of a eukaryotic ON riboswitch. If an aptamer is placed in a 5′ untranslated region (5′ UTR), the aptamer–ligand complex should prevent the ribosome from being loaded or proceeding on the mRNA: this is an OFF riboswitch (*vide infra*). Therefore, it has been a challenge to obtain a translation initiation-regulating eukaryotic ON riboswitch, though eukaryotic ON riboswitches driven by mRNA cleavage or degradation have been engineered through the use of an allosteric self-cleaving ribozyme (Ogawa, 2009; Win & Smolke, 2007), an endogenous splicing mechanism (Culler, Hoff, & Smolke, 2010), or the phenomenon of nonsense-mediated mRNA decay (Endo, Hayashi, Inoue, & Saito, 2013). However, I have recently succeeded in constructing mRNA cleavage-free ON (and OFF) riboswitches functioning even in a eukaryotic translation system by harnessing an internal ribosome entry site (IRES), which allows the ribosome to enter at the middle of an mRNA transcript (Ogawa, 2011, 2012). In addition, I established a method for the rational design of these IRES-based riboswitches using only the sequence information of *in vitro*-selected aptamers, the details of which are described elsewhere (Ogawa, 2011, 2012, 2014). Therefore, an arbitrary ligand-dependent eukaryotic riboswitch can be constructed in a relatively simple way with the corresponding well-minimized *in vitro*-selected aptamer. The mechanism of IRES-based riboswitches is very similar to that of the translation initiation-regulating bacterial riboswitches: in the case of the ON riboswitches, the IRES is deactivated by an upstream sequence (anti-IRES) complementary to a crucial part of the IRES in the absence of the ligand, while the ligand binds to the aptamer domain placed in advance of the anti-IRES to dehybridize the duplex between the IRES and the anti-IRES, thereby activating IRES-mediated translation (Ogawa, 2011, 2014). In other words, the IRES-based riboswitches share a common feature with most of bacterial riboswitches: they require hybridization switches to activate (or suppress) the gene expression. In contrast, in eukaryotic OFF riboswitches that are artificially constructed with aptamers inserted into the 5′ UTR as described above, the aptamer–ligand complex itself down-regulates the gene expression, in principle without any hybridization switches (Grate & Wilson, 2001; Harvey, Garneau, & Pelletier, 2002; Saito et al., 2010; Suess et al., 2003; Werstuck & Green, 1998). Therefore, they have a major advantage in terms of a low energy requirement for regulating the gene expression. In addition, they can be very easily prepared by simply inserting one or more aptamers into the 5′ UTR, without the need for a specialized device.

Although, a rigid and robust structure such as an aptamer–ligand complex on mRNA generally inhibits the ribosomal progression in this way, if we could use a translation mechanism wherein a rigid structure conversely activates ribosomal movement, it might be fairly easy to design a hybridization switch-free ON riboswitch. In fact, I have recently succeeded in construction of such a riboswitch, by harnessing a eukaryote-specific "ribosomal shunt" that meets the above conditions (Ogawa, 2013). In this chapter, a facile method for engineering this novel type of eukaryotic ON riboswitch by using a cell extract is described.

2. A EUKARYOTIC TRANSLATION MECHANISM REQUIRING A RIGID mRNA STRUCTURE FOR RIBOSOMAL PROGRESSION

To promote the ribosomal progression by complex formation between a ligand and the corresponding aptamer on mRNA, it is necessary to identify a natural mechanism by which a rigid structure on mRNA plays an important role in activating movement of the ribosome. Ideally, the mechanism would require a simple stem-loop structure that can be replaced with an aptamer–ligand complex. In eukaryotic translation systems, internal ribosome entry and ribosomal shunt have been floated as candidates. An

IRES for the former generally has some stem-loops within its relatively long sequence to form an active higher order structure for translation initiation (Pfingsten, Costantino, & Kieft, 2006). Nonetheless, it is highly possible to render the precisely folded active structure inactive if an aptamer–ligand complex is substituted for one of the stem-loops. In contrast, the requirement of a rigid stem-loop for the latter is considered to be due mainly to the rigidity (Pooggin, Hohn, & Fütterer, 2000), which renders it more potential for the exchange of the complex.

Ribosomal shunt is an atypical mechanism of ribosomal movements that some viruses (cauliflower mosaic virus (CaMV), rice tungro bacilliform virus (RTBV), adenovirus, and so on) use in eukaryotic cells (Pooggin, Ryabova, He, Fütterer, & Hohn, 2006). Although the mechanism is slightly different between these viruses, a stable rigid structure on mRNA is basically imperative for ribosomal shunt. Among the known viral mRNAs on which ribosomal shunt occurs, the CaMV 35S RNA has been the most well studied. On this mRNA, the 40S ribosome that has just finished translation of the 5′ terminal short open reading frame (sORF) bypasses, or shunts over, the rigid secondary structure lying downstream of sORF—without unwinding it—from the takeoff site (TS) to the landing site (LS), and then reinitiates translation of the downstream open reading frame (dORF) (Fig. 1). Although the original rigid structure right behind sORF is very large (~500 nt), stem section I (ssI, $\Delta G = -40$ kcal/mol) is sufficient for efficient ribosomal shunt

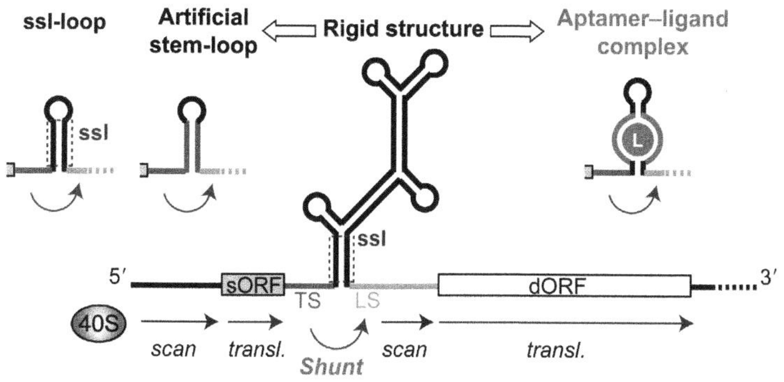

Figure 1 Schematic diagram of ribosomal shunt on the CaMV 35S mRNA. Thick lines and boxes represent noncoding regions and ORFs, respectively. The ribosomal movements are shown by single arrows and italics under the mRNA. The 60S ribosomal subunit is not shown here to simplify the diagram. The intergenic rigid structure required for shunting is replaceable with an ssI-loop (far left), a stable artificial stem-loop (second left), or an aptamer–ligand complex (right). The circle L represents the ligand. (See the color plate.)

(Ogawa, 2013). In addition, an artificial hairpin structure as stable as ssI can be substituted for it (Ogawa, 2013), so that it should also be possible to substitute an aptamer–ligand complex. Thus, the CaMV 35S RNA is chosen here as a foundation for hybridization switch-free eukaryotic riboswitches (other potential viral mRNAs may be available as alternatives, depending on the experimenter's circumstances, though detailed preliminary analyses of their mechanisms are recommended).

3. CHOICE OF A TRANSLATION SYSTEM FOR ENGINEERING ARTIFICIAL RIBOSWITCHES

An important step in the engineering of riboswitches is determining the translation system that will be used for their optimization and characterization. This, of course, depends on how the optimized riboswitch will be used. Although many applications are envisioned by virtue of the potential of riboswitches, they can be roughly divided into two types, *in vitro* and *in vivo* applications. As to the former, attention is currently focused on applications in cell-free synthetic biology (Hodgman & Jewett, 2012). In addition, cell-free riboswitches are expected to be useful tools in bioanalyses and diagnoses because they enable us to calculate the amount of a specific molecule (i.e., ligand) from that of the protein expressed from the regulated gene (Muranaka, Sharma, Nomura, & Yokobayashi, 2009; Ogawa, 2011, 2013; Ogawa & Maeda, 2007). In terms of the latter, it goes without saying that artificial riboswitches are promising as genetic tools (Nakahira, Ogawa, Asano, Oyama, & Tozawa, 2013). Furthermore, they have potential for use in the reprogramming of cellular functions and behaviors in synthetic biology or metabolic engineering applications (Liang, Bloom, & Smolke, 2011; Chang et al., 2012; Topp & Gallivan, 2010).

Normally, cell-free translation systems (i.e., cell extracts that include energy-regenerating components) should be used for optimizing riboswitches that will be applied to *in vitro* applications. Conversely, living cells are suitable for riboswitches that will be used in *in vivo* applications. Nonetheless, the former is useful even if artificial riboswitches are intended to be finally leveraged *in vivo*, because a riboswitch optimized in a certain cell extract is expected to function well in the related cells (or perhaps organisms) with minor corrections unless it is driven by mRNA cleavage (Ogawa, 2011). Cell-free systems also have advantages, because they reduce the time and effort needed to maintain cells and construct various transformants. They require for translation only *in vitro* transcripts (or DNA templates in some cases), which can be quickly and easily obtained as described below. Therefore, it is reasonable to use them for engineering riboswitches, no matter how the obtained riboswitch will finally be used.

Various types of cell extracts are available as cell-free translation systems (Endo & Sawasaki, 2006; Carlson, Gan, Hodgman & Jewett, 2012), depending on the mRNA selected as the foundation of the riboswitch. In the present case of the CaMV 35S RNA, a plant cell extract should be chosen for reason of its compatibility. A wheat germ extract (WGE) is used here on the grounds that ribosomal shunt is known to take place efficiently on CaMV 35S RNA-based mRNAs in WGE (Ogawa, 2013; Pooggin et al., 2006). Although WGE is commercially available from several companies, WGE from CellFree Sciences (Matsuyama, Japan) is recommended because it has almost no contamination of the suicide system directed against the translational machine, which usually does not exist in cells but invades when a cell wall is damaged (Madin, Sawasaki, Ogasawara, & Endo, 2000). Therefore, in the case that WGE is prepared in a laboratory, note that the wheat embryos should be extensively washed to remove the suicide system before preparing the extract, according to a protocol described elsewhere (Takai, Sawasaki, & Endo, 2010). If other viral mRNAs are chosen as the foundation of riboswitches, a compatible cell extract should be selected.

4. *IN VITRO*-SELECTED APTAMER FOR RIBOSOMAL SHUNT-MODULATING RIBOSWITCHES

A wide variety of RNA aptamers that specifically bind to their ligands have thus far been selected *in vitro* from randomized RNA libraries. Of course, not all of these *in vitro*-selected aptamers can be used for ribosomal shunt-modulating eukaryotic riboswitches in the form in which they were initially selected. To be useful in this context, an aptamer must bind tightly to its ligand under the conditions of the selected translation system (WGE here). It also needs to be well minimized; it is better to remove as many extra bases as possible. In addition, it is helpful to know the aptamer–ligand complex structure or at least which bases in the aptamer are responsible for the binding. The most important requirement to be fulfilled is that the aptamer has a loop that is not directly related to the ligand binding and thus can be cut into two fragments (a so-called split aptamer) as shown in Fig. 2A. This is because the aptamer must be split before being implanted in the mRNA to create a ribosomal shunt-modulating riboswitch (see Section 5).

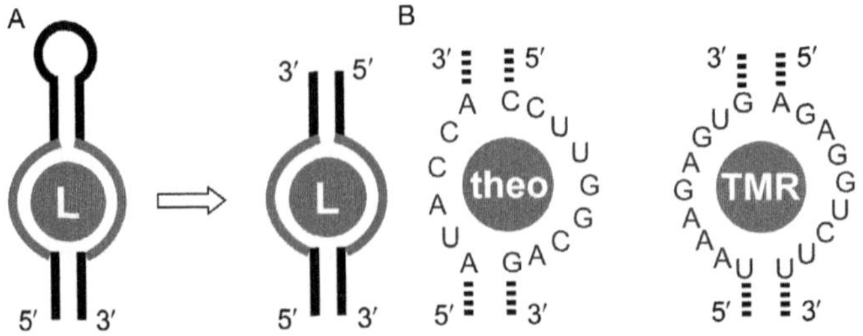

Figure 2 An ideal aptamer structure for ribosomal shunt-modulating riboswitches. (A) General structure of a splittable aptamer. (B) Examples of a split aptamer. A theophylline (theo)-binding aptamer (left) and a tetramethylrhodamine (TMR)-binding aptamer (right). Note that some base pairs above and below these aptamers are required for binding to their specific ligands.

Therefore, the ideal choice is an aptamer that strongly binds to the ligand in an internal loop under typical physiological conditions, such as a theophylline-binding aptamer (Jenison, Gill, Pardi, & Polisky, 1994) or a tetramethylrhodamine-binding aptamer (Carothers, Goler, Kapoor, Lara, & Keasling, 2010) (Fig. 2B).

If there is no desirable aptamer for a ligand of interest, it is necessary to select one using a well-established *in vitro* selection method (Hayashi & Nakatani, 2014; Murata & Sato, 2014; Ohuchi, 2014). In this case, to obtain an aptamer that fulfills the requirements described above, it is better to carefully design the initial library. Specifically, it is recommended that a fixed short stem-loop be inserted into the randomized region (Giver et al., 1993). In addition, the selection conditions (e.g., magnesium concentrations) should be adjusted to be similar to those of the translation system used for optimizing riboswitches (Carothers et al., 2010). After clarifying the sequences of *in vitro*-selected aptamers, minimization and characterization should be performed as described above. In particular, knowledge of the base(s) responsible for binding to the ligand is useful for constructing a negative control mRNA, which is necessary to confirm if the ligand–aptamer interaction on mRNA actually activates the downstream gene expression.

5. HOW TO IMPLANT THE SELECTED APTAMER INTO mRNA

Although it is entirely possible to use an aptamer–ligand complex to activate ribosomal shunt instead of the rigid secondary structure on the

CaMV 35S RNA in the presence of the ligand (i.e., in the ON state), it should also be considered how best to suppress the dORF expression in the absence of the ligand (i.e., in the OFF state). This is, of course, because the suppression in the OFF state is crucial for ON riboswitches to exert the high switching efficiency. As described above, in the case that a stable rigid structure lies downstream of sORF (actually, at a "*proper*" distance: *vide infra*), the 40S ribosome that has just finished translation of sORF can bypass (shunt over) the obstacle to LS (Fig. 1). In contrast, when there is no rigid structure in the intergenic region between sORF and dORF, the 40S after sORF translation scans for the next start codon and efficiently reinitiates translation of dORF (Ogawa, 2013). This means that dORF is likely to be well translated even in the OFF state if an aptamer, which is generally flexible without the ligand, is just replaced with the original rigid structure. Therefore, the scanning ribosome in the intergenic region must be stopped or trapped by some mechanism to suppress the gene expression in the OFF state.

For this purpose, a rigid stem-loop and/or a mimic start codon can be used. Although the former seems to be contradictory to the shunting mechanism, it generally inhibits the ribosomal progression as in the case of typical artificial OFF riboswitches for eukaryotic translation systems (see Section 1). Actually, ribosomal shunt occurs only when the distance between sORF and the rigid stem (Dss, i.e., the length of TS) is *proper*, as described above. When Dss is *improper*, the mode of ribosomal movement becomes not shunting but scanning, so that the 40S ribosome that has left sORF is inhibited by the stem structure. According to my previous results using WGE (Ogawa, 2013), the original 6-nt Dss is the most efficient for ribosomal shunt, while a Dss longer than 26 nt completely allows the 40S ribosome scan to proceed and stop at the rigid stem (Fig. 3A). Therefore, a split aptamer connected to a rigid stem-loop should be replaced with the original rigid structure so that the Dss in the OFF state (OFF-Dss) will be more than 26 nt and the Dss in the ON state (ON-Dss) will be 6 nt (Fig. 3B). In view of the fact that some stem-end base pairs longer than 3 bp are required for efficient shunting in the ON state (but stem-end base pairs that are too long will allow the ribosome to shunt even in the OFF state), the 5′ side of the split aptamer needs to be longer than 17 nt. If the selected aptamer does not meet this requirement, two split aptamers in tandem with some intervening base pairs (4–6 bp) can be used instead (Fig. 4A). In fact, the theophylline-dependent ON riboswitch that I optimized in WGE includes two split theophylline aptamers (Ogawa, 2013) (Fig. 4B). In regard to the rigid stem-loop to be connected to the split

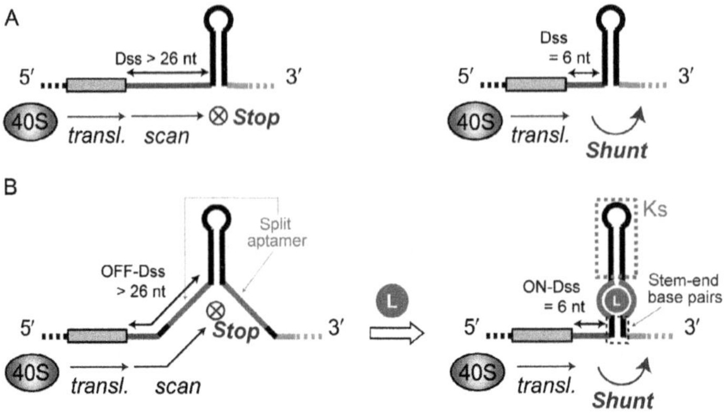

Figure 3 Design of ribosomal shunt-modulating riboswitches by using the mechanism for Dss-dependently switching the mode of ribosomal movement after sORF translation. (A) The effect of Dss on the ribosomal movement. The 40S ribosome that has finished translation of sORF scans TS and stops at the rigid stem when Dss is longer than 26 nt (left), while it efficiently shunts over the rigid stem when Dss is properly short (~6 nt) (right). (B) General design of ribosomal shunt-modulating riboswitches with a split aptamer (left, the OFF state; right, the ON state). (See the color plate.)

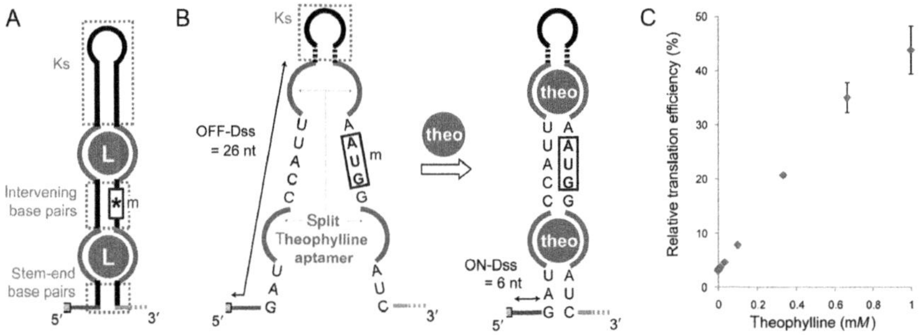

Figure 4 Ribosomal shunt-modulating riboswitches with two split aptamers in tandem. (A) General structure of the aptamer–Ks conjugate part in the ON state. The boxed star, m, represents the mimic start codon. (B) The optimized theophylline-dependent riboswitch (left, the OFF state; right, the ON state). (C) Theophylline dependence of the optimized riboswitch.

aptamer(s), a slightly modified Kozak stem (Ks, 5′ GCA AGC UGG GGC GCG UGG UGG CGG CUG CAG CCG CCA CCA CGC GCC CCA AGC UUG C 3′) is recommended (Pooggin et al., 2000). As to the stem-end base pairs, though the original ones (5′-GAU(AA)-3′/5′-(UU)AUC-3′) work efficiently, other base pairs may also do so (Fig. 4B).

The use of another candidate (mimic start codon: start codon upstream of the dORF for trapping the ribosome) in combination with Ks is also effective for suppressing translation in the OFF state (Ogawa, 2009, 2013). It should be placed at the 3′ side of the stem-end (or intervening) base pairs (Fig. 4B) to trap the 40S ribosome that has somehow unwound or shunted over Ks in the OFF state. Note that it is better that the corresponding in-frame stop codon lies in the dORF or 3′ UTR so that the 40S ribosome cannot reinitiate translation of dORF from the original start codon. In addition, if possible, the Kozak sequence (5′ A/GNNAUGG 3′, the start codon is underlined) should be used within the vicinity of the mimic start codon to efficiently trap the 40S (Kozak, 1989). If the translation efficiency in the OFF state is still higher than desired, the insertion of an additional mimic start codon into the loop of Ks is moderately effective (Ogawa, 2013). In this case, the translational frame (and the sequence around the start codon) should be taken care of as in the case of the first mimic start codon.

6. GENERAL DESIGN OF mRNAs WITH RIBOSOMAL SHUNT-MODULATING RIBOSWITCHES

The following parts are required for mRNAs (in this order): 5′ UTR, sORF, TS, an aptamer–Ks conjugate with stem-end (and intervening) base pairs, LS, dORF, and 3′ UTR.

6.1. 5′ Untranslated region

Although the original 5′ UTR can be used as is, a translational enhancer, E01, is recommended for efficient translation of sORF without the 5′ cap in WGE (Kamura, Sawasaki, Kasahara, Takai, & Endo, 2005). The sequence of E01 is 5′ G_{+1}AA CUC ACC UAU CUC CCC AAC ACC UAA UAA CAU UCA AUC ACU CUU UCC ACU AAC CAC CUA UCU ACA UCA CCA AGA UAU CAC UAG U 3′. The 3′ terminal nucleotide is fused directly to the sORF.

6.2. Short open reading frame

The original CaMV sORF sequence (5′ AUG UGU GAG UAG 3′) is available as is. Although it has been reported that use of the RTBV sORF (5′ AUG GCU CAG GUC AGU GAG UAG 3′) instead of the original increased the shunting efficiency (Pooggin et al., 2006), this was not observed in my experiments (Ogawa, 2013).

6.3. Takeoff site

As described above, the original 6-nt sequence (5′ UUC CCA 3′) is suitable for efficient ribosomal shunt. The 5′ terminal U and 3′ terminal A are fused directly to the sORF and the aptamer–Ks conjugate, respectively.

6.4. Aptamer–Ks conjugate

The design of this part is discussed in Section 5. Basically, only this part is variable, so all one has to do is to optimize this part, especially the sequence of stem-end (and intervening) base pairs, the position and the number of the mimic start codons, the direction of the selected split aptamer, and so on. For a positive control, the following sequence including a slightly mutated ssI sequence is recommended: 5′ *GAU AAG GGA AUU GUG GUU CUU AUA GGG UUU CGC UU*G CAA GCU GGG GCG CGU GGU GGC GGC UGC AGC CGC CAC CAC GCG CCC CAA GCU UGC *GUA UUU ACC CUA UAU ACC ACA AUG GCC CCU UAU C* 3′ (italics: a slightly mutated ssI sequence; underlined: Ks).

6.5. Landing site

Because the original LS sequence is not long enough for efficient reinitiation of dORF, it is better to lengthen it with a nonstructured RNA sequence that has no AUG triplet. The recommended LS sequence is 5′ *GAU UUA AAG AAA UAA UCC GCA UAA GCC CCC GCU UAA AAA AUU GGU AUC AGA GCC* GAA CUC ACC UAU CUC UCU ACA CAA AAC AUU UCC CUA CAU ACA ACU UUC AAC UUC CUA UUC AGU 3′ (italics: the original LS sequence; underlined: the potential stop codons for the mimic start codon(s)). When the mimic start codon is used in the aptamer–Ks conjugate, it should be placed such that it not be in-frame of these stop codons to avoid reinitiating translation of dORF in the OFF state (the introduction of mutation into the stop codons is an option). The 5′ terminal nucleotide and the 3′ terminal nucleotide are fused directly to the aptamer–Ks conjugate and the dORF, respectively.

6.6. Downstream open reading frame

In optimizing riboswitches, a reporter gene whose expression product can be easily detected is the best choice. To detect low expression efficiencies, the firefly luciferase gene is recommended, the sequence of which is described elsewhere.

6.7. 3′ Untranslated region

Although any sequence can be used as the 3′ UTR, a length of more than 800 nt is required for efficient translation. This is for inhibiting mRNA degradation. Nonetheless, polyadenylation is not necessary because it has almost no effect on the translation efficiency in WGE if the 3′ UTR is sufficiently long (Sawasaki, Ogasawara, Morishita, & Endo, 2002).

7. EXPERIMENTS

7.1. Construction of DNA templates for mRNAs

A protocol for constructing DNA templates that will be transcribed to mRNAs is briefly described below. All PCRs should be performed with a high-fidelity DNA polymerase such as PrimeSTAR MAX DNA polymerase (Takara Bio, Shiga, Japan). A typical reaction mixture for PCR is shown in Table 1. Refer to the manufacturer's protocol for the thermal cycler program.

(i) Construct a plasmid encoding a positive control mRNA, which is designed according to Section 6. A gene synthesis service is commercially available to easily obtain it. For *in vitro* transcription afterward, the bacteriophage SP6 promoter (5′ ATT TAG GTG ACA CTA TA 3′) should be added to the 5′ terminus of 5′ UTR, though it can also be introduced by PCR at the next step.

(ii) PCR-amplify the sequence for *in vitro* transcription (from the SP6 promoter to the 3′ UTR composed of more than 800 nt) by using the plasmid (as a template) and appropriate primers. An example of a forward primer is 5′ ATT TAG GTG ACA CTA TAG AA 3′ (the SP6 promoter is underlined). As to the reverse primer, refer

Table 1 Reaction mixture for PCR

Nuclease-free water	4 μL
PrimeSTAR MAX Premix	10 μL
2 μ*M* Forward primer	2 μL
2 μ*M* Reverse primer	2 μL
10–100 pg/μL Plasmid (1–100 pM ligation product)	2 μL
Total	20 μL

to the plasmid sequence for the design. Purification of the PCR product is not necessary for the next *in vitro* transcription.

(iii) Design some riboswitch candidates (specifically, the aptamer part including its vicinity) according to Section 5.

(iv) In regard to DNA templates for mRNAs with these potential candidates, first PCR-amplify the upstream segment (from the 5′ terminus to the loop of Ks) and the remaining downstream segment, separately, by using the plasmid and the following primers. The forward primer for the former segment and the reverse primer for the latter segment are the same as those in Step (ii). The reverse primer for the former should be 5′ NNN NNN <u>CTG CAG</u> CCG CCA CCA CGC GCC CCA GCT TGC—X_n [a sequence complementary to the 5′ side of the aptamer part including the stem-end (and intervening) base pairs]—TGG GAA CTA CTC ACA C 3′ (the *Pst*I site is underlined and N means any base). The forward primer for the latter should be 5′ NNN NNN <u>CTG CAG</u> CCG CCA CCA CGC GCC CCA AGC TTG C—Y_m [the 3′ side of the aptamer part]—GAT TTA AAG AAA TAA TCC 3′. Incidentally, the 5′ terminal N_6 bases in these primers are helpful in allowing *Pst*I to recognize its specific site. Of course, other restriction enzyme sites can be used instead if the *Pst*I site is improper for some reason (e.g., it appears in the aptamer, dORF, or 3′ UTR). After PCRs, mix both products and purify them with a commercially available spin column kit.

(v) Digest these two segments in the mixture with *Pst*I according to the manufacturer's protocol and then purify them.

(vi) Ligate them with a commercially available ligation kit according to the manufacturer's protocol.

(vii) Separate the ligation reaction mixture on a 1.0–1.5% agarose gel and purify the desired ligation product from the gel with a commercially available spin column kit. Note that a maximum of five bands should be observed: two reactants (the upstream segment and the downstream segment) and three ligated products between two upstream segments, two downstream segments, or each segment. To cut off only the desired one (i.e., the last) without contamination of the others, separate these bands completely with a long electrophoresis time.

(viii) PCR-amplify the ligation product for *in vitro* transcription as in Step (ii).

(ix) Store the PCR products at approximately −20 °C until use.

Table 2 Reaction mixture for *in vitro* transcription

Nuclease-free water	1.75 μL
10 × Transcription buffer	1 μL
NTP mix at 25 m*M* each	2 μL
100 m*M* DTT	1 μL
1 U/μL RNase inhibitor	0.25 μL
Enzyme solution	1 μL
PCR product	3 μL
Total	10 μL

7.2. *In vitro* transcription

A protocol for preparing 5′ cap- and 3′ poly(A)-free mRNAs is shown below. It should be noted that both the 5′ cap and poly(A) are not necessarily required for efficient translation in WGE (Sawasaki et al., 2002).

(i) Transcribe the DNA templates with an SP6-Scribe Standard RNA IVT Kit (CELLSCRIPT, Madison, WI) in a 10-μL reaction volume. Specifically, prepare a reaction mixture as in Table 2 and incubate it at 37 °C overnight.

(ii) Add 0.5 μL of the DNase included in the kit to the mixture, and then further incubate at 37 °C for 15 min.

(iii) Purify the transcribed mRNA with an RNeasy MinElute Cleanup Kit (QIAGEN, Tokyo, Japan) according to the manufacturer's protocol. Use 14 μL of nuclease-free water for the elution of mRNA from the column in order to sufficiently concentrate the mRNA.

(iv) Quantify the purified mRNA with a UV spectrometer and adjust the concentration to 3 pmol/μL with nuclease-free water.

(v) Aliquot the mRNA into smaller volumes and store them at −80 °C until use.

7.3. *In vitro* translation in WGE

As described above, a protocol for use with the WEPRO1240 Expression Kit (CellFree Sciences) is introduced here. Note that it is different from the manufacturer's protocol.

Table 3 Reaction mixture for *in vitro* translation in WGE

Nuclease-free water	4.6–y μL
x μM Ligand	y μL
3 μM mRNA	1 μL
1 mg/mL Creatine kinase	0.4 μL
WEPRO1240	2 μL
5 × SUB-AMIX	2 μL
Total	10 μL

(i) The following preparation should be performed beforehand by using materials included in the kit: Add 125 μL each of kit solutions S1, S2, S3, and S4 in a sequential fashion to 500 μL of nuclease-free water to prepare 1 mL of 5 × SUB-AMIX, and aliquot it into smaller volumes. Dilute 20 μL of 20 mg/mL creatine kinase with 380 μL of nuclease-free water to 1 mg/mL, and aliquot it into smaller volumes. Aliquot 1 mL of kit solution WEPRO1240 (this is WGE) into smaller volumes. Store them at −80 °C until use.

(ii) Prepare a reaction mixture with or without the ligand as in Table 3 and incubate it at 26 °C for a fixed period of time (typically 1 h).

(iii) Place the mixture on ice to stop the reaction.

7.4. Characterization of riboswitches

(i) Measure the amount of reporter protein expressed from dORF by the corresponding method.

(ii) Compare the translation efficiency to that of the positive control. Incidentally, the relative translation efficiencies of the optimized theophylline-dependent riboswitch were 3.0% without theophylline and 43.8% with 1 mM theophylline, when dORF encoding the firefly luciferase gene was used (Fig. 4C). The ON/OFF efficiency, which is the ratio of the translation efficiency in the presence of a certain concentration of the ligand to that in its absence, is also useful for evaluating the switching ability of a riboswitch. That of the abovementioned theophylline-dependent riboswitch was 14.4 at 1 mM theophylline.

(iii) If the relative activity is not satisfactory, redesign the candidates and recharacterize them.
  (a) In the case of high activity in the OFF state
  - Make sure that OFF-Dss is longer than 26 nt.
  - Insert the mimic start codon(s) according to Section 5.
  - Change the position or the surrounding base(s) of the mimic start codon(s).
  - Change the sequence or shorten the length of the stem-end (and/or intervening) base pairs.

  (b) In the case of low activity in the ON state
  - Determine whether the ligand affects translation by using the positive control.
  - Change the direction of the aptamer.
  - Change the sequence or extend the length of the stem-end (and/or intervening) base pairs.
  - Confirm the switching activity of the already optimized theophylline-dependent riboswitch (Fig. 4B) under the same conditions.

(iv) If a desirable riboswitch is obtained, confirm that the ligand–aptamer interaction on mRNA actually turns on the dORF expression by using a negative control mRNA (see Section 4).

8. CONCLUSION

As we have seen, it is very straightforward to engineer ribosomal shunt-modulating eukaryotic ON riboswitches by using a cell-free translation system. The design is simple and does not require any calculations, unlike that of IRES-based eukaryotic riboswitches (Ogawa, 2011), thanks to a mechanism that does not require hybridization switching. In addition, the energy necessary for switching between ON and OFF states should be lower than that of hybridization switch-based riboswitches, so that the ribosomal shunt-modulating riboswitches are expected to show a higher switching efficiency (Ogawa, 2013). Therefore, these switches should be useful tools in cell-free synthetic biology and cell-free biotechnology (Hodgman & Jewett, 2012). Moreover, the optimized riboswitches also have the potential to function in compatible cells. In applying the ribosomal shunt-modulating riboswitches derived from the CaMV 35S RNA *in vivo*, it is recommended that the CaMV-encoded TAV protein, which is known to promote ribosomal shunt, be added or coexpressed (see the following report for details, Pooggin et al., 2000). It is my hope that ribosomal shunt-modulating riboswitches constructed by the method described herein will prove highly effective in a range of applications and studies.

ACKNOWLEDGMENTS

This work was supported by MEXT KAKENHI Grant number 24119510 and JSPS KAKENHI Grant number 25708027.

REFERENCES

Breaker, R. R. (2011). Prospects for riboswitch discovery and analysis. *Molecular Cell, 43*, 867–879.

Carlson, E. D., Gan, R., Hodgman, C. E., & Jewett, M. C. (2012). Cell-free protein synthesis: Applications come of age. *Biotechnology Advances, 30*, 1185–1194.

Carothers, J. M., Goler, J. A., Kapoor, Y., Lara, L., & Keasling, J. D. (2010). Selecting RNA aptamers for synthetic biology: Investigating magnesium dependence and predicting binding affinity. *Nucleic Acids Research, 38*, 2736–2747.

Chang, A. L., Wolf, J. J., & Smolke, C. D. (2012). Synthetic RNA switches as a tool for temporal and spatial control over gene expression. *Current Opinion in Biotechnology, 23*, 679–688.

Culler, S. J., Hoff, K. G., & Smolke, C. D. (2010). Reprogramming cellular behavior with RNA controllers responsive to endogenous proteins. *Science, 330*, 1251–1255.

Davidson, M. E., Harbaugh, S. V., Chushak, Y. G., Stone, M. O., & Kelley-Loughnane, N. (2013). Development of a 2,4-dinitrotoluene-responsive synthetic riboswitch in *E. coli* cells. *ACS Chemical Biology, 8*, 234–241.

Ellington, A. D., & Szostak, J. W. (1990). *In vitro* selection of RNA molecules that bind specific ligands. *Nature, 346*, 818–822.

Endo, K., Hayashi, K., Inoue, T., & Saito, H. (2013). A versatile cis-acting inverter module for synthetic translational switches. *Nature Communications, 4*, 2393.

Endo, Y., & Sawasaki, T. (2006). Cell-free expression systems for eukaryotic protein production. *Current Opinion in Biotechnology, 17*, 373–380.

Giver, L., Bartel, D., Zapp, M., Pawul, A., Green, M., & Ellington, A. D. (1993). Selective optimization of the Rev-binding element of HIV-1. *Nucleic Acids Research, 21*, 5509–5516.

Grate, D., & Wilson, C. (2001). Inducible regulation of the *S. cerevisiae* cell cycle mediated by an RNA aptamer–ligand complex. *Bioorganic and Medicinal Chemistry, 9*, 2565–2570.

Harvey, I., Garneau, P., & Pelletier, J. (2002). Inhibition of translation by RNA-small molecule interactions. *RNA, 8*, 452–463.

Hayashi, G. & Nakatani K. (2014). Development of photoswitchable RNA aptamer–ligand complexes. In J.M. Walker (Series Ed.) & A. Ogawa (Vol. Ed.), *Methods in molecular biology: Vol. 1111. Artificial riboswitches* (pp. 29–40). New York: Springer.

Hodgman, C. E., & Jewett, M. C. (2012). Cell-free synthetic biology: Thinking outside the cell. *Metabolic Engineering, 14*, 261–269.

Jenison, R. D., Gill, S. C., Pardi, A., & Polisky, B. (1994). High-resolution molecular discrimination by RNA. *Science, 263*, 1425–1429.

Kamura, N., Sawasaki, T., Kasahara, Y., Takai, K., & Endo, Y. (2005). Selection of 5′-untranslated sequences that enhance initiation of translation in a cell-free protein synthesis system from wheat embryos. *Bioorganic & Medicinal Chemistry Letters, 15*, 5402–5406.

Kozak, M. (1989). Context effects and inefficient initiation at non-AUG codons in eucaryotic cell-free translation systems. *Molecular and Cellular Biology, 9*, 5073–5080.

Liang, J. C., Bloom, R. J., & Smolke, C. D. (2011). Engineering biological systems with synthetic RNA molecules. *Molecular Cell, 43*, 915–926.

Lynch, S. A., & Gallivan, J. P. (2009). A flow cytometry-based screen for synthetic riboswitches. *Nucleic Acids Research, 37*, 184–192.

Madin, K., Sawasaki, T., Ogasawara, T., & Endo, Y. (2000). A highly efficient and robust cell-free protein synthesis system prepared from wheat embryos: Plants apparently contain a suicide system directed at ribosomes. *Proceedings of the National Academy of Sciences of the United States of America, 97*, 559–564.

Muranaka, N., Sharma, V., Nomura, Y., & Yokobayashi, Y. (2009). Efficient design strategy for whole-cell and cell-free biosensors based on engineered riboswitches. *Analytical Letters, 42*, 108–122.

Murata, A., & Sato, S. (2014). *In vitro* selection of RNA aptamers for a small-molecule dye. In J. M. Walker (Series Ed.) & A. Ogawa (Vol. Ed.), *Methods in molecular biology: Vol. 1111. Artificial riboswitches* (pp. 17–28). New York: Springer.

Nakahira, Y., Ogawa, A., Asano, H., Oyama, T., & Tozawa, Y. (2013). Theophylline-dependent riboswitch as a novel genetic tool for strict regulation of protein expression in cyanobacterium *Synechococcus elongatus* PCC 7942. *Plant & Cell Physiology, 54*, 1724–1735.

Nomura, Y., & Yokobayashi, Y. (2007). Reengineering a natural riboswitch by dual genetic selection. *Journal of the American Chemical Society, 129*, 13814–13815.

Ogawa, A. (2009). Biofunction-assisted sensors based on a new method for converting aptazyme activity into reporter protein expression with high efficiency in wheat germ extract. *ChemBioChem, 10*, 2465–2468.

Ogawa, A. (2011). Rational design of artificial riboswitches based on ligand-dependent modulation of internal ribosome entry in wheat germ extract and their applications as label-free biosensors. *RNA, 17*, 478–488.

Ogawa, A. (2012). Rational construction of eukaryotic OFF-riboswitches that downregulate internal ribosome entry site-mediated translation in response to their ligands. *Bioorganic & Medicinal Chemistry Letters, 22*, 1639–1642.

Ogawa, A. (2013). Ligand-dependent upregulation of ribosomal shunting. *ChemBioChem, 14*, 1539–1543.

Ogawa, A. (2014). Rational design of artificial ON-riboswitches. In J. M. Walker (Series Ed.) & A. Ogawa (Vol. Ed.), *Methods in molecular biology: Vol. 1111. Artificial riboswitches* (pp. 165–181). New York: Springer.

Ogawa, A., & Maeda, M. (2007). Aptazyme-based riboswitches as label-free and detector-free sensors for cofactors. *Bioorganic & Medicinal Chemistry Letters, 17*, 3156–3160.

Ohuchi, S. (2014). Identification of RNA aptamers against recombinant proteins with a hexa-histidine tag. In J. M. Walker (Series Ed.) & A. Ogawa (Vol. Ed.), *Methods in molecular biology: Vol. 1111. Artificial riboswitches* (pp. 41–56). New York: Springer.

Pfingsten, J. S., Costantino, D. A., & Kieft, J. S. (2006). Structural basis for ribosome recruitment and manipulation by a viral IRES RNA. *Science, 314*, 1450–1454.

Pooggin, M. M., Hohn, T., & Fütterer, J. (2000). Role of a short open reading frame in ribosome shunt on the *Cauliflower mosaic virus* RNA leader. *The Journal of Biological Chemistry, 275*, 17288–17296.

Pooggin, M. M., Ryabova, L. A., He, X., Fütterer, J., & Hohn, T. (2006). Mechanism of ribosome shunting in *Rice tungro bacilliform pararetrovirus*. *RNA, 12*, 841–850.

Roth, A., & Breaker, R. R. (2009). The structural and functional diversity of metabolite-binding riboswitches. *Annual Review of Biochemistry, 78*, 305–334.

Saito, H., Kobayashi, T., Hara, T., Fujita, Y., Hayashi, K., Furushima, R., et al. (2010). Synthetic translational regulation by an L7Ae-kink-turn RNP switch. *Nature Chemical Biology, 6*, 71–78.

Sawasaki, T., Ogasawara, T., Morishita, R., & Endo, Y. (2002). A cell-free protein synthesis system for high-throughput proteomics. *Proceedings of the National Academy of Sciences of the United States of America, 99*, 14652–14657.

Suess, B., Hanson, S., Berens, C., Fink, B., Schroeder, R., & Hillen, W. (2003). Conditional gene expression by controlling translation with tetracycline-binding aptamers. *Nucleic Acids Research, 31*, 1853–1858.

Takai, K., Sawasaki, T., & Endo, Y. (2010). Practical cell-free protein synthesis system using purified wheat embryos. *Nature Protocols, 5*, 227–238.

Topp, S., & Gallivan, J. P. (2010). Emerging applications of riboswitches in chemical biology. *ACS Chemical Biology, 5*, 139–148.

Tuerk, C., & Gold, L. (1990). Systematic evolution of ligands by exponential enrichment: RNA ligands to bacteriophage T4 DNA polymerase. *Science, 249*, 505–510.

Werstuck, G., & Green, M. R. (1998). Controlling gene expression in living cells through small molecule-RNA interactions. *Science, 282*, 296–298.

Wieland, M., & Hartig, J. S. (2008). Artificial riboswitches: Synthetic mRNA-based regulators of gene expression. *ChemBioChem, 9*, 1873–1878.

Win, M. N., & Smolke, C. D. (2007). A modular and extensible RNA-based gene-regulatory platform for engineering cellular function. *Proceedings of the National Academy of Sciences of the United States of America, 104*, 14283–14288.

Wittmann, A., & Suess, B. (2012). Engineered riboswitches: Expanding researchers' toolbox with synthetic RNA regulators. *FEBS Letters, 586*, 2076–2083.

CHAPTER SEVEN

Live-Cell Imaging of Mammalian RNAs with Spinach2

Rita L. Strack, Samie R. Jaffrey[1]
Department of Pharmacology, Weill Medical College, Cornell University, New York, New York, USA
[1]Corresponding author: e-mail address: srj2003@med.cornell.edu

Contents

Abstract

The ability to monitor RNAs of interest in living cells is crucial to understanding the function, dynamics, and regulation of this important class of molecules. In recent years, numerous strategies have been developed with the goal of imaging individual RNAs of interest in living cells, each with their own advantages and limitations. This chapter provides an overview of current methods of live-cell RNA imaging, including a detailed discussion of genetically encoded strategies for labeling RNAs in mammalian cells. This chapter then focuses on the development and use of "RNA mimics of GFP" or Spinach technology for tagging mammalian RNAs and includes a detailed protocol for imaging 5S and CGG_{60} RNA with the recently described Spinach2 tag.

1. INTRODUCTION

Our knowledge of the diverse roles of RNA molecules in every aspect of gene expression has expanded tremendously in recent years. To

Methods in Enzymology, Volume 550
ISSN 0076-6879
http://dx.doi.org/10.1016/bs.mie.2014.10.044

understand better the abundance, dynamics, and localization of individual RNAs, tools for labeling and imaging RNAs are necessary to study their spatiotemporal regulation. Recently, several major independent advances in imaging technology have allowed for RNAs to be monitored in living cells.

Many powerful RNA imaging techniques involve the use of exogenously added probes, such as labeled antisense probes (Molenaar, Abdulle, Gena, Tanke, & Dirks, 2004; Molenaar et al., 2001) and molecular beacons (Bao, Tsourkas, & Santangelo, 2004; Bratu, Cha, Mhlanga, Kramer, & Tyagi, 2003; Nitin, Santangelo, Kim, Nie, & Bao, 2004; Tyagi & Alsmadi, 2004). These techniques have the advantage of targeting native, nonengineered RNAs. However, they are limited by the ability to deliver probe into cells without perturbing the biological system (Santangelo, Alonas, Jung, Lifland, & Zurla, 2012). However, several fully genetically encoded systems for studying RNA in living cells have been developed. These include the MS2-GFP imaging system, the Pumilio-split-FP imaging system, and the use of "RNA mimics of GFP" to tag RNAs of interest. For the purposes of this chapter, these genetically encoded systems will be discussed in detail.

The MS2-GFP system for imaging tagged RNAs in living cells provided the first genetically encoded system for labeling endogenous RNAs (Brodsky & Silver, 2002; Fusco et al., 2003). In this system, an RNA of interest is tagged with up to 24 copies of the MS2 RNA hairpin and coexpressed with the MS2 coat protein tagged with GFP (Fig. 1A) (Fusco et al., 2003; Shav-Tal et al., 2004). Because MS2-GFP binds to the MS2 hairpin as a high-affinity dimer, the tagged RNAs become labeled with up to 48 GFP molecules, allowing for imaging of even individual mRNAs in living cells. This technique was recently used to study the trafficking of individual β-actin mRNAs in the neuronal cells of a living mouse (Park et al., 2014).

Although powerful, this technique has several drawbacks. First, the MS2-GFP fusion protein is fluorescent throughout the cell, which can lead to high background. To mitigate this problem, MS2-GFP can be fused with a nuclear localization signal, so that the excess unbound MS2-GFP concentrates in the nucleus, lowering cytoplasmic background fluorescence. However, this prevents nuclear and cytoplasmic RNAs from being imaged in a single experiment. In addition, the presence of up to 48 nuclear localization signals tethered to the tagged RNA can lead to mislocalization and confound the interpretation of trafficking behaviors of tagged RNAs (Tyagi, 2009).

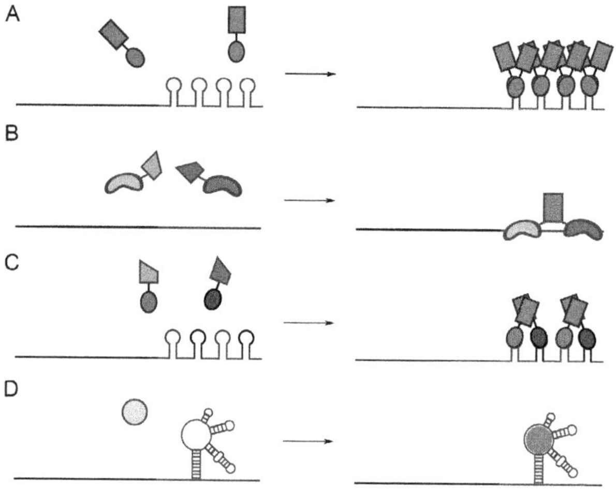

Figure 1 Genetically encoded strategies for labeling RNA in living cells. (A) The MS2-GFP imaging system involves labeling an mRNA of interest (black line) with up to 24 copies of the MS2 RNA hairpin in the 3′ UTR (gray line). When this tagged RNA is coexpressed with the MS2 coat protein fused to GFP, these hairpins are bound by MS2-GFP as dimers, leading to tagging by up to 48 MS2-GFP molecules. (B) The PUM-HD split-FP imaging system involves engineering two PUM-HDs to distinct 8-nt regions in an RNA of interest. Each PUM-HD is then fused to half of a split fluorescent protein. When both PUM-HDs are bound to the target RNA, fluorescence complementation occurs and the RNA is labeled with a single fluorescent protein. (C) The MS2/PP7 split-FP imaging system involves tagging an RNA with both the MS2 RNA hairpin (gray hairpin) and the PP7 RNA hairpin (blue hairpin) in alternation. MS2 and PP7 coat proteins are each tagged with complementary halves of an FP. When both are bound, complementation occurs, leading to labeling of RNAs. (D) Imaging an RNA of interest by Spinach tagging involves fusing Spinach to either the 5′ or 3′ end of an RNA and incubating cells with a dye that is nonfluorescent in solution. Only when the dye is bound to the tagged RNA does the Spinach–dye complex become fluorescent, specifically labeling the tagged RNA. (See the color plate.)

To bypass some of these problems, a similar system was developed using split fluorescent proteins fused to Pumilio homology domains (PUM-HDs) (Ozawa, Natori, Sato, & Umezawa, 2007). PUM-HDs are site-specific RNA-binding proteins that can be engineered to bind an 8-nt target of interest. In this case, the authors fused two halves of a fluorescent protein to two different PUM-HDs that were targeted to adjacent 8-nt sites on an RNA. Only when both proteins bound were the two halves of the fluorescent protein brought into close proximity to allow for bimolecular fluorescence complementation and fluorescence signal (Fig. 1B). This technique has several advantages. The first is that unbound proteins are nonfluorescent, reducing background fluorescence. The second is that PUM-HDs can be engineered to target native RNAs. A primary disadvantage is that each RNA of interest is only labeled with a single fluorescent protein, which does not allow imaging of individual RNA molecules. A second disadvantage is that PUM-HDs must be engineered every time a new RNA is targeted.

The Singer laboratory recently developed a new technology that weds the advantages of the MS2-GFP system and this PUM-HD system, in which two different RNA-binding proteins (the MS2 coat protein and the PP7 coat protein) are each tagged with a different half of a fluorescent protein (Wu, Chen, & Singer, 2014). Then, these two fusion proteins are coexpressed with an RNA containing both MS2 and PP7 binding sites in the 3′ UTR. Only when both fusion proteins have bound do the two fluorescent protein halves interact to become fluorescent, nearly eliminating background fluorescence (Fig. 1C). However, this technique still requires the large insertion of sequence into an RNA which may alter folding, function, or regulation.

A different approach for imaging RNA in living cells is based on appending RNA tags that directly confer fluorescence to RNAs of interest (Paige, Wu, & Jaffrey, 2011; Strack, Disney, & Jaffrey, 2013). Here, "RNA mimics of GFP" are used to tag RNAs for live-cell imaging (Fig. 1D). A detailed discussion of the development of Spinach and its derivative Spinach2 follows, along with a detailed protocol for imaging RNA in mammalian cells.

2. DEVELOPING SPINACH, AN RNA MIMIC OF GFP

In order to image RNA dynamics in living cells, we sought to develop an RNA tag that would be analogous to the GFP imaging system for protein localization in live cells. RNA aptamers had previously been described that can bind fluorescent dyes, which could potentially be used for labeling RNAs (Holeman, Robinson, Szostak, & Wilson, 1998). However, the fluorescence of these dyes was high regardless of whether they were bound to RNA, leading to high cellular background fluorescence. One exception was an aptamer that could conditionally activate a dye called malachite green (Babendure, Adams, & Tsien, 2003). In solution, malachite green is nonfluorescent, but when bound to a cognate aptamer RNA, it became brightly fluorescent. However, malachite green binds to cellular components which also induce its fluorescence, resulting in high background fluorescence levels (Paige et al., 2011).

To harness the power of an activatable dye, it is essential to use a dye that does not get activated by the diverse cellular constituents in the cell. Previous studies on GFP revealed that under protein denaturing conditions the chromophore was absorbant but not fluorescent, indicating that the folded protein modifies the structure of the chromophore to allow for fluorescence (Ward & Bokman, 1982). For this reason, we synthesized the GFP chromophore and several derivatives (Paige et al., 2011). We found these to be strongly absorbant, but essentially nonfluorescent in solution. Moreover, when cells were treated with these dyes, nominal cell-induced fluorescence was observed (Paige et al., 2011).

SELEX was then carried out against these dyes immobilized on agarose beads. Following SELEX, eluted RNAs were screened for their ability to induce fluorescence. The brightest RNA–dye pair was an RNA that binds 3,5-difluoro-4-hydroxybenzylidene imidazolinone (DFHBI), a fluorophore that resembles the GFP fluorophore but contains two aromatic fluorine atoms. This RNA, designated Spinach, is approximately 80% as bright as wild-type GFP and half as bright as enhanced GFP (EGFP) when bound to DFHBI (Fig. 2) (Paige et al., 2011).

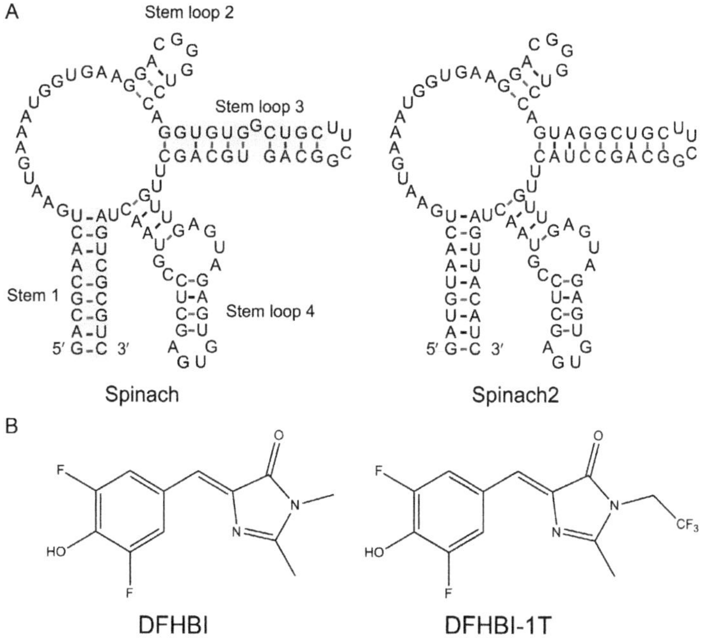

Figure 2 Schematic representations of the aptamers and dyes. (A) The mFOLD-predicted secondary structures of Spinach (left) and Spinach2 (right) are shown. The green (light gray in print version) highlighted region shows the regions that contain mutations. (B) DFHBI (left) and DFHBI-1T (right).

Spinach–DFHBI fluorescence was then tested in bacterial cells. When *Escherichia coli* cells overexpressing Spinach were incubated in media containing DFHBI, the cells were brightly fluorescent, indicating that Spinach was properly folded in this context and that DFHBI was able to enter the bacterial cells (Paige et al., 2011). Next, the utility of Spinach in tagging a mammalian RNA was examined. 5S RNA was tagged at the 3′ end with Spinach and expressed from its native promoter in HEK 293-T cells. Here, diffuse cytoplasmic signal was observed for 5S-Spinach that was relocalized to stress granules after sucrose treatment, as expected for untagged 5S (Paige et al., 2011). These experiments demonstrated that Spinach–DFHBI could be used for live-cell imaging.

3. IMAGING WITH SPINACH2, A SUPERFOLDING VARIANT OF SPINACH

Although Spinach could be used to tag 5S RNA, the exposure times necessary for imaging revealed that 5S-Spinach was substantially dimmer than expected, suggesting that Spinach might not perform well as a tag in mammalian systems. Moreover, experiments tagging other RNAs demonstrated that Spinach was not bright enough to be imaged, although tagged RNA was readily detected by fluorescence *in situ* hybridization (Strack et al., 2013).

We reasoned that several factors could affect the fluorescence of Spinach in cells, including ion concentration, temperature, the overall percentage folded, and the amount that is folded in the context of tagged RNA. Testing these parameters showed that even under ideal *in vitro* conditions, only ~30% of Spinach is folded (Strack et al., 2013). Melting curve analysis also showed that more than 50% of Spinach is unfolded at 37 °C, the temperature routinely used for imaging. Based on these results, we carried out systematic mutagenesis of Spinach to improve folding and thermostability. The result was Spinach2, which folds twice as well as Spinach *in vitro* and is substantially brighter than Spinach at 37 °C (Strack et al., 2013) (Fig. 2). Moreover, Spinach2 retains a high percent folding rate in diverse sequence contexts, unlike Spinach. These mutations seem to affect only the percentage of Spinach2 that folds properly, as the extinction coefficient, quantum yield, and K_D for dye binding were unchanged (Strack et al., 2013).

Spinach2 was then used to tag the noncoding RNAs 5S and 7SK. In both cases, Spinach2 was markedly brighter than Spinach (Strack et al., 2013). In addition, Spinach2 was used to label a "toxic" model RNA associated with

Fragile X tremor/ataxia syndrome (FXTAS). When the 5′ UTR of *FMR1* contains 55–200 CGG repeats, the resulting RNA forms aggregates (Dombrowski et al., 2002; Fu et al., 1991; Hagerman et al., 2001; Tassone, Iwahashi, & Hagerman, 2004) that are thought to cause toxicity by sequestration of nuclear proteins involved with essential cellular processes such as microRNA biogenesis (Sellier et al., 2013) and splicing (Sellier et al., 2010). For our work, Spinach2 was used to label a model transcript containing 60 CGG repeats in order to image the dynamic nature of these RNAs in living cells.

Although Spinach2 provides a simple and convenient approach to image RNA in living cells, the technology is currently limited to labeling highly abundant RNAs and is not yet suited for imaging individual RNAs. The remainder of this chapter will provide a detailed protocol for imaging 5S-Spinach2 and CGG_{60}-Spinach2 in mammalian cells and will provide a general framework for tagging any RNA of interest with Spinach2.

3.1 Tagging an RNA of interest with Spinach2

Our work tagging RNAs with Spinach and Spinach2 has revealed that the presence of an RNA scaffold sequence greatly enhances imaging in cells (Strack et al., 2013). For this reason, we routinely tag RNAs with Spinach2 in the context of a modified human $tRNA^{Lys}$ scaffold (Ponchon & Dardel, 2007), in which Spinach2 is inserted into a truncated anticodon stem loop. This scaffold has been shown to enhance the folding of Spinach and was essential for its stability in bacterial cells (Paige et al., 2011). Both 5S and CGG_{60} were also successfully imaged in cells when fused to tRNA–Spinach2.

The folding of both the target RNA and the Spinach2 tag may be affected by the position of insertion. For this reason, we typically generate RNAs with both 5′ and 3′ tRNA–Spinach2 tags and test them for fluorescence *in vitro* as well as in cells. Spinach2-tagged RNAs that retain the fluorescence of Spinach2 *in vitro* are then tested for performance in live cells.

3.2 Testing a tagged RNA for Spinach2 fluorescence *in vitro*

1. Design primers to amplify your tagged RNA of interest by PCR. The forward promoter should contain the T7 promoter sequence and two additional Gs (5′-TAATACGACTCACTATAGGG-3′) directly upstream of the target RNA sequence.

2. Carry out PCR with these primers using your tagged constructs as template. Also amplify tRNA–Spinach2 as a positive control and untagged target RNA as a negative control. Run the products on an agarose gel and gel purify the PCR products to ensure the correct band is used to template transcription.
3. Carry out *in vitro* transcription using the Ampliscribe transcription kit (Ambion) with a total transcription volume of 10 μl. Incubate the reaction at 37 °C for 4–16 h. Following incubation, treat reaction with 1 μl DNase included in the kit to remove any residual template at 37 °C for 1 h.
4. Purify RNA using ammonium acetate precipitation. Add 1 μl GlycoBlue coprecipitant (Life Technologies) and 10 μl 5 *M* ammonium acetate to each 10 μl reaction. Incubate on ice for 15–30 min. Spin at 20,000 × *g* at 4 °C for 15 min. Remove supernatant. Carefully wash pellet with ice-cold 70% ethanol. Spin at 20,000 × *g* at 4 °C for 5 min. Remove supernatant carefully and do not disturb the pellets. Allow the tubes to sit open on the bench for 10–20 min or until the pellet is dry. Resuspend pellet in 50 μl RNase-free water. Store samples on ice when in use and at −20 °C for long-term stability.
5. Measure the RNA concentration using an absorbance spectrophotometer such as a NanoDrop. To convert the concentration from ng/μl to molarity, simply divide the value in ng/μl by the molecular weight (MW) in kDa. For example, if 300 ng/μl Spinach2 RNA is measured, this value is divided by the MW of Spinach2, which is 31.3 kDa to obtain a concentration of 9.6 μ*M*.
6. Measure fluorescence of RNA–fluorophore complex. Prepare DFHBI (or DFHBI-1T) as a 40 m*M* stock in DMSO (10 mg/ml) and dilute to 100 μ*M* in H_2O immediately before use. The 40 m*M* stock can be stored indefinitely at 4 °C in the dark. Prepare a 200 μl solution containing 1 μ*M* RNA and 10 μ*M* DFHBI (Lucerna) in a buffer with a final concentration of 40 m*M* K-HEPES, pH 7.4, 100 m*M* KCl, and 1 m*M* $MgCl_2$. Smaller or larger volumes may be needed based on the sample volume appropriate for the fluorimeter being used. Sample can be measured immediately after mixing in a fluorimeter. To measure fluorescence, use an excitation wavelength of 470-nm light and an emission wavelength between 500 and 520 nm.
7. Calculate the fluorescence emission intensity of the tagged samples relative to Spinach2 control.

This method of measuring the fluorescence of tagged constructs *in vitro* ensures that Spinach2 is folded properly in the context of the tagged

RNA. Using this method, we were able to observe that the original Spinach exhibits misfolding depending on the identity of the flanking sequences, while Spinach2 generally folds robustly regardless of the identity of adjacent sequences. In our experience, tagged RNAs that exhibit less than 50% of the brightness of Spinach2 alone, on a mole per mole basis, are typically not likely to exhibit fluorescence in *in vivo* imaging experiments. If this occurs, changing the position of the tRNA–Spinach2 tag within the RNA may improve signal and should be attempted before further imaging is pursued. For 5S-Spinach2 and CGG_{60}-Spinach2, the signals were at least 70% as high as untagged Spinach2.

3.3 Expressing 5S RNA in HEK 293-T cells

1. Obtain HEK 293-T cells (ATCC #CRL-11268) and grow to 80–90% confluence in Dulbecco's modified Eagle's medium (DMEM) + 10% fetal bovine serum (FBS) + penicillin/streptomycin.
2. Prior to splitting cells for imaging, treat two wells of a 24-well glass-bottom dish (Mat Tek Corp # P24G-1.5-13F) with 300 μl of poly-L-lysine (PLL) per well for 2 h at 37 °C. Remove PLL and wash wells with 1 ml ddH_2O followed by UV sterilization for 5 min. Plating cells directly onto coated glass coverslips allows for direct detection of live cells on an inverted widefield microscope. PLL coating followed by treatment with laminin (next step) provides a substrate that HEK 293-T cells can adhere to more readily than uncoated glass.
3. Add 100 μl of 10 μg/ml laminin (Trevigen, Inc., Cat # 3400-010-01) to each well and incubate for 2 h at 37 °C. Remove laminin and wash with 1 ml sterile ddH_2O.
4. Remove media from 10 cm dish of HEK 293-T cells by aspiration. Replace with 10 ml prewarmed, 37 °C DMEM + 10% FBS without antibiotics and create a cell suspension by repeated pipetting. For imaging, we routinely use DMEM that contains high glucose, L-glutamine, and sodium pyruvate for robust cell growth (Life Technologies, Cat # 11995-065). Media may contain or lack phenol red. Prior to transfection, cells are grown without additional antibiotic, as specified in the FugeneHD manual.
5. Seed both wells in the 24-well coated dish with 40,000 HEK 293-T cells in 500 μl DMEM + 10% FBS without antibiotic. Allow cells to adhere and grow for 24 h. After 24 h, cells should be ~25% confluent.
6. Transfect cells using a 3:1 ratio of DNA to FugeneHD (Promega, Cat # E2311). In the first well, transfect cells with pAV5S-lambda (negative

control). In the second well, transfect cells with pAV5S-Spinach2. For transfections, add 25 μl water to each of two tubes. Next, add 600 μg of appropriate plasmid to each tube and mix. Next, add 1.65 μl FugeneHD to each tube and mix immediately by vigorously flicking the tube. Collect liquid at the bottom by brief centrifugation and incubate for 5–10 min at room temperature.

7. Add 25 μl of transfection mixture in droplets over the appropriate well of cells. Mix by gently pipetting up and down five times using a 1000 μl pipette set on 400 μl pipetting volume, being careful not to detach cells.
8. Twelve- to 24-h posttransfection, replace media with 500 μl DMEM + 10% FBS + penicillin/streptomycin and allow cells to grow for 24 h.
9. Image cells 48 h posttransfection (see below for details).

This protocol should ensure robust expression of 5S-Spinach2 in HEK 293-T cells. The relatively low confluence of cells at the time of transfection is important for optimal imaging conditions, because after 48 h of growth following transfection, they reach only 50–80% confluence. At this cell density, individual cells that are spread out nicely along the glass bottom are readily imaged, and cellular death and autofluorescence are relatively low. Parallel experiments using a control plasmid expressing a fluorescent protein can be used to control for transfection efficiency.

5S-Spinach2 can, in principle, be imaged in any mammalian cell type that is readily transfected. However, we routinely image this construct in HEK 293-T cells due to their robust expression levels. Highest expression will be observed in cell lines that express the T antigen, as the pAV5S vector uses the SV40 origin of replication. 5S-Spinach expression can also be further increased by induction with sucrose, as described previously, which enhances the transcription of 5S and relocalizes the 5S-Spinach2 signal to punctate stress granules (Paige et al., 2011).

3.4 Expressing CGG_{60}-Spinach2 in COS-7 cells

1. Grow COS-7 cells (ATCC CRL-1651) to 80–90% confluence in DMEM + 10% FBS + penicillin/streptomycin. Ideally, the COS-7 cells used for imaging will be no more than 5–15 passages.
2. Resuspend COS-7 cells. Incubate COS-7 cells with 2 ml TrypLE (Life Technologies, Cat # 12604-013) at 37 °C for 3–5 min. Resuspend cells in 8 ml prewarmed, 37 °C DMEM + 10% FBS without antibiotics and create a cell suspension by trituration.

3. Plate four wells in the 24-well dish with 80,000 COS-7 cells in 500 μl DMEM + 10% FBS without antibiotic. Allow cells to adhere and grow for 24 h. After 24 h, cells should be ~40% confluent.
4. Transfect cells using a 3:1 ratio of DNA to FugeneHD. In well 1, transfect cells with $pCGG_{60}$ (negative control). In well 2, transfect $pCGG_{60}$-Spinach2. In well 3, cotransfect $pCGG_{60}$ and pCDNA-m Cherry-Sam68. In well 4, cotransfect $pCGG_{60}$-Spinach2 and pCDNA-m Cherry-Sam68. For transfections, add 25 μl water to each of two tubes. Next, add 600 μg of appropriate plasmid to each tube and mix. For the cotransfections, add 400 μg of the appropriate $pCGG_{60}$ vector and 200 μg of pCDNA-m Cherry-Sam68. Next, add 1.65 μl FugeneHD to each tube and mix immediately by vigorously flicking the tube. Collect liquid at the bottom by brief centrifugation and incubate for 5–10 min at room temperature.
5. Add 25 μl of transfection mixture in droplets over the appropriate well of cells. Mix by gently pipetting up and down five times using a 1000 μl pipette set on 400 μl pipetting volume, being careful not to resuspend cells.
6. Twelve- to 24-h posttransfection, replace media with 500 μl DMEM + 10% FBS + penicillin/streptomycin and allow cells to grow for 12–24 h.
7. Image cells 24–48 h posttransfection (see below for details).

This protocol should allow for robust expression of CGG_{60}-Spinach2 in ~10–40% of COS-7 cells. It is important to note that most cell types do not form aggregates after expression of CGG_{60} RNA (Sellier et al., 2010). CGG_{60} aggregates can be seen in COS-7 and differentiated PC12 cells, but not other common cell lines such as HeLa and HEK 293-T cells. This indicates that CGG_{60} RNA aggregation is regulated by cellular pathways. Coexpression of mCherry-Sam68 serves as a positive control for CGG_{60} RNA expression and aggregate formation (Sellier et al., 2010; Strack et al., 2013). Typically, if the CGG_{60} RNA has aggregated, then the mCherry-Sam68 signal will be predominantly localized to intranuclear aggregates. However, if the CGG_{60} RNA is not expressed or is not aggregated, then the mCherry-Sam68 signal is diffusely nucleoplasmic, and not in foci. In these cells, CGG aggregate formation has not occurred and therefore will not be imaged.

CGG_{60} aggregates have been shown to grow larger over time (Sellier et al., 2010). This was also observed for CGG_{60}-Spinach2 aggregates (Strack et al., 2013). At 24-h posttransfection, cells are highly heterogeneous, yet typically contain 10–20 small–medium CGG_{60}-Spinach2 foci per nucleus. By 72-h posttransfection, cells may contain fewer than five very large foci. It is important to note that CGG_{60} aggregates appear to be cytotoxic to COS-7 cells, which can reduce the number of CGG_{60}-Spinach2-expressing cells after transfection. Additionally, passage of these cells is extremely difficult, with almost no positively transfected cells being observed after passaging.

4. FLUORESCENCE IMAGING OF SPINACH2-TAGGED RNAs

All imaging of Spinach2 was carried out using a conventional widefield microscope (Nikon TE2000 epifluorescence microscope) with a CoolSnap HQ2 CCD camera through a 60 × oil objective (Plan Apo 1.4 NA) housed entirely in an environmental chamber that regulates temperature and CO_2 levels. Spinach2 fluorescence is captured through a FITC/EGFP filter set (480/40 ex and 535/50 em), and mCherry fluorescence is captured through a Texas Red filter set (560/55 ex and 645/75 em).

Spinach2 is nonfluorescent in the absence of dye. Until recently, the best dye for imaging was DFHBI (Lucerna) (Paige et al., 2011; Strack et al., 2013). However, testing of novel derivatives of DFHBI revealed that another dye, DFHBI-1T (Lucerna) (Fig. 2), is optimally suited for mammalian imaging (Song, Strack, Svensen, & Jaffrey, 2014). DFHBI-1T has three primary advantages. The first is that Spinach2–DFHBI-1T is nearly twice as bright as Spinach2–DFHBI *in vitro*. The second is that Spinach2–DFHBI-1T is red-shifted relative to Spinach2–DFHBI, allowing for a better spectral overlap with the GFP filter sets, which enhances the overall observed fluorescence signal. The third is that DFHBI-1T itself, when incubated with cells, shows even lower background signals than DFHBI, allowing for greater sensitivity when imaging Spinach2–DFHBI-1T. When quantified in living cells, the signal for Spinach2–DFHBI-1T was greater than 1.6-fold higher than that of Spinach2–DFHBI (Song et al., 2014) (Fig. 3). For these reasons, DFHBI-1T can be useful to improve the detection limits when imaging Spinach2-tagged RNAs in both bacterial and mammalian cells.

Spinach2–DFHBI is known to photobleach rapidly in some situations (Han, Leslie, Fei, Zhang, & Ha, 2013). However, due to the nature of the RNA–dye complex, this photobleaching is reversible as the bleached dye molecule is unbound and replaced with an unbleached dye molecule. Thus, in time-course imaging where images are taken at least 10 s apart, Spinach–DFHBI and Spinach2–DFHBI-1T signal is remarkably resistant to photobleaching.

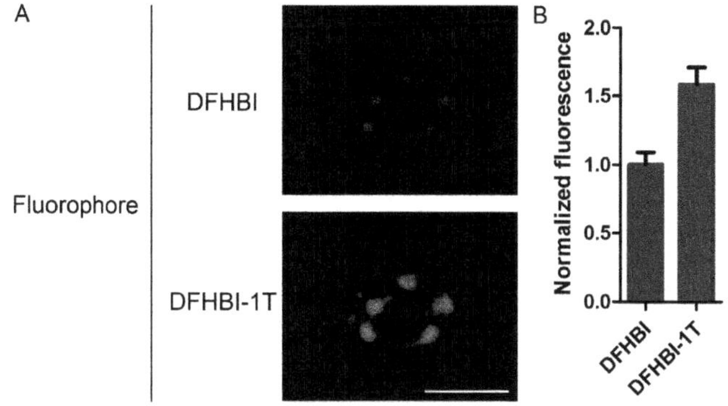

Figure 3 Spinach2–DFHBI-1T is brighter than Spinach2–DFHBI in live cells. (A) COS-7 cell expressing CGG_{60}-Spinach2. The cell was imaged using a widefield microscope with EGFP filter sets and a 100-ms exposure time. Cells were first incubated with 20 μ*M* DFHBI and imaged. Media was then exchanged with media lacking dye for 30 min to remove DFHBI. This media was then supplemented with 20 μ*M* DFHBI-1T for 30 min. Spinach2–DFHBI-1T images were collected following this incubation. (B) Quantification of green fluorescence signal from Spinach2–DFHBI and Spinach2–DFHBI-1T in living cells. Scale bar, 20 μm. *Images were used with permission from Song et al., 2014.* (See the color plate.)

4.1 Imaging 5S-Spinach2

1. Create a 10× stock of imaging solution containing 200 m*M* Na–HEPES, pH 7.4, 50 m*M* $MgSO_4$, and 200 μ*M* DFHBI-1T.
2. Thirty minutes prior to imaging add 50 μl imaging solution directly into media of both wells and mix by gently swirling the plate. Incubate the plate at 37 °C for at least 30 min to allow the dye to enter the cells.
3. Using your imaging software, take images of negative control cells transfected with pAV5S-lambda using 1-s exposure times. Using linear adjustment, subtract background from these images such that any green autofluorescence or residual fluorescence from DFHBI-1T is eliminated. Save these background adjustment values.
4. Image cells expressing 5S-Spinach2 using 1-s exposure times. Carry out background subtraction using the background adjustment values obtained for negative control cells in step 3. All resulting green fluorescence represents 5S-Spinach2.

This imaging protocol will produce images of 5S-Spinach2 in HEK 293-T cells (Fig. 4). However, it should be noted that 5S-Spinach2 signal when diffuse throughout the nucleus and cytoplasm is not easily observed by the naked eye. Indeed, optimal images are obtained using 0.5- to 1-s exposure times and background subtraction. Cotransfection with a plasmid expressing mCherry can be used to determine which cells are transfected

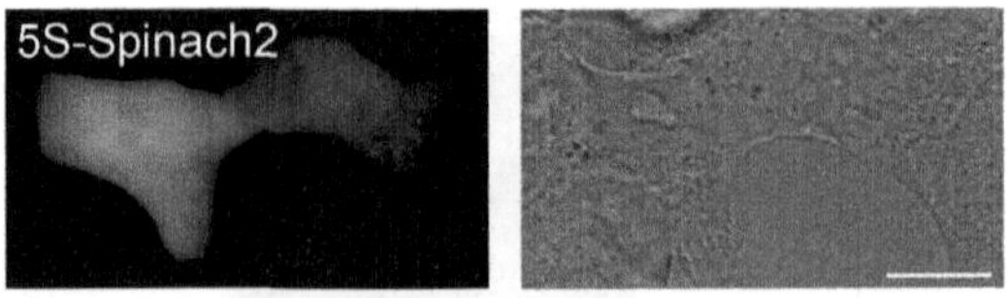

Figure 4 HEK-293T cells expressing 5S-Spinach2. Cells were incubated with 20 μ*M* DFHBI for 30 min prior to imaging. Shown are green fluorescence (left) and differential interference contrast (DIC, right) images. Fluorescence image was collected by widefield microscopy with EGFP filter sets with a 1-s exposure time. Scale bar, 10 μ*M*. (See the color plate.)

with pAV5S–Spinach2; however, the green fluorescence signal is noticeably lower in cotransfected cells. If cotransfection is used, it must also be carried out for the control plasmid to ensure proper background subtraction.

To ensure specificity of signal, cells can be imaged before and 30 min after addition of DFHBI-1T. In addition, after imaging with DFHBI-1T, imaging media can be replaced with media lacking dye. Thirty minutes after dye removal, Spinach2 fluorescence will disappear. Finally, the 5S promoter is induced by treatment with 600 m*M* sucrose for 30 min and sucrose treatment causes 5S-Spinach to relocalize to stress granules (Paige et al., 2011). This relocalization to distinct foci can also be observed for 5S-Spinach2 and can provide additional confirmation that the fluorescent signal reflects Spinach2.

4.2 Imaging CGG_{60}-Spinach2

1. Create a 10× stock of imaging solution containing 200 m*M* Na–HEPES, pH 7.4, 50 m*M* $MgSO_4$, and 200 μ*M* DFHBI-1T.
2. Thirty minutes prior to imaging add 50 μl imaging solution directly into media of both wells and mix by gently swirling the plate. Incubate the plate at 37 °C for 30 min to allow for dye to enter the cells.
3. Using your imaging software, take images of negative control cells transfected with $pCGG_{60}$ using 100-ms exposure times. Using linear adjustment, subtract background from these images such that any green autofluorescence or residual fluorescence from DFHBI-1T is eliminated. Save these background adjustment values. Repeat for cells coexpressing CGG_{60} and mCherry-Sam68 to subtract any contribution of mCherry signal in the green channel.
4. Image cells expressing CGG_{60}-Spinach2. Due to the brightness of the foci, imaging should be carried out using 100-ms exposure times. Carry out background subtraction using the background adjustment values

obtained for negative control cells in step 3. All resulting green fluorescence represents CGG_{60}-Spinach2.

5. For cells coexpressing mCherry-Sam68, image cells using the Texas Red filter set using 100- to 400-ms exposure times.

CGG_{60}-Spinach2 will form large foci that colocalize with mCherry-Sam68 (Fig. 5) in COS-7 cells. Unlike 5S-Spinach2, which is only faintly visible, CGG_{60}-Spinach2 is clearly visible by the naked eye through the microscope, as ~150–200 copies of CGG_{60}-Spinach2 are present in a typical aggregate (Strack et al., 2013). CGG_{60}-Spinach2 aggregates are highly heterogeneous (Fig. 6) and can change dramatically over time (Strack et al., 2013).

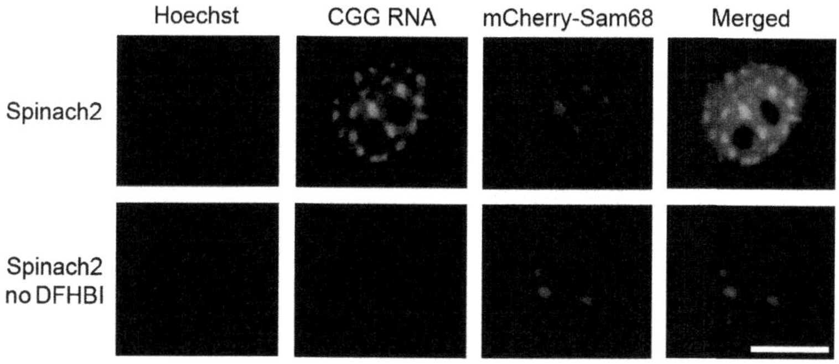

Figure 5 CGG_{60}-Spinach2 colocalizes with mCherry-Sam68. Shown are images of COS-7 nuclei containing CGG_{60} aggregates labeled with Spinach2–DFHBI or mCherry-Sam68. Spinach signal was collected by widefield microscopy with EGFP filter sets with a 100-ms exposure time. mCherry signal was collected using a Texas Red filter set and 200-ms exposure time. CGG_{60}-Spinach2 and mCherry-Sam68 are shown with and without DFHBI. Scale bar, 20 μm. (See the color plate.)

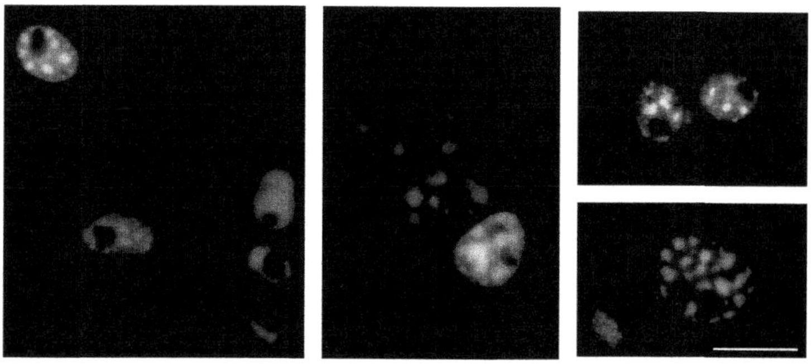

Figure 6 Examples of COS-7 nuclei with CGG_{60}-Spinach2 aggregates. Aggregates are highly heterogeneous and range in both size and number. Some representatives are shown. Images were collected by widefield microscopy with EGFP filter sets with 50- to 200-ms exposure times. (See the color plate.)

5. IMAGING OTHER RNAs USING SPINACH2

Spinach2 can, in principle, be used to tag any RNA of interest for live-cell imaging. However, to date, the most successful imaging applications have involved tagging abundant and relatively long-lived RNAs (Strack et al., 2013). This is largely due to each RNA molecule being tagged by only a single Spinach2 molecule. Thus, with the current technology, the most likely RNAs to be tagged and imaged using Spinach2 include highly expressed noncoding RNAs, abundant viral RNAs, and other trinucleotide-repeat containing RNAs that form intranuclear aggregates.

The challenge of imaging rare or even individual RNAs using Spinach2 is currently being addressed. An obvious method for enhancing signal would be to attach "cassettes" of 2–20 Spinach2 molecules to an RNA of interest. Studies of this nature are currently underway in order to optimize the sequence, length, and number of Spinach2 tags necessary to image lower abundance RNAs in living cells.

In addition, new RNA–dye complexes with unique features, including brighter, better folding green fluorescent complexes as well as RNA–dye complexes in the yellow to far-red regions are being developed. Such studies will enhance RNA tagging in living cells and allow for imaging multiple RNAs in the same cell simultaneously.

ACKNOWLEDGMENTS

We thank K.Y. Wu and J.S. Paige for their role in developing the original protocols for imaging Spinach in mammalian cells. This work was supported by US National Institutes of Health NINDS NS010249 (S. R. J.) and NIGMS F32 GM106683 (R. L. S.).

REFERENCES

Babendure, J. R., Adams, S. R., & Tsien, R. Y. (2003). Aptamers switch on fluorescence of triphenylmethane dyes. *Journal of the American Chemical Society*, *125*, 14716–14717.

Bao, G., Tsourkas, A., & Santangelo, P. J. (2004). Engineering nanostructured probes for sensitive intracellular gene detection. *Mechanics & Chemistry of Biosystems*, *1*, 23–36.

Bratu, D. P., Cha, B. J., Mhlanga, M. M., Kramer, F. R., & Tyagi, S. (2003). Visualizing the distribution and transport of mRNAs in living cells. *Proceedings of the National Academy of Sciences of the United States of America*, *100*, 13308–13313.

Brodsky, A. S., & Silver, P. A. (2002). Identifying proteins that affect mRNA localization in living cells. *Methods*, *26*, 151–155.

Dombrowski, C., Lévesque, S., Morel, M. L., Rouillard, P., Morgan, K., & Rousseau, F. (2002). Premutation and intermediate-size FMR1 alleles in 10572 males from the general population: Loss of an AGG interruption is a late event in the generation of fragile X syndrome alleles. *Human Molecular Genetics*, *11*, 371–378.

Fu, Y. H., Kuhl, D. P., Pizzuti, A., Pieretti, M., Sutcliffe, J. S., Richards, S., et al. (1991). Variation of the CGG repeat at the fragile X site results in genetic instability: Resolution of the Sherman paradox. *Cell*, *67*, 1047–1058.

Fusco, D., Accornero, N., Lavoie, B., Shenoy, S. M., Blanchard, J. M., Singer, R. H., et al. (2003). Single mRNA molecules demonstrate probabilistic movement in living mammalian cells. *Current Biology*, *13*, 161–167.

Hagerman, R. J., Leehey, M., Heinrichs, W., Tassone, F., Wilson, R., Hills, J., et al. (2001). Intention tremor, parkinsonism, and generalized brain atrophy in male carriers of fragile X. *Neurology*, *57*, 127–130.

Han, K. Y., Leslie, B. J., Fei, J., Zhang, J., & Ha, T. (2013). Understanding the photophysics of the spinach-DFHBI RNA aptamer-fluorogen complex to improve live-cell RNA imaging. *Journal of the American Chemical Society*, *135*, 19033–19038.

Holeman, L. A., Robinson, S. L., Szostak, J. W., & Wilson, C. (1998). Isolation and characterization of fluorophore-binding RNA aptamers. *Folding & Design*, *3*, 423–431.

Molenaar, C., Abdulle, A., Gena, A., Tanke, H. J., & Dirks, R. W. (2004). Poly(A)+ RNAs roam the cell nucleus and pass through speckle domains in transcriptionally active and inactive cells. *The Journal of Cell Biology*, *165*, 191–202.

Molenaar, C., Marras, S. A., Slats, J. C., Truffert, J. C., Lemaître, M., Raap, A. K., et al. (2001). Linear 2′ O-methyl RNA probes for the visualization of RNA in living cells. *Nucleic Acids Research*, *29*, E89–E99.

Nitin, N., Santangelo, P. J., Kim, G., Nie, S., & Bao, G. (2004). Peptide-linked molecular beacons for efficient delivery and rapid mRNA detection in living cells. *Nucleic Acids Research*, *32*, e58.

Ozawa, T., Natori, Y., Sato, M., & Umezawa, Y. (2007). Imaging dynamics of endogenous mitochondrial RNA in single living cells. *Nature Methods*, *4*, 413–419.

Paige, J. S., Wu, K. Y., & Jaffrey, S. R. (2011). RNA mimics of green fluorescent protein. *Science*, *333*, 642–646.

Park, H. Y., Lim, H., Yoon, Y. J., Follenzi, A., Nwokafor, C., Lopez-Jones, M., et al. (2014). Visualization of dynamics of single endogenous mRNA labeled in live mouse. *Science*, *343*, 422–424.

Ponchon, L., & Dardel, F. (2007). Recombinant RNA technology: The tRNA scaffold. *Nature Methods*, *4*, 571–576.

Santangelo, P. J., Alonas, E., Jung, J., Lifland, A. W., & Zurla, C. (2012). Probes for intracellular RNA imaging in live cells. *Methods in Enzymology*, *505*, 383–399.

Sellier, C., Freyermuth, F., Tabet, R., Tran, T., He, F., Ruffenach, F., et al. (2013). Sequestration of DROSHA and DGCR8 by expanded CGG RNA repeats alters microRNA processing in fragile X-associated tremor/ataxia syndrome. *Cell Reports*, *3*, 869–880.

Sellier, C., Rau, F., Liu, Y., Tassone, F., Hukema, R. K., Gattoni, R., et al. (2010). Sam68 sequestration and partial loss of function are associated with splicing alterations in FXTAS patients. *The EMBO Journal*, *29*, 1248–1261.

Shav-Tal, Y., Darzacq, X., Shenoy, S. M., Fusco, D., Janicki, S. M., Spector, D. L., et al. (2004). Dynamics of single mRNPs in nuclei of living cells. *Science*, *304*, 1797–1800.

Song, W., Strack, R. L., Svensen, N., & Jaffrey, S. R. (2014). Plug-and-play fluorophores extend the spectral properties of Spinach. *Journal of the American Chemical Society*, *136*, 1198–1201.

Strack, R. L., Disney, M. D., & Jaffrey, S. R. (2013). A superfolding Spinach2 reveals the dynamic nature of trinucleotide repeat RNA. *Nature Methods*, *10*, 1219–1224.

Tassone, F., Iwahashi, C., & Hagerman, P. J. (2004). FMR1 RNA within the intranuclear inclusions of fragile X-associated tremor/ataxia syndrome (FXTAS). *RNA Biology*, *1*, 103–105.

Tyagi, S. (2009). Imaging intracellular RNA distribution and dynamics in living cells. *Nature Methods*, *6*, 331–338.

Tyagi, S., & Alsmadi, O. (2004). Imaging native beta-actin mRNA in motile fibroblasts. *Biophysical Journal, 87*, 4153–4162.

Ward, W. W., & Bokman, S. H. (1982). Reversible denaturation of Aequorea green-fluorescent protein: Physical separation and characterization of the renatured protein. *Biochemistry, 21*, 4535–4540.

Wu, B., Chen, J., & Singer, R. H. (2014). Background free imaging of single mRNAs in live cells using split fluorescent proteins. *Scientific Reports, 4*, 3615.

CHAPTER EIGHT

In Vitro Analysis of Riboswitch–Spinach Aptamer Fusions as Metabolite-Sensing Fluorescent Biosensors

Colleen A. Kellenberger*, Ming C. Hammond*,†,1
*Department of Chemistry, University of California, Berkeley, California, USA
†Department of Molecular & Cell Biology, University of California, Berkeley, California, USA
1Corresponding author: e-mail address: mingch@berkeley.edu

Contents

Abstract

The development of fluorescent biosensors has been motivated by the interest to monitor and measure the levels of specific metabolites in live cells in real time. Common approaches include fusing a protein-based receptor to fluorescent proteins or synthesizing a small molecule reactive probe. Natural metabolite-sensing riboswitches also have been used in reporter-based systems that take advantage of ligand-dependent regulation of downstream gene expression. More recently, it has been shown that RNA-based fluorescent biosensors can be generated by fusing a riboswitch aptamer to the *in vitro* selected Spinach aptamer, which binds a cell-permeable and conditionally fluorescent molecule. Here, we describe methods to design, prepare, and analyze riboswitch–Spinach aptamer fusion RNAs for ligand-dependent activation of fluorescence *in vitro*. Examples of procedures to measure fluorescence activation, ligand

Methods in Enzymology, Volume 550
ISSN 0076-6879
http://dx.doi.org/10.1016/bs.mie.2014.10.045

binding selectivity and affinity, and binding kinetics are given for a cyclic di-GMP-responsive biosensor. The relative ease of *in vitro* RNA synthesis and purification should make this method accessible to other researchers interested in developing riboswitch-based fluorescent biosensors.

1. INTRODUCTION

Many soluble signaling molecules dynamically regulate cellular physiology in response to changes in environmental conditions. Thus, observing correlations between the concentration of a specific signaling molecule and input stimuli or output responses is an important step toward deconvoluting often complex signal transduction pathways. Furthermore, there is growing appreciation that small molecule signals function in a spatially and temporally resolved manner inside cells (Newman, Fosbrink, & Zhang, 2011). The development of tools to tackle outstanding questions in the cell biology and *in vivo* biochemistry of small molecule signals or metabolites is an exciting and rapidly advancing research area.

An ideal method for tracking the dynamics of a small molecule in cells is through development of a biosensor, which is a receptor that specifically recognizes a small molecule ligand and provides a dynamic and measurable output in response to changing concentrations of the ligand. Most common biosensors are protein-based fluorescent turn-on or FRET sensors (Newman et al., 2011). While proteins are advantageous because they are genetically encodable and naturally specific for the metabolite of interest, it is often difficult to find a suitable receptor and to engineer a fusion construct that shows a large ligand-dependent change in signal. Another powerful approach toward detecting small molecule flux is the development of small molecule reactive probes (Liu, He, & Guo, 2013). However, small molecule probes with specific chemical reactivity toward a target analyte must be chemically synthesized and must be able to permeate through the cell membrane (Domaille, Zeng, & Chang, 2010; Liu et al., 2013). They are often designed to react irreversibly with the target analyte, although this is not always the case.

As an alternative, riboswitches naturally possess many of the traits required of biosensors. These RNAs are naturally evolved to recognize a specific metabolite and upon ligand binding, undergo a conformational change (Serganov & Nudler, 2013). Riboswitches have previously been used as small molecule reporters by placing a reporter gene, such as GFP or luciferase, under the control of the riboswitch (Fowler, Brown, & Li, 2010). While these reporters are metabolite specific and genetically encodable, they lack the dynamic response of a biosensor, as the lifetime of the protein reporter does not correspond to the concentration of the metabolite.

Recently, Jaffrey and coworkers reported the development of the Spinach aptamer (Paige, Wu, & Jaffrey, 2011) that was evolved to bind the cell-permeable molecule 3,5-difluoro-4-hydroxybenzylidene imidazolinone (DFHBI) and to elicit fluorescence. The crystal structure of Spinach bound to DFHBI has recently been solved and has resulted in a revision of the secondary structure model (Fig. 1; Deigan Warner et al., 2014; Huang et al., 2014). Spinach can be engineered to fluoresce in response to metabolite binding through replacement of the second stem-loop of Spinach with other RNA aptamers (Kellenberger, Wilson, Sales-Lee, & Hammond, 2013; Nakayama, Luo, Zhou, Dayie, & Sintim, 2012; Paige, Nguyen-Duc,

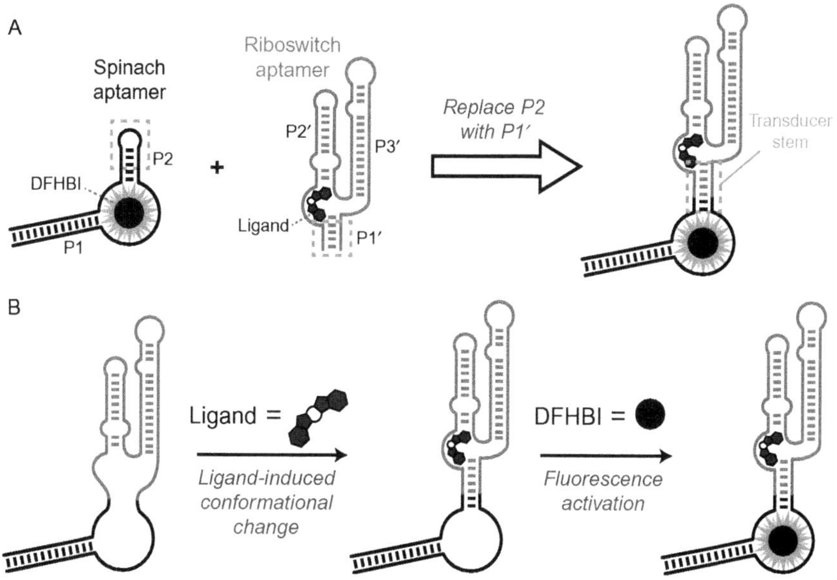

Figure 1 Design of Spinach-based biosensor and model of fluorescence activation. (A) The riboswitch aptamer is inserted into the Spinach scaffold in place of the P3 stem of Spinach. The P1′ stem of the Vc2 riboswitch aptamer, which is stabilized upon binding of ligand, serves as the transducer stem between the riboswitch and Spinach aptamers. (B) Binding of ligand to the riboswitch aptamer induces a conformational change in the transducer stem that enables the Spinach aptamer to bind DFHBI and elicit fluorescence.

Song, & Jaffrey, 2012). Binding of Spinach to DFHBI is dynamic and dependent on the ligand-bound state of the riboswitch aptamer, thus providing a direct fluorescent read-out of fluctuating metabolite concentrations. Furthermore, in comparison to protein-based fluorescent biosensors, it is more tractable to use RNA-based biosensors as both *in vivo* and *in vitro* sensors of metabolites. While use of a protein biosensor *in vitro* requires extensive purification of the protein from cells and any bound metabolites, RNA-based biosensors can be readily synthesized through *in vitro* transcription and directly used for analyses.

Following this strategy, here we describe the development of an RNA-based biosensor for the bacterial signaling molecule, cyclic di-GMP (c-di-GMP), by fusing Spinach to the Vc2 aptamer, a GEMM-I class riboswitch (Kellenberger et al., 2013). Binding of c-di-GMP to Vc2 stabilizes the transducer stem to Spinach, thus enabling Spinach to bind DFHBI and elicit fluorescence (Fig. 1). The development of this biosensor provides a sensitive method for studying the regulation of c-di-GMP, which is an important bacterial second messenger involved in the regulation of bacterial biofilms and pathogenesis (Römling et al., 2013).

1.1. General equipment

Micropipettor
Multichannel micropipettor
Vortex mixer
Microcentrifuge with temperature control
Millipore water filter with a BioPak unit

1.2. General materials

Micropipettor tips
Filter-tip micropipettor tips
1.5 mL microcentrifuge tubes
Sterile filter units
dNTPs: dATP, dCTP, dGTP, dTTP
rNTPs: ATP, CTP, GTP, UTP
Magnesium chloride ($MgCl_2$)
4-(2-hydroxyethyl)-1-piperazineethanesulfonic acid (HEPES)
Sucrose
Bromophenol blue
Xylene cyanol

SDS
Tris base
Boric acid
Sodium chloride (NaCl)
Potassium chloride (KCl)
EDTA, pH 8.0
Sodium carbonate (Na_2CO_3)
Acetic acid
T7 RNA polymerase
Inorganic pyrophosphatase
Ethanol (EtOH)
Sodium hydroxide (NaOH)
Acetic Acid
3,5-difluoro-4-hydroxybenzylidene imidazolinone (DFHBI)
Dimethyl sulfoxide (DMSO)
Cyclic di-GMP
National Diagnostics Sequagel mix
GE healthcare G-25 spin purification column
Dithiothreitol (DTT)
Glycerol
Double-distilled H_2O (ddH_2O, from Millipore water filtration system, 18.2 $M\Omega$ at 25 °C)

2. DESIGN AND PREPARATION OF AN RNA-BASED FLUORESCENT BIOSENSOR

For the fluorescent biosensor to function, ligand binding to the riboswitch aptamer, which is the domain that specifically recognizes a small molecule, must lead to a change in fluorescence, preferably a turn-on of fluorescence. This is affected by fusing the riboswitch aptamer to the Spinach aptamer in such a way that ligand binding enhances the affinity for DFHBI (Fig. 1). Formation of the Spinach P2 stem is required for binding of DFHBI, so connecting a riboswitch aptamer at this location creates a "transducer" stem that is affected by ligand binding (Paige et al., 2012). While others have screened artificial transducer stems and found some functional sequences (Nakayama et al., 2012; Paige et al., 2012), we recommend using the natural P1 stem of a riboswitch aptamer, in this case Vc2 (labeled as P1′ in Fig. 1), as a starting point for biosensor design. There are two main reasons, the first being that many riboswitches utilize conformational changes involving the P1 stem to modulate gene expression (Blount, Wang, Lim, Sudarsan, & Breaker, 2007; Mandal & Breaker, 2004), so the natural sequence has evolved to perform the function of a transducer. Secondly, the natural P1 stem has evolved as part of the riboswitch fold, whereas artificial stems may perturb the folding pathway or stability of the RNA structure. For example, Vc2-Spinach biosensors incorporating artificial P1 stems showed drastic loss of ligand binding affinity (Nakayama et al., 2012).

In the case that one selects the natural P1 stem to serve as the transducer, there are still some design choices as to optimal length, inclusion or deletion of non-base-paired nucleotides, and mutations to increase or decrease stem stability. We typically screen several different construct designs varying these types of parameters to identify a working biosensor sequence. In our experience, even relatively modest changes such as comparing 4-nt versus 5-nt length transducer stems can have a large impact on the extent of ligand-induced fluorescence activation. These effects from modifying the transducer stem are not only due to altering the thermodynamic stability of the stem, because the three-dimensional arrangement of the two aptamers are likely to be important as well. An A-type RNA double helix has an average rotation of 37.4° per base-pair (Cruse et al., 1994), so a difference of one base-pair changes the relative orientation of the two aptamers and may lead to misfolding due to steric clash between folded domains. A long transducer stem would increase separation between the two aptamers but may be overstabilized, leading to higher background fluorescence.

The process of transducer stem optimization is similar to the need to screen different linker regions in the construction of FRET-based protein biosensors (Zhou, Herbst-Robinson, & Zhang, 2012). One major benefit is that *in vitro* transcription of RNA-based biosensors from a DNA template, followed by PAGE gel purification, is a straightforward procedure. Furthermore, as riboswitch aptamer sequences are usually 50–200 nt in length and the Spinach aptamer is 73 nt in length, full-length DNA templates of the riboswitch–Spinach aptamer fusions can be purchased as single-stranded oligonucleotides from commercial vendors or assembled by the overlapping PCR method (Nelson & Fitch, 2011).

2.1. Preparation of DNA templates of riboswitch–Spinach aptamer fusion

Shown below is the sequence of the WT Vc2-Spinach biosensor construct. The capitalized nucleotides indicate the 5′ and 3′ portions of the Spinach aptamer (black in Fig. 1) and the lowercase nucleotides represent the Vc2 riboswitch (blue (gray in the print version) in Fig. 1). The transducer stem, consisting of the top four base-pairs of the natural Vc2 P1′ stem and containing a single nucleotide bulge, is underlined. Bolded regions anneal to the universal forward and reverse primers used in PCR that generates and amplifies a double-stranded template (from a single-stranded commercial oligonucleotide) and introduces the T7 promoter sequence. To design another biosensor, insert the desired aptamer sequence in place of the lowercase WT Vc2 sequence. This procedure works best for aptamers with a pairing stem that comprises the 5′ and 3′ ends of the aptamer.

WT Vc2-Spinach sequence: 5′-**GACGCGACTGAATGAAATGG** TGAAGGACGGGTCCAcacgcacagggcaaaccattcgaaagagtgggacgcaaagcctc cggcctaaaccagaagacatggtaggtagcggggttaccgatgTTGTTGAGTAGAGTGT **GAGCTCCGTAACTAGTCGCGTC**-3′

Universal forward primer (underlined is minimal T7 promoter with additional bases upstream to enhance polymerase binding and initiation in lower caps): 5′-ccaagTAATACGACTCACTATAGACGCGACTGAAT GAAATGG

Universal reverse primer: 5′ GACGCGACTAGTTACGGAGCTC

Generating the construct through overlapping extension PCR from shorter oligonucleotides is particularly useful if one wants to test a series of mutations or changes to the same biosensor, or if the biosensor constructs are greater than 200 nt. In the procedure described below, universal primers that anneal to the flanking Spinach sequences are used to generate a double-stranded template and introduce the T7 promoter sequence.

Notes and tips:

- T7 RNA polymerase prefers to initiate transcription at G nucleotides (Kuzmine, Gottlieb, & Martin, 2003). The Spinach aptamer starts with a single G nucleotide, but for riboswitch–Spinach fusion constructs that do not start with a G, one should add 1–2 G nucleotides immediately after the TATA in the promoter sequence.
- If interested in testing alternate transducer stems, design constructs that include altered lengths of the natural riboswitch P1 stem.
- Another DFHBI-binding aptamer sequence, called Spinach2, has been developed that demonstrates enhanced folding stability and brightness at 37 °C (Strack, Disney, & Jaffrey, 2013). To use Spinach2, simply replace the Spinach sequence in the biosensor design (nucleotide changes are bold underlined).

WT Vc2-Spinach2 sequence: 5′-GA**T**G**TA**AC**T**GAATGAAATGGTGA AGGACGGGTCCAcacgcacagggcaaaccattcgaaa gagtgggacgcaaagcctccggcc taaaccagaagacatggtaggtagcggggttaccgatgTTGTTGAGTAGAGTGTGAGC TCCGTAACT**A**GT**TA**C**A**TC-3′

2.1.1 Equipment and materials

PCR thermal cycler
Agarose gel electrophoresis equipment
Phusion polymerase with HF buffer
dNTPs
Gel extraction or PCR clean-up kit

2.1.2 Procedure for generating DNA template for transcription

To prepare a 100-μL PCR reaction, combine the following in a thin-walled PCR tube:

20 μL 5 × Phusion HF buffer
10 μL 2 m*M* each dNTP
1 μL 40 μ*M* forward primer
1 μL 40 μ*M* reverse primer
1 μL 10–100 ng DNA template
76 μL ddH_2O
1 μL Phusion HF polymerase (add last)

The following is a standard thermocycler protocol that has been used for the WT Vc2-Spinach construct: Initial denaturation 98 °C for 1 min; 35 cycles of 98 °C 5–10 s, 66 °C for 20 s, 72 °C for 15 s, final extension 72 °C for 5 min. After the reaction, an aliquot of the PCR reaction is analyzed on an agarose gel. The PCR products are purified either via commercial gel extraction or PCR clean-up kits following the manufacturer's protocol and the product can be eluted in ddH_2O or the provided elution buffer.

Notes and tips:

- The NEB Website has a web form for estimating annealing temperatures for Phusion reactions (https://www.neb.com/tools-and-resources/interactive-tools/tm-calculator).

2.2. Preparation of RNAs by *in vitro* transcription

This step describes how to transcribe RNA *in vitro*, purify the RNA by polyacrylamide gel electrophoresis (PAGE), and accurately quantify the RNA concentration using a hydrolysis assay (Wilson, Cohen, Wang, & Hammond, 2014) (Estimated duration: 1 day).

2.2.1 Materials and equipment

UV/Vis spectrophotometer
Heat block (37, 70, 95 °C)
Polyacrylamide gel electrophoresis equipment

UV fluorescent TLC plate
Short wave UV light source
Razor blades
Microcentrifuge tube rotator
National Diagnostics Sequagel UreaGel system
Inorganic phosphatase
T7 RNA polymerase (NEB, 50 U/μL)
Ammonium persulfate (APS)
Tetramethylethylenediamine (TEMED)

Buffers and solutions that can be prepared in advance are listed here, whereas solutions that should be freshly prepared are described in the procedure. Prepare all solutions using ddH_2O water.

10× Transcription buffer

Component	Final concentration (m*M*)	Stock	Amount
Tris–HCl, pH 8.0	400	1 *M*	4 mL
$MgCl_2$	200	1 *M*	2 mL
DTT	100	1 *M*	1 mL
Spermidine	20	500 m*M*	400 μL

Add water to 10 mL, sterile filter, store at −20 °C.

Crush soak buffer

Component	Final concentration (m*M*)	Stock (*M*)	Amount (mL)
Tris–HCl, pH 7.5	10	1	10
NaCl	200	5	40
EDTA, pH 8.0	1	0.5	2

Add water to 1 L, autoclave, store at RT.

10× Tris–Borate–EDTA (TBE) buffer

Component	Final concentration (m*M*)	Stock	Amount
Tris	900	–	108.99 g
Boric acid	900	–	55.64 g
EDTA, pH 8.0	10	0.5 *M*	20 mL

Adjust pH to 8.3, add water to 1 L, sterile filter, store at RT.

2× Urea gel loading buffer (ULB)

Component	Final concentration	Stock	Amount
Sucrose	20%	–	2 g
Bromophenol Blue	0.05%	–	5 mg
Xylene cyanol	0.05%	–	5 mg
SDS	0.1%	10%	100 μL
TBE	1×	10×	1 mL
Urea	~18 *M*	–	11 g

Add water to 10 mL, stir for 2 h at room temperature, decant the solution from any undissolved urea, filter to sterilize, and store at 4 °C.

1× Tris–EDTA (TE) buffer

Component	Final concentration (m*M*)	Stock (*M*)	Amount (μL)
Tris–HCl, pH 7.5	10	1	100
EDTA, pH 8.0	1	0.5	20

Add water to 10 mL, sterile filter, store at RT.

Inorganic pyrophosphatase solution

Component	Final concentration	Stock	Amount
Glycerol	25%	50%	100 μL
HEPES–KOH, pH 7.5	50 m*M*	1 *M*	10 μL
DTT	0.5 m*M*	200 m*M*	0.5 μL
H_2O	–	–	89.5 μL
Inorganic pyrophosphatase	0.5 U/μL	–	100 U

Aliquot and store at −20 °C.

10× Sodium carbonate buffer

Component	Final concentration	Stock	Amount
Na_2CO_3	500 m*M*	–	0.53 g
EDTA	10 m*M*	0.5 *M*	200 μL

Adjust pH to 7.0, add water to 10 mL, sterile filter, store at −20 °C.

2.2.2 Procedures

Step 1: Transcription of RNA (estimated duration: 3 h)

1.1 To prepare a 50 μL transcription reaction, combine the following in a 1.5 mL microcentrifuge tube:
   5 μL 10 × transcription buffer
   8 μL 25 m*M* rNTPs
   1–32 μL DNA template (1 μg)
   1 μL inorganic pyrophosphatase
   ddH_2O to 46 μL
   4 μL T7 RNA polymerase (add last)

1.2 Incubate reaction at 37 °C for 3 h.

1.3 Quench the reaction with 50 μL 2 × ULB. Place on ice if not immediately loading on the gel for purification. It is also possible to store the RNA overnight at − 20 °C until ready to purify by PAGE.

Step 2: Purification of RNA (estimated duration: 4 h to 1 day)

2.1 To purify the desired RNA product away from abortive transcripts and unreacted NTPs, first prepare a 6% polyacrylamide gel. For sample volumes of 100–200 μL, we prepare a 28 cm long by 16.5 cm wide and 1.5 mm thick gel with eight sample lanes. Set up the electrophoresis equipment and fill buffer reservoirs with 1 × TBE buffer.

2.2 Load quenched reaction on polyacrylamide gel using a narrow-tip pipette and run the gel at 25 W. Run for ~1.5 h, or until the RNA has migrated 1/3–1/2 of the distance of the gel as indicated by the distance traveled by the xylene cyanol and bromophenol blue dyes, which migrate similar to 106 and 26 nt length oligonucleotides, respectively (Sambrook, Fritsch, & Maniatis, 1989).

2.3 To isolate the RNA, first peel off one glass plate from the gel and cover the exposed side with plastic wrap. If more than one RNA is being purified, be sure to label the order of the lanes at this time on the plastic wrap. Flip the gel and repeat for the other side, so that the gel is encased in plastic wrap.

2.4 To visualize the RNA by UV shadowing, place the wrapped gel on a fluorescent TLC plate and hold a UV light over the gel in a dark location. While the TLC plate should fluoresce, nucleotides absorb UV light and are revealed as dark, nonfluorescent bands in the gel. Typically the amount of DNA template used is too low to be visualized by this method, but RNA and NTPs are visible. Quickly mark the location of the RNA band, either by outlining the band or marking dots on the four corners of the band, with a marker to minimize RNA damage from UV exposure.

2.5 Using a clean razor blade or other cutting utensil, excise the outlined RNA band, transfer it to a fresh glass plate, and dice the gel piece into ~1 mm cubes. Transfer the diced gel pieces into a fresh 1.5 mL microcentrifuge tube and add 450 μL crush soak buffer.

2.6 To extract the RNA, incubate gel pieces in crush soak buffer on a rotator for either 2 h at RT or overnight at 4 °C.

Step 3: Precipitation of RNA (estimated duration: 2 h)

3.1 To collect the RNA that has been extracted by the crush soak solution, centrifuge the sample briefly at 4500 rcf for 30 s at 4 °C then transfer the supernatant into a fresh 1.5 mL microcentrifuge tube. Use a narrow-tip pipette to avoid transferring any gel pieces. To precipitate the RNA, add 1 mL of ice-cold 100% ethanol. Mix the solution well by inverting the tube and store at −20 °C for at least 1 h.

3.2 Pellet the RNA by spinning at 13,200 rpm for 20 min at 4 °C in a microcentrifuge. Decant or remove the supernatant using a micropipettor and let the RNA pellet dry in open air for ~30 min or until all ethanol has evaporated.

3.3 Resuspend the pellet in 30 μL 1 × TE buffer. For a transcription reaction as described in Step 2, this typically results in an RNA solution of ~10 μ*M* concentration. Aliquot the RNA solution for storage at −20 °C to prevent degradation from multiple freeze-thaw cycles.

Step 4: Quantification of RNA concentration (estimated duration: 2 h)

4.1 Determine the approximate RNA concentration by measuring the A_{260} on a UV/Vis spectrophotometer. For structured RNAs like riboswitches and these biosensor constructs, the nearest-neighbor method (Kallansrud & Ward, 1996) overestimates the extinction coefficients because it does not take into account hypochromicity from base-pairing (Wilson et al., 2014).

4.2 To determine the accurate RNA concentration, perform the neutral pH thermal hydrolysis assay (Wilson et al., 2014). In this method, the RNA is hydrolyzed to individual NMPs, which eliminates the hypochromic effect, then the extinction coefficient is calculated as a summation of the extinction coefficients of NMPs in proportion to their representation in the sequence. Briefly, dilute an aliquot of the RNA stock to a starting A_{260} ~ 10 a.u. and prepare the following reaction in a 0.5-mL thin-walled microcentrifuge tube:

2 μL diluted RNA stock
2 μL 10× Na_2CO_3 buffer
16 μL ddH_20

Incubate the reaction at 95 °C for 1.5 h, cool to RT. Measure the A_{260} with a UV/Vis spectrophotometer and calculate the RNA concentration of the reaction using the following formula:

$$c_{\text{RNA}} = \frac{A_{260,\text{hydrolyzed}}}{b \sum_{i}^{\text{A,C,U,G}} n_i \varepsilon_{260,i}}$$

where c is the RNA concentration, b is the path length, i is the nucleotide identity (A, C, G, or U), n_i is the frequency of nucleotide i present in the sequence, and ε_i is the molar extinction coefficient of the nucleotide i (Wilson et al., 2014). Multiply by the dilution factor to determine the stock RNA concentration. For a more detailed protocol, please see the methods paper by Wilson et al.

Notes and tips:

- To prevent RNAse contamination, autoclave or filter sterilize all solutions using a 0.22 μm filter. Keep samples containing RNA on ice to prevent breakdown. Use filter tips when pipetting RNA samples to avoid contamination.
- In Step 1.1, multiple transcription reactions can be set up in parallel, only limited by the quantity that can be purified on the PAGE gel. The reactions can be scaled up in quantity, depending on researcher needs or problems with transcription yield.
- If too many RNAs are being screened to purify by PAGE, it is possible to use a commercial RNA clean-up kit, such as the RNA Clean & Concentrator kit (Zymo Research). In this case, do not quench the reaction with ULB as described in Step 1.3. Replace Steps 2 and 3 with the manufacturer's protocol. However, it is advisable to check for abortive transcripts and to confirm the purity of the RNA through other means, such as denaturing agarose gel electrophoresis.
- In Step 2.1, to reduce band distortion and to remove unreacted ammonium persulfate from the gel, pre-run the gel for ~30 min prior to sample loading.
- In Step 2.2, to prevent contamination between samples, it is advisable to load samples in every other lane.
- In Step 2.5, to avoid contamination, do not reuse the cutting implement between different RNA samples.

- In Step 2.5, the efficiency of RNA extraction from the gel pieces may be increased by performing one freeze-thaw cycle prior to addition of crush soak buffer.

3. DETERMINATION OF LIGAND SELECTIVITY AND AFFINITY OF BIOSENSOR BY FLUORESCENCE ACTIVATION

A key functional test of any designed biosensor is whether fluorescence activation is observed in the presence of the ligand versus without ligand. Furthermore, fluorescence must be selectively turned on by the ligand of interest versus other related compounds that might be present in the cell. Finally, the apparent dissociation constant (K_d) of the biosensor–ligand complex should be within the physiologically relevant range, if the purpose is to measure *in vivo* concentrations of a metabolite or signaling molecule. Note that the apparent K_d is a simplified estimate of ligand binding affinity since the biosensor has to bind both ligand and DFHBI to elicit fluorescence. For this reason, the measurements are performed with saturating concentrations of DFHBI. Since temperature and magnesium concentrations can have dramatic effects on the stabilities of RNA structures, we recommend measuring the ligand selectivity and K_d at 37 °C and 3 m*M* Mg^{2+}, which is closer to cellular conditions than standard assay conditions.

Nevertheless, care needs to be taken in terms of interpreting these *in vitro* results as being fully reflective of cellular fluorescence measurements. While the maximum fluorescence activation *in vitro* for a given biosensor generally corresponds to the signal-to-noise observed *in vivo*, the *in vitro* measurement is performed under conditions such that the biosensor is at limiting concentrations relative to ligand, which is not always the case in the cell. In addition, the *in vitro* binding buffer is an inadequate proxy for the complex cellular milieu, so the binding affinity of the biosensor *in vivo* may be different. Nevertheless, we have found that the selectivity and relative binding affinities of different biosensor constructs *in vivo* are accurately reflected by the *in vitro* measurements (Kellenberger et al., 2013).

3.1. Determination of ligand selectivity

3.1.1 Additional materials and equipment

Fluorescent plate reader with heating control
Corning Costar 96-well black plate
c-di-GMP (Axxora, LLC; Farmingdale, NY)
Other ligand analogs for selectivity experiments.

Buffers and solutions that can be prepared in advance are listed here, whereas solutions that should be freshly prepared are described in the procedure. Prepare all solutions using ddH_2O water.

2× Renaturing buffer

Component	Final concentration (m*M*)	Stock (*M*)	Amount
HEPES–KOH, pH 7.5	80	1	800 μL
$MgCl_2$	6	1	60 μL
KCl	250	2	1.25 mL

Add water to 10 mL, sterile filter, store at RT.

DFHBI stock (Lucerna; New York, NY)

Component	Final concentration	Stock	Amount
DFHBI	20 m*M*	–	1 mg
DMSO	–	–	198 μL

Aliquot in amber tubes and store at −20 °C.

3.1.2 Procedures

Step 5: Preparation of binding reaction components (estimated duration: 20 min)

5.1 To determine the fluorescent activation of the biosensor for different ligands, prepare 1 m*M* ligand stocks (e.g., c-di-GMP) for a final concentration of 100 μ*M* in the binding reactions.

5.2 To prepare the binding reactions, first a master mix containing buffer components and DFHBI is freshly prepared:

4 μL 1 *M* HEPES, pH 7.5
6.25 μL 2 *M* KCl
0.3 μL 1 *M* $MgCl_2$
10 μL 100 μ*M* DFHBI
59.45 μL ddH_2O.

For analyzing additional or replicate samples, multiply these volumes by the number of samples plus one. The master mix given above accounts for 80 μL of the 100 μL final reaction volume, in which the remaining volume will be from RNA and ligand solutions added separately. The final concentration of components of the binding reaction will be 40 m*M* HEPES, pH 7.5, 125 m*M* KCl, 3 m*M* $MgCl_2$, 10 μ*M* DFHBI, 100 n*M* RNA, and 100 μ*M* ligand.

5.3 To renature the RNA biosensor before addition to the binding reaction, prepare a 2 μ*M* stock solution of RNA by diluting in ddH_2O, then add an equal volume of 2 × renaturation buffer to make a 1 μ*M* stock solution of RNA in renaturation buffer. Prepare enough volume to add 10 μL of this RNA solution to each binding reaction for a final concentration of 100 n*M*. Heat the renaturation reaction to 70 °C for 3 min, then cool at ambient temperature for 5 min.

Step 6: Perform binding reactions and measure fluorescence activation (estimated duration: 2.5 h)

6.1 Add 10 μL of ligand stock or ddH_2O for the "no ligand" control to 1–3 replicate wells in a 96-well black plate followed by 80 μL master mix solution. A multichannel pipet may be used; however, do not reuse tips between different wells to prevent cross-contamination. To compare the sample fluorescence to the background fluorescence of DFHBI without RNA, prepare 1–3 replicate wells with 20 μL ddH_2O + 80 μL master mix.

6.2 Add 10 μL renatured RNA to each binding reaction, excluding "no RNA" control. Mix reactions well and check that there are no bubbles in or on the side of each well. To remove any bubbles, either use a sterile pipet tip to pop or mix with the solution, or centrifuge the plate in a microcentrifuge equipped with a 96-well plate holder for 30 s at 500 rcf.

6.3 To measure fluorescence, load the plate in a fluorescence plate reader and monitor fluorescence using the following parameters:

Excitation: 460 nm
Emission: 500 nm
Total time: 2 h
Read time: every 5 min
Temperature: constant 37 °C

6.4 To analyze the data, choose a time point where the fluorescence reading has equilibrated. To determine ligand selectivity, graph fluorescence intensity for each ligand versus no ligand control, subtracting from each sample the background fluorescence of an empty well or a well containing buffer without DFHBI, which have approximately equal fluorescence. We do not subtract the fluorescence of DFHBI without RNA, as this represents the real background in the experiment

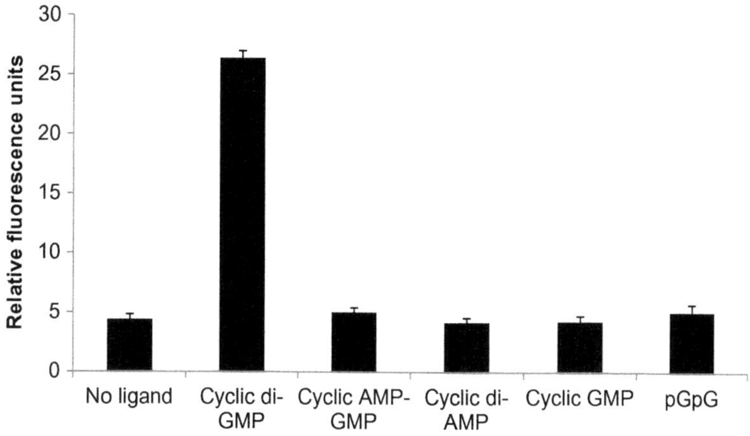

Figure 2 Selectivity measurements for the WT Vc2-Spinach aptamer. The WT Vc2-Spinach aptamer demonstrates selectivity for cyclic di-GMP. WT Vc2-Spinach (100 n*M*) elicits fluorescence from DFHBI (10 μ*M*) selectively in response to cyclic di-GMP (100 μ*M*) but not other related ligands (100 μ*M*). Error bars represent the standard deviation between three independent replicates carried out at 37 °C, 3 m*M* Mg^{2+}.

and would misrepresent the fold activation of the biosensor. Representative results for the Vc2-Spinach biosensor are shown in Fig. 2.

Notes and tips:

- In the procedure described, a single high concentration (100 μ*M*) of ligands is analyzed. If the biosensor has a very low dissociation constant (K_d) for binding its cognate ligand (high affinity), it may appear to have a fluorescence response to other ligands. In case this is observed, an intermediate (e.g., 50-fold lower) versus high ligand concentration may be tested, or determination of ligand affinity should be performed following the procedure described below.
- It is suggested that at least two technical replicates should be performed for each sample on a plate to account for systematic error. Furthermore, at least two independent replicates of the experiment should be performed to account for sample variability. Independent experiments should be prepared from different master mixes, although preparing samples on different days also helps account for other sources of variability.
- In Step 6.1, it is recommended to add ligand first before the master mix as this makes it easier to keep track of which samples have already been added. However, this order does require that micropipette tips be changed between additions of master mix to prevent contamination.

The reverse order, where master mix is added before ligand, is also fine and allows micropipette tips to be reused.

- In Step 6.3, fluorescence readings are measured at 37 °C and the binding reactions prepared in Step 5.2 contain 3 mM $MgCl_2$ in order to model biosensor activity under physiological conditions. For initial screening purposes, the binding reactions may be performed at room temperature with 10 mM $MgCl_2$ instead, which typically increases the fluorescence activation and ligand binding affinity of the biosensor.
- Use of higher concentrations of the RNA biosensor would also increase fluorescence signal. Thus, it is important to determine the RNA concentration accurately (see Step 4).
- If high background fluorescence is observed without ligand, one may consider modifying the construct to destabilize the transducer stem. To increase fluorescence activation, one may consider using the Spinach2 sequence or the alternate fluorescent dyes DFHBI-1T (Lucerna), which has an improved fluorescence excitation and emission profile (Song, Strack, Svensen, & Jaffrey, 2014; Strack et al., 2013).
- In Step 6.4, keep in mind that different biosensor constructs may demonstrate different binding kinetics. For example, Vc2-Spinach takes 40 min to fully equilibrate with ligand, but other biosensors display full activation within 5 min (Kellenberger et al., 2013, unpublished results).

3.2. Determination of ligand affinity

A protocol similar to the one described above for determination of ligand selectivity (Steps 5 and 6) can be used to measure the affinity of the biosensor for a specific ligand, with the following changes:

1. The RNA biosensor concentration must be below the expected K_d to measure the dissociation constant accurately (Hall & Kranz, 2007). Thus, we typically use a concentration of RNA that is approximately 10-fold lower than the expected K_d. For example, in Step 5.3, the final concentration for the 30 nM (Kd ~ 230 ± 50 nM). It is advisable to determine the lower limit of detection for the fluorescence activation over background.
2. In Step 5.1, in place of different ligand solutions, use different concentrations of the ligand of interest that span both sides of the expected K_d value to obtain a full binding curve. For example, to determine the binding affinity of WT Vc2-Spinach to c-di-GMP, a no ligand control and 12 serial dilutions of c-di-GMP were made that ranged from final concentrations of 100 μM to 100 pM in ~3-fold increments.

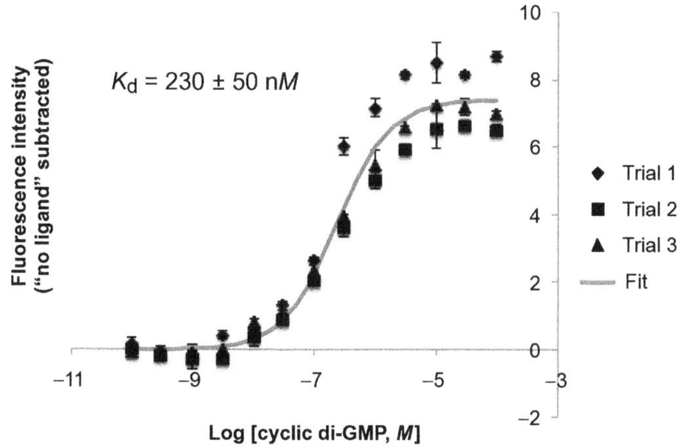

Figure 3 Affinity measurement of the WT Vc2-Spinach aptamer for c-di-GMP. WT Vc2-Spinach (30 n*M*) binds cyclic di-GMP with an apparent K_d of 230 ± 50 n*M* in the presence of saturating DFHBI (10 μ*M*). Trials from three independent replicates of experiments at 37 °C and 3 m*M* Mg^{2+} are shown.

3. Compile the fluorescence intensity for each ligand concentration from a timepoint where binding has equilibrated. Subtract the background of the "no ligand" sample from each data point and fit the data to an appropriate binding model using an Excel spreadsheet or commercial software to calculate the K_d value. At least three independent replicates of the experiment should be performed to account for sample variability, and the average K_d value with standard deviation is reported. Representative results for the Vc2-Spinach biosensor are shown in Fig. 3.

4. DETERMINATION OF BINDING KINETICS OF BIOSENSOR

For a fluorescent biosensor to be used to study the dynamics of a target metabolite or signaling molecule, it must respond within the time range of the cellular process being studied. Furthermore, it is important to characterize both ligand association and dissociation, as one advantage of fluorescent biosensors versus reporter systems is the ability to monitor signal production and breakdown dynamically. Again, we recommend measuring binding kinetics at 37 °C and 3 m*M* Mg^{2+} to mimic physiological conditions. The measured activation and deactivation rates will also depend on the concentrations of the biosensor, ligand of interest, and DFHBI.

Since fluorescence activation involves a ternary complex, one may not easily derive rate constants for the individual steps from the kinetic experiments described. Our proposed binding model assumes DFHBI can only bind to a specific conformation of the biosensor (RNA*) that is normally higher in energy than the ground-state conformation (Fig. 4). Thus, background fluorescence is low because there is very little partitioning to the RNA*•DFHBI complex. For example, we observe <5% fluorescence change for Vc2-Spinach + DFHBI versus DFHBI alone samples, which implies that very little RNA*•DFHBI is present under these conditions. Upon addition of ligand, ligand binding favors the conformational change to RNA*•L, which results in fluorescence activation of DFHBI by formation of the ternary complex. This model does not take into account the *cis* and *trans* forms of DFHBI, which have been shown to bind Spinach with slightly different affinities and to be thermally reversible (Wang et al., 2013).

For the activation experiments described below, the biosensor is added directly to a mixture of cyclic di-GMP and excess DFHBI (10 μ*M*). Because the amount of RNA*•DFHBI is low, presumably we are observing the combined steps of ligand binding, RNA conformational change, and

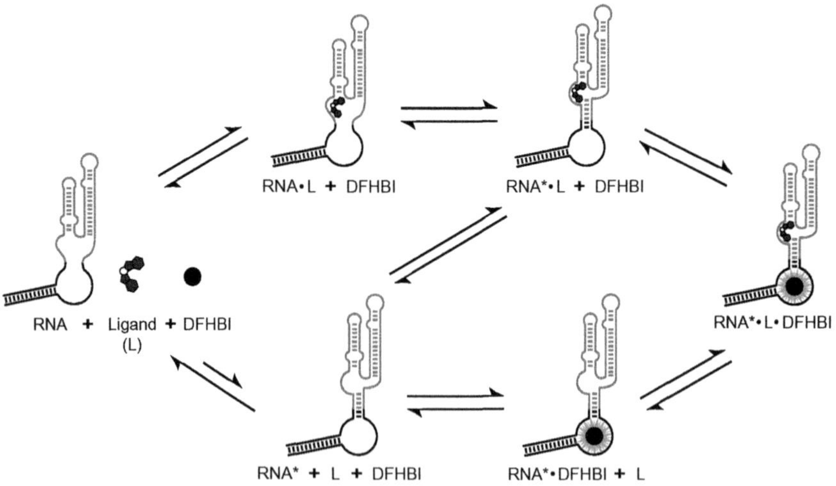

Figure 4 Association of RNA, ligand, and DFHBI to form the fluorescent ternary complex. Before the stable ternary complex forms, ligand binds to the RNA, RNA presumably undergoes a conformational change to RNA*, and DFHBI binds to RNA*. Predicted pathways of complex formation are depicted. (See the color plate.)

DFHBI binding to form the fluorescent ternary complex. For the deactivation experiments described below, the RNA is first equilibrated with excess ligand and DFHBI to form the fluorescent ternary complex RNA*•L•DFHBI. The binding reaction then is split and passed through two gel fitration columns that have been preequilibrated with either buffer containing DFHBI + ligand or DFHBI alone. In the latter sample, absence of free ligand in the reaction solution should move the overall equilibrium to the left, although the kinetic parameters are complex and reflect two separate pathways for fluorescence deactivation.

4.1. Determination of activation rate

A protocol similar to the one described above for determination of ligand selectivity (Steps 5 and 6) can be used to measure biosensor binding kinetics. Begin by preparing the same ligand dilution and master mix as in Steps 5.1 and 5.2. To avoid fluctuations in fluorescence due to temperature, the most important aspect of the activation experiment is that the ligand/master mix and RNA solutions are preequilibrated at 37 °C before fluorescence measurement. To ensure that all solutions are warmed to 37 °C and to minimize RNA degradation, first carry out Step 6.1 followed by Step 5.3 with the modifications below, then proceed with the remaining steps in order.

1. In Step 6.1, after addition of ligand or ddH_2O for "no ligand" sample to the reaction, prewarm the solution to 37 °C in the plate reader.
2. In Step 5.3, after preparing the 1 μ*M* stock solution of RNA in renaturation buffer, heat the RNA to 70 °C for 3 min, then incubate at 37 °C until added to the plate. To minimize RNA degradation, do not perform this step until the reaction solutions containing DFHBI and ligand are equilibrated at 37 °C.
3. In Step 6.2, add 10 μL RNA to each binding reaction using a multichannel pipettor, and mix reactions quickly by pipetting up and down. Immediately proceed to the next step, which is taking fluorescence readings.
4. In Step 6.3, use the following parameters:
   Excitation: 460 nm
   Emission: 500 nm
   Total time: 2 h
   Read time: every 1–2 min
   Temperature: constant 37 °C

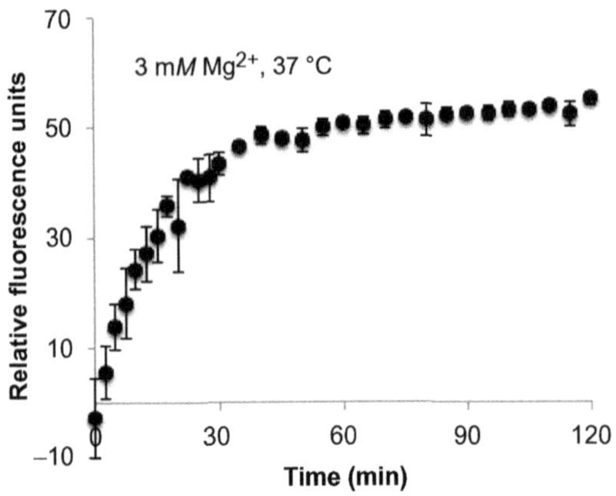

Figure 5 Measurement of the activation rate of WT Vc2-Spinach. WT Vc2-Spinach (100 n*M*) fully binds to DFHBI (10 μ*M*) and cyclic di-GMP (100 μ*M*) within 40 min at 37 °C and 3 m*M* Mg^{2+}. Error bars represent the standard deviation of two replicate samples.

5. In Step 6.4, plot the fluorescence intensity over time to determine time required for full ligand association. Representative results for the Vc-Spinach biosensor are shown in Fig. 5.

Notes and tips:

- If unexpected fluorescence variation is observed in the first few minutes, this is likely due to fluctuation in temperature from the sample. Ensure that both the RNA and the ligand mixture are preequilibrated at 37 °C before mixing.

4.2. Determination of deactivation rate

To perform the deactivation experiment, free ligand in the sample will be removed by passing half of an equilibrated binding reaction through a size-exclusion column that has been preloaded with DFHBI and buffer but no c-di-GMP. The free c-di-GMP in the sample will be retained by the column, whereas ligand-bound RNA and free RNA, should elute from the column along with preloaded DFHBI and buffer. As a control, the other half of the sample is passed through a size-exclusion column that has been preloaded with c-di-GMP, DFHBI, and buffer.

4.2.1 Additional materials and equipment

G-25 columns.

4.2.2 Procedure

Step 7: Preparation of binding reaction (estimated duration: 1.5 h)

7.1 Prepare the following two exchange buffers, one with and the other without ligand, and prewarm the solution to 37 °C. The amounts shown should be sufficient for 3 column volumes each (~1.5 mL).

60 μL 1 *M* HEPES, pH 7.5
93.75 μL 2 *M* KCl
4.5 μL 1 *M* $MgCl_2$
150 μL 100 μ*M* DFHBI
150 μL 1 m*M* ligand (e.g., c-di-GMP) or ddH_2O
1041.75 μL ddH_2O

7.2 Renature the RNA before adding to the binding reaction by preparing a 2 μ*M* stock solution of RNA in ddH_2O, then add an equal volume of 2 × renaturation buffer to make a 1 μ*M* stock solution of RNA in renaturation buffer. Prepare enough volume to add 21 μL of this RNA solution to each binding reaction for a final concentration of 100 n*M*. Heat the renaturation reaction to 70 °C for 3 min, then cool at ambient temperature for 5 min.

7.3 Prepare the following binding reaction (210 μL total volume) in one well of a black 96-well plate:

8.4 μL 1 *M* HEPES, pH 7.5
13.13 μL 2 *M* KCl
0.63 μL 1 *M* $MgCl_2$
21 μL 100 μ*M* DFHBI
21 μL 1 m*M* ligand (e.g., c-di-GMP)
21 μL 1 μ*M* renatured WT Vc2-Spinach RNA
124.84 μL ddH_2O

The final concentrations of components of the binding reaction are 40 m*M* HEPES, pH 7.5, 125 m*M* KCl, 3 m*M* $MgCl_2$, 10 μ*M* DFHBI, 100 n*M* RNA, and 100 μ*M* ligand.

7.4 Incubate the reaction at 37 °C in the fluorescent plate reader. Monitor fluorescence using the parameters given in Step 6.3 until the reaction has fully equilibrated. For the sample with Vc2-Spinach, this takes approximately 1 h.

Step 8: Perform buffer exchange and measure fluorescence deactivation (estimated duration: 2 h)

8.1 First prepare two G-25 spin columns, one equilibrated with ligand and one without, by saturating each of the columns with 3 column volumes (~500 μL) of the exchange buffer. Briefly, load 500 μL buffer on each column and centrifuge the spin columns at 730 rcf for 1 min at 37 °C. Discard the flow through and repeat this buffer exchange two more times. Place the column onto a clean collection tube.

8.2 Once the binding reaction from Step 7.4 is equilibrated and the G-25 columns are prepared, immediately load 100 μL of the binding reaction onto each of the G-25 spin columns and centrifuge at 730 rcf for 1 min to elute. Due to size exclusion, the ligand-bound RNA and RNA will flow through, whereas the free ligand will be retarded in the gel matrix.

8.3 Immediately transfer each of the flow-through solutions to clean and separate wells of the 96-well plate and record fluorescence with the following parameters:

Excitation: 460 nm
Emission: 500 nm
Total time: 2 h
Read time: every 1–2 min
Temperature: constant 37 °C

8.4 To analyze the data, plot the fluorescence intensity over time for each sample. Representative results are shown in Fig. 6.

Notes and tips:

- The manufacturer notes that G-25 column should not be left dry for a long period of time. Proceed directly from Step 8.1 to 8.2.

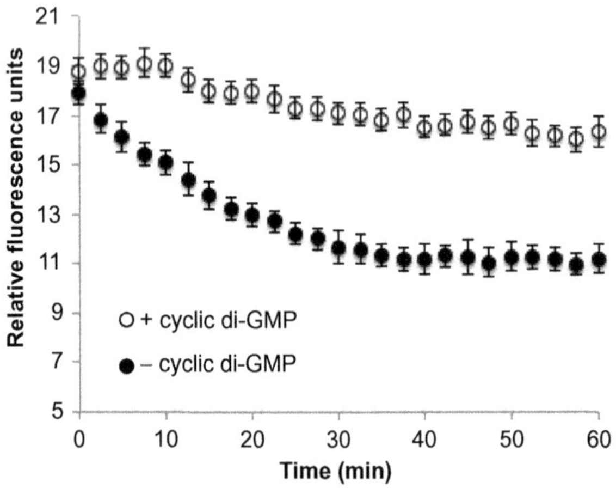

Figure 6 Measurement of the deactivation rate of WT Vc2-Spinach. WT Vc2-Spinach RNA (100 n*M*) fully deactivates in response to loss of ligand within 40 min at 37 °C and 3 m*M* Mg^{2+}. Some loss of ligand is observed in the sample passed through a column buffer exchanged with cyclic di-GMP, indicating retention of cyclic di-GMP or DFHBI on the column. Error bars represent the standard deviation of two replicate samples.

- Steps 8.2 and 8.3 should be performed quickly. If the deactivation rate is too fast or if the above procedure proves problematic, consider other strategies to examine the deactivation of the biosensor, such as enzymatic degradation of the metabolite or favoring ligand dissociation by dilution.
- To avoid fluctuations in fluorescence due to differential elution and retention between the two columns, consider eluting the RNA sample into another 100 μL of the exchange buffer in Step 8.2. This helps to ensure that the buffer components stay the same between samples.
- If possible, use a saturating concentration of ligand in the binding reaction and an excess of ligand when equilibrating the spin column. This will reduce variations in fluorescence due to small amounts of ligand retained in the column.

REFERENCES

Blount, K. F., Wang, J. X., Lim, J., Sudarsan, N., & Breaker, R. R. (2007). Antibacterial lysine analogs that target lysine riboswitches. *Nature Chemical Biology, 3*, 44–49.

Cruse, W. B., Saludjian, P., Biala, E., Strazewski, P., Prangé, T., & Kennard, O. (1994). Structure of a mispaired RNA double helix at 1.6-Å resolution and implications for the prediction of RNA secondary structure. *Proceedings of National Academy of Sciences of United States of America, 91*, 4160–4164.

Deigan Warner, K., Chen, M. C., Song, W., Strack, R. L., Thorn, A., Jaffrey, S. R., et al. (2014). Structural basis for activity of highly efficient RNA mimics of green fluorescent protein. *Nature Structural and Molecular Biology, 21*, 658–663.

Domaille, D. W., Zeng, L., & Chang, C. J. (2010). Visualizing ascorbate-triggered release of labile copper within living cells using a ratiometric fluorescent sensor. *Journal of the American Chemical Society, 132*, 1194–1195.

Fowler, C. C., Brown, E. D., & Li, Y. (2010). Using a riboswitch sensor to examine coenzyme B12 metabolism and transport in *E. coli*. *Chemistry & Biology, 17*, 756–765.

Hall, K. B., & Kranz, J. K. (2007). Nitrocellulose filter binding for determination of dissociation constants. *Methods in Molecular Biology, 118*, 1–10.

Huang, H., Suslov, N. B., Li, N.-S., Shelke, S. A., Evans, M. E., Koldobskaya, Y., et al. (2014). A G-quadruplex-containing RNA activates fluorescence in a GFP-like fluorophore. *Nature Chemical Biology, 10*, 686–691.

Kallansrud, G., & Ward, B. (1996). A comparison of measured and calculated single- and double-stranded oligodeoxynucleotide extinction coefficients. *Analytical Biochemistry, 236*, 134–138.

Kellenberger, C. A., Wilson, S. C., Sales-Lee, J., & Hammond, M. C. (2013). RNA-based fluorescent biosensors for live cell imaging of second messengers cyclic di-GMP and cyclic AMP-GMP. *Journal of the American Chemical Society, 135*, 4906–4909.

Kuzmine, I., Gottlieb, P. A., & Martin, C. T. (2003). Binding of the priming nucleotide in the initiation of transcription by T7 RNA polymerase. *The Journal of Biological Chemistry, 278*, 2819–2823.

Liu, Z., He, W., & Guo, Z. (2013). Metal coordination in photoluminescent sensing. *Chemical Society Reviews, 42*, 1568–1600.

Mandal, M., & Breaker, R. R. (2004). Adenine riboswitches and gene activation by disruption of a transcription terminator. *Nature Structural & Molecular Biology, 11*, 29–35.

Nakayama, S., Luo, Y., Zhou, J., Dayie, T. K., & Sintim, H. O. (2012). Nanomolar fluorescent detection of c-di-GMP using a modular aptamer strategy. *Chemical Communications, 48*, 9059–9061.

Nelson, M. D., & Fitch, D. H. A. (2011). Molecular methods for evolutionary genetics. In V. Orgogozo, & M. V. Rockman (Eds.), *Vol. 772. Methods in molecular biology* (pp. 459–470). New York, NY: Humana Press.

Newman, R. H., Fosbrink, M. D., & Zhang, J. (2011). Genetically encodable fluorescent biosensors for tracking signaling dynamics in living cells. *Chemical Reviews, 111*, 3614–3666.

Paige, J. S., Nguyen-Duc, T., Song, W., & Jaffrey, S. R. (2012). Fluorescence imaging of cellular metabolites with RNA. *Science, 335*, 1194.

Paige, J. S., Wu, K. Y., & Jaffrey, S. R. (2011). RNA mimics of green fluorescent protein. *Science, 333*, 642–646.

Römling, U., Galperin, M. Y., & Gomelsky, M. (2013). Cyclic di-GMP: The first 25 years of a universal bacterial second messenger. *Microbiology and Molecular Biology Reviews, 77*, 1–52.

Sambrook, J., Fritsch, E. F., & Maniatis, T. (1989). *Molecular cloning: A laboratory manual.* Cold Spring Harbor, NY: Cold Spring Harbor Press.

Serganov, A., & Nudler, E. (2013). A decade of riboswitches. *Cell, 152*, 17–24.

Song, W., Strack, R. L., Svensen, N., & Jaffrey, S. R. (2014). Plug-and-play fluorophores extend the spectral properties of Spinach. *Journal of the American Chemical Society, 136*, 22–25.

Strack, R. L., Disney, M. D., & Jaffrey, S. R. (2013). A superfolding Spinach2 reveals the dynamic nature of trinucleotide repeat-containing RNA. *Nature Methods, 10*, 1219–1224.

Wang, P., Querard, J., Maurin, S., Nath, S. S., Le Saux, T., Gautier, A., et al. (2013). Photochemical properties of Spinach and its use in selective imaging. *Chemical Science, 4*, 2865–2873.

Wilson, S. C., Cohen, D. T., Wang, X. C., & Hammond, M. C. (2014). A neutral pH thermal hydrolysis method for quantification of structured RNAs. *RNA, 20*, 1153–1160.

Zhou, X., Herbst-Robinson, K. J., & Zhang, J. (2012). Visualizing dynamic activities of signaling enzymes using genetically encodable FRET-based biosensors from designs to applications. *Methods in Enzymology, 504*, 317–340.

CHAPTER NINE

Using Spinach Aptamer to Correlate mRNA and Protein Levels in *Escherichia coli*

Georgios Pothoulakis*,†, Tom Ellis*,†,1
*Centre for Synthetic Biology and Innovation, Imperial College London, South Kensington Campus, London, United Kingdom
†Department of Bioengineering, Imperial College London, South Kensington Campus, London, United Kingdom
[1]Corresponding author: e-mail address: t.ellis@imperial.ac.uk

Contents

Abstract

In vivo gene expression measurements have traditionally relied on fluorescent proteins such as green fluorescent protein (GFP) with the help of high-sensitivity equipment such as flow cytometers. However, fluorescent proteins report only on the protein level inside the cell without giving direct information about messenger RNA (mRNA) production. In 2011, an aptamer termed Spinach was presented that acts as an RNA mimic of GFP when produced in *Escherichia coli* and mammalian cells. It was later shown that coexpression of a red fluorescent protein (mRFP1) and the Spinach aptamer, when included into the same gene expression cassette, could be utilized for parallel *in vivo* measurements of mRNA and protein production. As accurate characterization of

Methods in Enzymology, Volume 550
ISSN 0076-6879
http://dx.doi.org/10.1016/bs.mie.2014.10.047

component biological parts is becoming increasingly important for fields such as synthetic biology, Spinach in combination with mRFP1 provide a great tool for the characterization of promoters and ribosome binding sites. In this chapter, we discuss how live-cell imaging and flow cytometry can be used to detect and measure fluorescence produced in *E. coli* cells by different constructs that contain the Spinach aptamer and the *mRFP1* gene.

1. INTRODUCTION

With the emergence of fields such as synthetic biology, which relies on biological systems exhibiting modularity to some degree and as such can be described using engineering principles, accurate characterization of gene expression control elements has become very important (Purnick & Weiss, 2009). Selection of different biological parts (such as promoters and ribosome binding sites) for gene networks and metabolic pathways relies on quantitative and qualitative data provided from several sources (Ellis, Wang, & Collins, 2009; Oyarzun & Stan, 2012; Temme, Hill, Segall-Shapiro, Moser, & Voigt, 2012). *In vivo* gene expression measurements traditionally relied on fluorescent proteins such as the green fluorescent protein (GFP) that link protein expression to engineered devices using high-sensitivity equipment such as flow cytometers and plate readers. Fluorescent proteins are limited by the fact that they report only on protein production, which is the end result of the gene expression process. Since gene expression also involves messenger RNA (mRNA) conversion from DNA before proteins are produced, fluorescent proteins are unable to provide information about control elements that are involved during this process.

In summer 2011, researchers at the Cornell University presented for the first time a RNA motif termed Spinach that, when bound to a specific fluorophore, could mimic GFP in *Escherichia coli* and mammalian cells (Paige, Wu, & Jaffrey, 2011). The Spinach aptamer fluoresces when expressed in the presence of the constructed fluorophore molecule, 3,5-difluoro-4-hydroxybenzylidene imidazolinone (DFHBI), which is a derivative of the HBI fluorophore of GFP. The complexes formed between the RNA and the fluorophore have an excitation maximum at 460 nm and an emission maximum at 510 nm. Since then, a new version of the aptamer called Spinach2 was developed countering Spinach's thermal instability and propensity for misfolding issues that lead to dimmer signals (Strack, Disney, & Jaffrey, 2013). In addition, compatible fluorophores that extend the spectral properties of Spinach (Song, Strack, Svensen, & Jaffrey, 2014) as well as Spinach-based sensors that can be used to image intracellular metabolites and proteins in living bacterial cells have also been described (Strack, Song, & Jaffrey, 2014).

The focus of this chapter is to describe how the Spinach aptamer, when coexpressed with a fluorescent protein under the same expression cassette, can be used to simultaneously measure transcription and translation of mRNAs in bacterial cells, providing a valuable tool for gene expression characterization. For these experiments, a described set of constructed plasmid vectors that carry specific promoters and ribosome binding sites are used (Pothoulakis, Ceroni, Reeve, & Ellis, 2013). Flow cytometry and live-cell imaging is used to measure the fluorescence output of Spinach and the mRFP1 protein under different scenarios. The parts selected for the construction of these vectors are based on the behavior of the original Spinach aptamer.

2. PARTS SELECTION AND PLASMID CONSTRUCTION

The first step involves the selection of the biological parts that need to be characterized and the construction of the final vectors to be transformed in *E. coli* cells. It has already been shown that the Spinach aptamer and the *mRFP1* gene can be used to characterize strong inducible promoters as well as ribosome binding sites designed using the Salis Lab RBS calculator (https://salis.psu.edu/software/; Salis, Mirsky, & Voigt, 2009). Moreover, it was shown that maximum green fluorescence output is obtained when the aptamer within the tRNA scaffold is placed after the *mRFP1* gene (Pothoulakis et al., 2013). Naturally, parts selection should not be limited only to the examples given here, as long as the promoters selected are able to drive significant Spinach production. Different fluorescent proteins are also a possibility but are yet to be explored in this context.

In this study, the constructs consisted of either the T7 or the T5 promoter. The T7 bacteriophage promoter was originally used by the Jaffrey Lab in the pET28c-Spinach plasmid vector and has a strong output. The T5 bacteriophage promoter, which according to our experiments is significantly weaker than the T7, also maintains a relatively high output (Pothoulakis et al., 2013). With the inclusion of a gene expressing a fluorescent protein marker, characterization of the ribosome binding site sequences can also be accomplished by dual measurement of transcription and translation.

For all constructs used here, the pET28c-Spinach vector was used as a backbone. The pET28c-Spinach plasmid can be obtained by the Samie Jaffrey Lab (Cornell University) and contains the Spinach aptamer inside a tRNA scaffold that protects the aptamer from the cellular RNases (Ponchon & Dardel, 2007). All other vectors shown here are derived from that plasmid and are designed by the Ellis Lab. These constructs carry both the Spinach aptamer and the *mRFP1* gene under the same promoter (either the T7 or the T5) and a variety of synthetic ribosome binding sequences designed using the Salis Lab RBS calculator (with the exception of RBS1).

To insulate the individual parts from local folding and ribosome binding, in all constructs random, synthetic 30 nucleotide spacer sequences are placed between the promoter, the RBS-mRFP1 region, and the Spinach aptamer (Pothoulakis et al., 2013). These 30mer DNA sequences were generated using a Monte Carlo Algorithm written to generate sequences with 50% GC content (Casini et al., 2014).

The *mRFP1* gene for the pGPR plasmids was PCR-amplified from part BBa_E1010 of the Registry of Standard Biological Parts (http://partsregistry.org) using specific primers that introduce the necessary overhangs for Gibson Assembly (Gibson et al., 2009). Assembly is performed as described by Daniel Gibson on his paper. All constructs and their characteristics are shown in Table 1.

3. *E. coli* STRAIN SELECTION

Since the T7 promoter system needs a nonnative T7 RNA polymerase to work, the BL21 (DE3) *E. coli* strain that carries the λ DE3 lysogen is required for protein expression. In this strain, the T7 RNA polymerase gene is controlled by LacI, which can lead to leaky expression even in the absence of the allolactose mimic IPTG (isopropyl β-D-1-thiogalactopyranoside) which acts as an inducer. The pLysS and pLysE versions of BL21 (DE3) cells reduce the amount of leakage but create induction problems and are not recommended. An alternative method of overnight culturing with the addition of glucose and lower temperature to reduce leaky expression is shown here. The same strain is used (but not required) for T5 promoter expression. The T5 promoter does not require phage RNA polymerase to function, instead using native polymerases.

Table 1 Plasmid vector list and construct characteristics

Vector name	Promoter	*mRFP1* gene	RBS name	RBS DNA sequence	Predicted translation initiation rate[a]
pET28c-Spinach	T7[b]	Not present	N/A	N/A	N/A
pGPR01	T7[b]	Present	RBS1	GCAATTTTAAGGAGGTAACT	–
pGPR05	T7[b]	Present	R1K	TTATCCTTCCCTAATCCGAAAGGCATCCCTACGGT	938.51
pGPR06	T7[b]	Present	R9K	TCGATCGCAAACGAATCGCCAGTAGGGTAAGAGT	9166.64
pGPR07	T7[b]	Present	R148K	ACACAGAAAGCAAAAACATTAAGGGGGTAATA	148,124 (MAX)
pGPR08	T5	Present	RBS1	GCAATTTTAAGGAGGTAACT	–
pGPR09	T5	Present	R38K	CCACCGCCCGGTATATTAAAATTAGGAGGTTAAT	38,394
pGPR10	T5	Present	R148K	ACACAGAAAGCAAAAACATTAAGGGGGTAATA	148,124 (MAX)
pGPS01	T5	Not present	N/A	N/A	N/A

[a]Predicted translation initiation rates as given by the Salis Lab RBS Calculator using the forward engineering mode.
[b]The original pET28c vector carries a *lac* operator sequence immediately downstream of the T7 promoter. That sequence was deleted during the construction of the pET28c-Spinach vector by the Jaffrey Lab. As a result, in our constructs, the *lac* operator is also omitted.

4. CULTURING AND INDUCING *E. coli* CELLS

Culturing and induction conditions were formed based on the characteristics of the T7 and T5 promoter systems and the induction conditions suggested by the Jaffrey Lab (http://www.jaffreylab.org/Pages/SpinachPlasmids.aspx). Both promoter systems are induced by IPTG. Usually, in the T7 inducible system, the *lac* operator can be found in both the upstream promoter driving the T7 RNA Polymerase expression and in the T7 promoter itself (although in our constructs it is omitted). In the T5 system, a *lac* operator is also adjacent to the T5 promoter and no alterations were made. Cells are cultured at 30 °C in rich media containing glucose to minimize expression leakiness of the T7 promoter:

1. BL21 (DE3) competent cells are transformed with the pET28c-Spinach, pGPR, or pGPS plasmids following the appropriate transformation protocol provided by the supplier and are plated on LB plates containing kanamycin.
2. Using a loop, pick a colony from a plate and inoculate a round-bottom culture tube containing 3 mL LB + kanamycin media supplemented with 1% glucose. Cells are grown overnight in a shaking incubator at 30 °C/225 rpm overnight. We found that 1% glucose in combination with the lower temperature greatly minimizes expression leakage.
3. Following the O/N growth, 150 μL of culture is diluted in round-bottom culture tubes containing 3 mL of LB + kanamycin media and grown at 37 °C/225 rpm until at $OD_{600} = 0.4$.
4. IPTG is then added to 1 m*M* final concentration. Incubation with shaking is continued at 37 °C/225 rpm for 2 h.

5. CORRELATING mRNA AND PROTEIN PRODUCTION USING FLOW CYTOMETRY

We found that the best way to correlate mRNA and protein production in living cells is to use flow cytometry. It enables multiparametric analysis of the physical and fluorescence characteristics of thousands of cells individually in a few seconds. Other methods, such as fluorescent plate reader measurement, give only a general idea of the whole cell population behavior without taking into account possible subpopulations. For our experiments, we used a modified two-laser Becton–Dickinson FACScan flow cytometer using the settings shown in Table 2. Make sure that your equipment is properly calibrated before use especially for *E. coli* usage.

Table 2 Two-color Becton–Dickinson FACScan flow cytometer settings for Spinach and mRFP1 analysis

Number of cells measured	10,000 Events inside the gate or 60 s
Spinach measurements	Excitation: 488-nm laser Detection: FL1 band-pass filter (530 nm)
mRFP1 measurements	Excitation: 561-nm laser Detection: FL5 band-pass filter (610 nm)
Voltage	FSC: E01; SSC: 375; FL1: 890; FL5: 850 (Rainbow)
Threshold	FSC: 52; SSC: 150; FL1: 52; FL5: 52
Mode	FSC: Log; SSC: Log; FL1: Log; FL5: Log
Compensation	No compensation

5.1 Prepping cells for analysis

1. To measure fluorescence, after the 2-h IPTG induction, 10 μL of culture was diluted into a 1.5-mL tube containing 90 μL LB media.
2. DFHBI (Lucerna Technologies) is added to 200 μ*M* final concentration (usually 1 μL from a 20-m*M* stock). Samples are incubated at 37 °C/ 225 rpm for 10 min, placed on ice for 10 min, and then taken for analysis. Usually, by lowering the temperature, RNA folding capability increases and the secondary structures of the tRNA scaffold/Spinach aptamer complex can be formed easier.
3. From each tube, 2 μL are transferred in round-bottom polystyrene tubes containing 1 mL of water. For large-scale experiments, 96-well plates can be used but sample volume must be adjusted accordingly.

5.2 Calibrating and setting up the flow cytometer

Before running the samples, the flow cytometer must be configured properly using the FACScan's control software Cellquest (see Table 2). Due to the small size of bacterial cells, it is very important to choose the correct settings and gate properly. For Spinach detection, we are using a 488-nm laser for excitation detecting through a 530-nm band-pass filter (FL1). Red fluorescence was excited with a 561-nm laser and 610 nm filter (FL5). FL5 is controlled by the add-on software "Rainbow":

1. To view your data, a dot plot of FSC versus SSC and a dot plot of FL1 (green fluorescence) and FL5 (red fluorescence) must be created. It is also recommended to include separate histograms of FL1 and FL5 as seen in Fig. 1.
2. Appropriate voltages for the flow cytometer's photomultiplier tubes (PMTs) must be set. For BL21 (DE3), we found that the following settings are appropriate: FSC at E01, SSC at 375 V, FL1 at 890 V, and FL5 at 850 V. Threshold of SSC at 150 V is needed.
3. Set a gate based on the FSC/SSC dot plot trying to exclude small particles, dead cells, or doublets. It is recommended to create a fairly large gate that stays constant between measurements so that the results are comparable. In this case, we use a gate between 10^1 and 10^2 arbitrary units for both FSC and SSC.
4. You can set measurements to stop after a certain period of time or until a certain number of events has been reached inside or outside the gate. Usually, for significant results, at least 5000 events within the gate are needed. In this case, we gate 10,000 cells, with a cutoff of 60 s (if fewer than 10,000 are collected after 1 min, then collection stops).

5.3 Analysis of flow data and presentation

There are several commercially available programs that can be used for data analysis and presentation. Apart from FACScan's control software CellQuest (BD biosciences), we also recommend FlowJo (Tree Star), which enables large-scale analysis with powerful analysis tools. Free programs for flow cytometric analysis include Cyflogic (http://www.cyflogic.com) and Flowing Software (http://www.flowingsoftware.com), which provide basic analysis tools.

6. CORRELATING mRNA AND PROTEIN PRODUCTION USING FLUORESCENCE MICROSCOPY

For a more straightforward correlation between Spinach and mRFP1 expression, live-fluorescence imaging can be used to assay *E. coli* cells. Using this method, photos of individual cells can be taken and analyzed for green and red fluorescence signal coming from the aptamer and the protein, respectively. Photos are taken through a 60 × CPI60 objective mounted

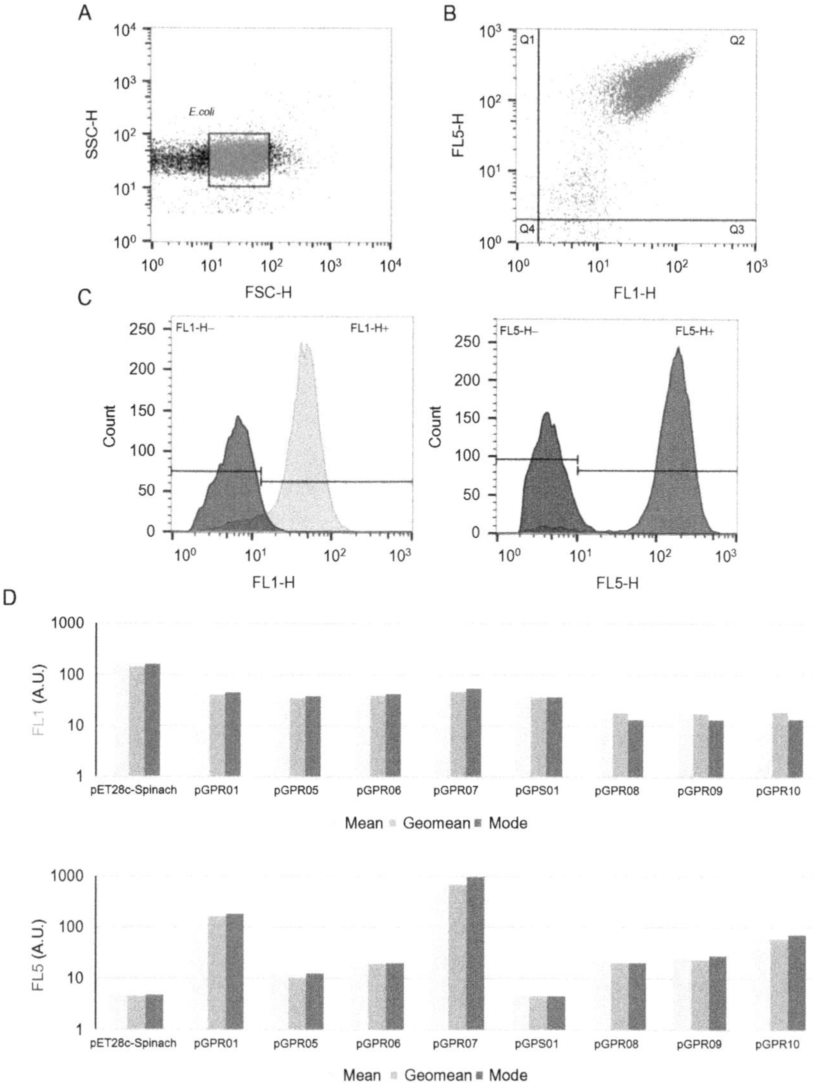

Figure 1 Flow cytometry analysis process for *E. coli* strains expressing Spinach and/or mRFP1 under the control of different promoters and ribosome binding sites. (A–D) Flow cytometry analysis of BL21(DE3) cells carrying the pGPR01 plasmid. (A) Area plot of forward scatter weight (FSC-H) versus side scatter height (SSC-H). Analyzed population (10,000 cells) is gated and shown in orange (gray in the print version). (B) Area plot of FL1-H (green fluorescence) versus FL5-H (red fluorescence) showing only the gated population of (A). Cells showing no fluorescence are seen in the lower left corner. FL1 and FL5 values below 2×10^{0} arbitrary units (subpopulations Q1, Q3, and Q4) are excluded. (C) Individual FL1-H and FL5-H histograms of the Q2 subpopulation of B (shown with light colors (light gray shades)). Dark-colored (dark gray shades) *(Continued)*

on a Nikon Eclipse Ti inverted microscope, with live cells imaged on 1% agarose pads in standard M9-Glucose media supplemented with 1 m*M* thiamine hydrochloride and 0.2% casamino acids (Sambrook & Russell, 2001).

6.1 Agarose pads

Agarose pads are broadly used for live-cell imaging of several cell types as they greatly improve the quality of images acquired by immobilizing living cells and increasing focus. That is important especially for bacterial cells which are very small, hard to focus, and mobile. Agarose pads are usually made using colorless media (so that the image is as clear as possible) and agarose. In this case, we prepare 1–1.2% agarose pads in M9:

1. In a 1.5-mL tube, measure 10–12 mg of agarose and add 1 mL of M9-glucose media.
2. Using a heat block or water bath, heat up the tube at 90 °C for 10–15 min while vortexing every 2 min.
3. Let the tube to gradually cool down to 55–60 °C.
4. Make 1-mm thick agarose pads by adding 40–50 μL of the M9-agarose between two glass slides separated by another slide or tape (≈1 mm thick). Apply weight on top of the slides and let them cool in the fridge for 20–30 min.
5. Remove the top slide and using a scalpel cut the pads in half (or more) and transfer them on a clean class slide.

6.2 Prepping cells for analysis

1. After the 2-h IPTG induction, 100 μL of culture was transferred into a 1.5-mL tube.
2. DFHBI (Lucerna Technologies) is added to 200 μ*M* final concentration (usually 1 μL from a 20-m*M* stock). Samples are incubated at 37 °C/225 rpm for 10 min, placed on ice for 10 min, and then taken for analysis.
3. Pipette 1–3 μL of each sample on top of the agarose pad. Slides are left to dry for a few minutes (not completely though) before getting covered with a coverslip.

Figure 1—Cont'd histograms represent control samples where DFHBI was not added. Black lines show the range of the positive (FL1-H+ and FL5-H+) and negative (FL1-H− and FL5-H−) populations based on the 95th percentile of the control populations. (D) Green (FL1) and red (FL5) fluorescence of *E. coli* BL21(DE3) cells expressing mRNA from nine different vectors (pET28c-Spinach, pGPR01, pGPR05-10, and pGPS01). Data show mean, geometric mean, and modal fluorescence of one biological replicate as determined by FlowJo (Tree Star).

6.3 Setting up the fluorescence microscope

1. Place the slide upside down.
2. Use the 10 × objective under the bright-field or phase filter to locate the cells.
3. Zoom-in using the 60 × CPI60 objective and adjust the base so that the cells are focused. Oil is required for the 60 × CP160 objective.
4. Without moving the slide, switch between the GFP and RFP channels. In this experiment, excitation, emission filters, and exposure time were, respectively, 480 nm, 535 nm, 4000 ms for the GFP channel (Spinach) and 532 nm, 590 nm, 1000 ms for the Cy3 channel (mRFP1).

Phase, green (GFP) and red fluorescence (Cy3) images of BL21(DE3) *E. coli* cells carrying either the pET28c-Spinach or pGPR01 plasmid are shown in Fig. 2. Detailed settings are given in Table 3.

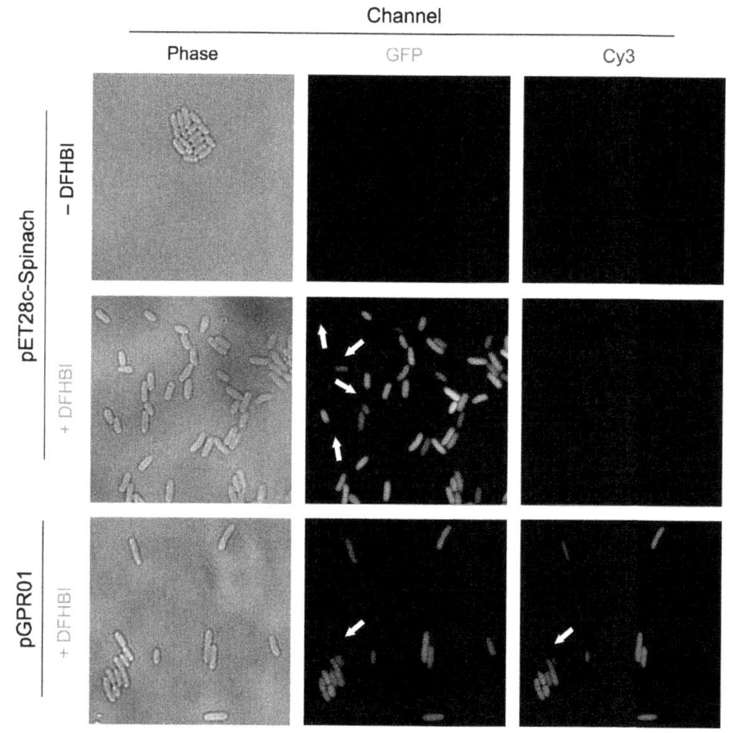

Figure 2 Live-cell fluorescence (GFP/Green and Cy3/Red) and phase images of BL21 (DE3) *E. coli* cells carrying either the pET28c-Spinach or pGPR01 plasmid in the presence or absence of 200 μ*M* DFHBI following IPTG induction. OFF-population cells not showing any fluorescence are highlighted with arrows. (See the color plate.)

Table 3 Nikon Eclipse Ti fluorescence microscope settings

Intermediate magnification	1.5 ×
Objectives	10 ×/0.30 CP160 Plan Fluor 60 ×/1.40 CP160 Plan Apo
Phase microscopy	Ph3
Spinach measurement (GFP channel)	Excitation: 480 nm Emission: 535 nm Exposure: 4000 ms Gain: 0
mRFP1 measurement (Cy3 channel)	Excitation: 532 nm Emission: 590 nm Exposure: 1000 ms Gain: 0
Frame size	1024 × 1024

6.4 Image analysis

By default, image files generated by NIS-elements Microscope Imaging Software (Nikon) are stored using the jp2 (JPEG2000) extension, which is not a widely recognized format. It is therefore preferable to convert the images using the NIS-Elements Imaging Software to another file type, such as tiff, that is recognized by the majority of specialized imaging programs. A very powerful Java-based program for image analysis is ImageJ (http://imagej.nih.gov/ij/), which was developed by the National Institutes of Health. With ImageJ it is easy to process and correlate fluorescence across different samples and identify negative populations.

7. SUMMARY

Measuring gene expression *in vivo* has been limited mainly to several fluorescent proteins such as GFP, which only report on the combined transcription and translational level. With the arrival of the Spinach aptamer, it is now possible to measure in *E. coli* the number of transcripts in the cell when the aptamer is included in the gene expression cassette. In combination with fluorescent proteins such as mRFP1, it is now possible to have parallel *in vivo* characterization of several biological parts using simple methods such as flow cytometry. The range of biological parts is limited to strong promoters and ribosome binding site sequence, although with the arrival of the new Spinach2 aptamer it may now be possible to measure weaker promoters as well.

Due to the very small size of the *E. coli* cells, for flow cytometry experiments it is important to calibrate the machinery and gate properly, always excluding events that are near the edges of the plots. For live-cell imaging, in order to avoid significant drawbacks in the quality of the acquired image, the creation of high-quality agar pads is strongly encouraged.

ACKNOWLEDGMENTS

The authors wish to thank Samie Jaffrey (Cornell University). Work at CSYNBI is supported by the UK Engineering and Physical Sciences Research Council (EPSRC), and B.R. is additionally co-funded by TMO Renewables Ltd.

REFERENCES

Casini, A., Christodoulou, G., Freemont, P. S., Baldwin, G. S., Ellis, T., & MacDonald, J. T. (2014). R2oDNA designer: Computational design of biologically neutral synthetic DNA sequences. *ACS Synthetic Biology, 3*, 525–528, 140211070819005.

Ellis, T., Wang, X., & Collins, J. J. (2009). Diversity-based, model-guided construction of synthetic gene networks with predicted functions. *Nature Biotechnology, 27*, 465–471.

Gibson, D. G., Young, L., Chuang, R.-Y., Venter, J. C., Hutchison, C. A., & Smith, H. O. (2009). Enzymatic assembly of DNA molecules up to several hundred kilobases. *Nature Methods, 6*, 343–345.

Oyarzun, D. A., & Stan, G. B. (2012). Synthetic gene circuits for metabolic control: Design trade-offs and constraints. *Journal of the Royal Society, Interface*. http://dx.doi.org/10.1098/rsif.2012.0671.

Paige, J. S., Wu, K. Y., & Jaffrey, S. R. (2011). RNA mimics of green fluorescent protein. *Science, 333*, 642–646.

Ponchon, L., & Dardel, F. (2007). Recombinant RNA technology: The tRNA scaffold. *Nature Methods, 4*, 571–576.

Pothoulakis, G., Ceroni, F., Reeve, B., & Ellis, T. (2013). The Spinach RNA aptamer as a characterization tool for synthetic biology. *ACS Synthetic Biology, 3*, 182–187, 130830130257008.

Purnick, P. E., & Weiss, R. (2009). The second wave of synthetic biology: From modules to systems. *Nature Reviews. Molecular Cell Biology, 10*, 410–422.

Salis, H. M., Mirsky, E. A., & Voigt, C. A. (2009). Automated design of synthetic ribosome binding sites to control protein expression. *Nature Biotechnology, 27*, 946–950.

Sambrook, J., & Russell, D. W. (2001). *Molecular cloning: A laboratory manual* (3rd ed.). New York: Cold Spring Harbor Laboratory Press, A2.2.

Song, W., Strack, R. L., Svensen, N., & Jaffrey, S. R. (2014). Plug-and-play fluorophores extend the spectral properties of Spinach. *Journal of the American Chemical Society, 136*, 1198–1201.

Strack, R. L., Disney, M. D., & Jaffrey, S. R. (2013). A superfolding Spinach2 reveals the dynamic nature of trinucleotide repeat-containing RNA. *Nature Methods, 10*, 1219–1224.

Strack, R. L., Song, W., & Jaffrey, S. R. (2014). Using Spinach-based sensors for fluorescence imaging of intracellular metabolites and proteins in living bacteria. *Nature Protocols, 9*, 146–155.

Temme, K., Hill, R., Segall-Shapiro, T. H., Moser, F., & Voigt, C. A. (2012). Modular control of multiple pathways using engineered orthogonal T7 polymerases. *Nucleic Acids Research, 40*, 8773–8781.

CHAPTER TEN

Monitoring mRNA and Protein Levels in Bulk and in Model Vesicle-Based Artificial Cells

Pauline van Nies, Alicia Soler Canton, Zohreh Nourian, Christophe Danelon[1]

Department of Bionanoscience, Kavli Institute of Nanoscience, Delft University of Technology, Delft, The Netherlands

[1]Corresponding author: e-mail address: c.j.a.danelon@tudelft.nl

Contents

Methods in Enzymology, Volume 550
ISSN 0076-6879
http://dx.doi.org/10.1016/bs.mie.2014.10.048

Abstract

With rising interest in utilizing cell-free gene expression systems in bottom-up synthetic biology projects, novel labeling tools need to be developed to accurately report the dynamics and performance of the biosynthesis machinery operating in various reaction conditions. Monitoring the transcription activity has been simplified by the Spinach technology, an RNA aptamer that emits fluorescence upon binding to a small organic dye. Recently, we tracked the fluorescence of Spinach-tagged messenger RNA (mRNA) and its translation product the yellow fluorescent protein (YFP), both synthesized in the protein synthesis using recombinant elements system from a DNA template. Building on our previous study, we describe here an improved Spinach reporter with modified flanking sequences that confer higher propensity for aptamer folding and, thus, enhanced fluorescence brightness. Hence, the kinetics of mRNA and YFP production could be simultaneously monitored with unprecedented sensitivity. A combination of methodologies, comprising RNA gel analysis, real-time quantitative polymerase chain reaction, absorbance measurements, and fluorescence correlation spectroscopy, was used to convert fluorescence intensity units into absolute concentrations of transcript and YFP translational product. Furthermore, we demonstrated that the new Spinach construct greatly enhanced mRNA detection when gene expression was confined inside self-assembled lipid vesicles. Therefore, we argue that this assay could be used to evaluate systematically the performance of transcription and translation in model vesicle-based artificial cells.

1. INTRODUCTION

Gene expression is a fundamental and ubiquitous cellular process that involves a succession of events. The DNA sequence information is transferred into a single-stranded messenger RNA (mRNA) that serves as a template for protein production. The quantitative relation between the levels of mRNA and protein is an essential feature that governs a variety of cellular behaviors, including cell-fate decisions (Kaufmann & van Oudenaarden, 2007; Maamar, Raj, & Dubnau, 2007; Ozbudak, Thattai, Kurtser, Grossman, & van Oudenaarden, 2002). Its knowledge is also crucial to understand the dynamics of natural and synthetic gene regulatory networks *in vivo* (Golding, Paulsson, Zawilski, & Cox, 2005; Kaern, Elston, Blake, & Collins, 2005; Pedraza & Paulsson, 2008).

In cell-free transcription–translation systems, inferring the dynamics of synthesized mRNA and protein is crucial to optimize gene circuit performance for further implementation into living cells (Noireaux, Bar-Ziv, & Libchaber, 2003; Shin & Noireaux, 2012; Siegal-Gaskins, Tuza, Kim, Noireaux, & Murray, 2014) or for the construction of genetically controlled artificial cell models based on lipid vesicles (Noireaux & Libchaber, 2004; Nomura et al., 2003; Nourian, Roelofsen, & Danelon, 2012). In these two examples, the temporal dynamics at which mRNA and protein are produced is as important as their final amount. To control the information flow underlying gene expression, it is therefore essential to track in real time the level of both messenger and protein (Karzbrun, Shin, Bar-Ziv, & Noireaux, 2011; Niederholtmeyer, Xu, & Maerkl, 2013; Rosenblum et al., 2012; Stögbauer, Windhager, Zimmer, & Rädler, 2012; van Nies et al., 2013).

The detection of synthesized protein output is facilitated by the use of fluorescent reporter proteins. However, simultaneous detection of the transcriptional and translational activities requires orthogonal fluorescence labeling tools of the RNA and protein. A widely used strategy to visualize RNA is to genetically fuse the target RNA to sequences that recruit green fluorescent protein (GFP)-tagged proteins (Bertrand et al., 1998; Tyagi, 2009) or hybridization probes (Sei-Iida, Koshimoto, Kondo, & Tsuji, 2000; Tyagi & Kramer, 1996). An alternative approach consists of using RNA aptamers that bind and activate the fluorescence of small organic dyes (Babendure, Adams, & Tsien, 2003; Paige, Wu, & Jaffrey, 2011). Inserting the aptamer sequence in either of the two untranslated regions (usually in the 3′-UTR) of the mRNA encoding for a fluorescent protein of different color enables to interrogate the kinetics of transcript and protein synthesis separately and simultaneously. For the first time, we designed and implemented such an aptamer-based dual gene expression assay in bulk reactions and inside self-assembled lipid vesicles (van Nies et al., 2013; Fig. 1). This strategy was recently exploited in *in vivo* (Pothoulakis, Ceroni, Reeve, & Ellis, 2014) and in cell-free measurements (Chizzolini, Forlin, Cecchi, & Mansy, 2014; Siegal-Gaskins et al., 2014).

In our previous study, we expressed Spinach (Paige et al., 2011), an RNA aptamer that forms a fluorescent complex upon binding with (*Z*)-4-(3,5-difluoro-4-hydroxybenzylidene)-1,2-dimethyl-1*H*-imidazol-5(4*H*)-one (DFHBI), to monitor transcription activity in a cell-free expression system (van Nies et al., 2013). Herein, we report on a variant of Spinach with modified flanking sequences that confer improved stability resulting in enhanced brightness in bulk and liposome-confined reactions. We characterize the Spinach fluorescence and validate its use in combination with a fluorescent protein reporter for quantitative monitoring of transcription

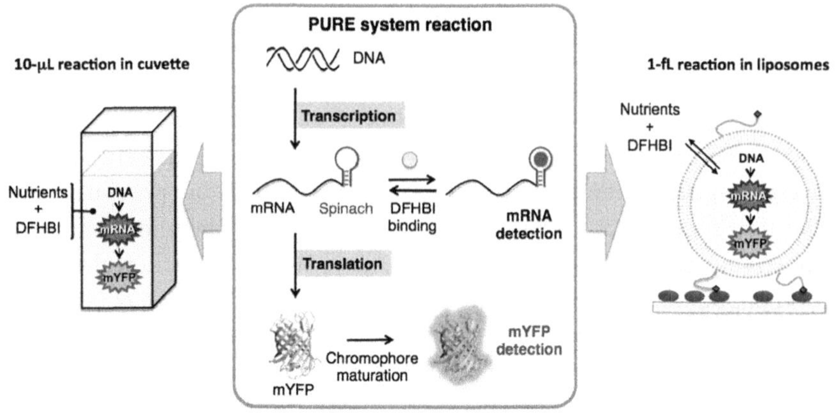

Figure 1 Schematic overview of our two-reporter assay for mRNA and protein levels in bulk (microtubes or cuvettes) and in liposome-confined gene expression reactions. The Spinach technology consisting of an RNA aptamer sequence that binds and turns the chromophore DFHBI into a fluorescent state was used to monitor transcription activity. (See the color plate.)

and translation activities. The experimental procedures described here are generally applicable to quantifying gene expression in different cell-free conditions.

2. THE "SPINACH TECHNOLOGY" FOR COMBINED DETECTION OF mRNA AND PROTEIN IN CELL-FREE EXPRESSION SYSTEMS

2.1 Lighting up RNA with Spinach

A series of RNA aptamers capable of binding derivatives of the GFP chromophore has recently been engineered for fluorescence imaging of cellular RNAs in live cells (Paige et al., 2011). The various aptamer–dye pairs fluoresce over a range of wavelengths that covers the entire visible spectrum. One of these aptamers, named Spinach owing to its bluish-green fluorescence, recognizes and activates DFHBI, a nontoxic and membrane permeable fluorophore, to form a particularly bright complex. A cellular RNA of interest, whose sequence is fused to the Spinach aptamer, endogenously expressed in mammalian cells or in bacteria, can thus be visualized when exposed to DFHBI. Therefore, the dynamics of Spinach-tagged RNA localization can be imaged as a response to various external stimuli or cellular contexts. More recently, a second generation of Spinach with improved brightness (Strack, Disney, & Jaffrey, 2013) and of GFP-like fluorophores with tailored spectral properties (Song, Strack, Svensen, & Jaffrey, 2014) have been developed, further expanding the toolbox of the Spinach technology. Additionally, Spinach can be fused to aptamers that bind to specific metabolites, which in turn modulates the Spinach fluorescence (Paige, Nguyen-Duc, Song, & Jaffrey, 2012; Strack, Song, & Jaffrey, 2014).

Detection of synthesized transcripts in cell-free expression systems is also simplified with the use of the Spinach tag. Indeed, DFHBI exhibits no nonspecific fluorescence activation by the transcription–translation system and it has no inhibitory effects to the reaction efficiency (van Nies et al., 2013). Moreover, due to its small size and favorable structural properties, DFHBI readily permeates across liposome membrane, which is another benefit for monitoring transcriptional activity in vesicle-based artificial cells (van Nies et al., 2013). In contrast, fluorescence resonance energy transfer-based oligonucleotide probes exhibited detrimental effects on the yield of translated proteins in our hands (van Nies et al., 2013).

Nevertheless, we noticed that the Spinach fluorescence signal consistently reports the kinetics of RNA production only during the first 90 min and then it deviates to give underestimated intensity values, a result we attributed to suboptimal folding stability of the aptamer at 37 °C van Nies et al., 2013). Here, we sought to design a new Spinach reporter with improved folding in cell-free gene expression systems in order to reliably monitor the complete transcription dynamics in real time.

2.2 Synthesis of DFHBI

DFHBI was synthesized as previously described (Paige et al., 2011) following a modified literature protocol (van Nies et al., 2013; Fig. 2). Whereas conversion of 3,5-difluoro-4-hydroxybenzaldehyde into (*Z*)-2,6- difluoro-((2-methyl-5-oxooxazol-4(5*H*)-ylidene)methyl)phenylacetate (**1**) was essentially performed as originally described, we found that the use of 4-dimethylaminopyridine as a catalyst was crucial for achieving high yield

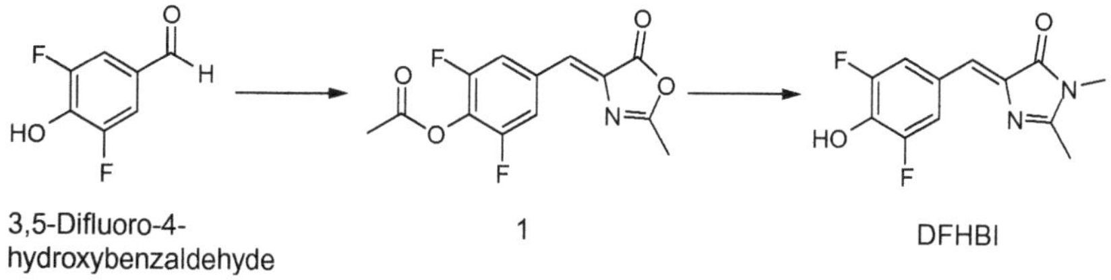

Figure 2 Two-step synthesis of the DFHBI fluorophore.

and selection conversion of (**1**) into DFHBI (van Nies et al., 2013). Synthesis details including reaction yields and spectroscopic data can be found in van Nies et al. (2013).

Throughout this study, the DFHBI synthesized in house was used. The commercially available DFHBI (Lucerna) was employed in a few experiments for comparison and no appreciable differences were noted (not shown).

2.3 Design and preparation of the DNA templates

A dual gene expression assay comprising two orthogonal fluorescent reporters for transcription and translation activities was developed to track the progression of mRNA and protein synthesis in real time and simultaneously. The Spinach aptamer sequence was included in the 3′-UTR of the yellow fluorescent protein (YFP) gene. Concurrent measurement of mRNA and protein levels is possible owing to the spectral compatibility between the Spinach and YFP fluorescence (Fig. 4A).

In our previous construct, denoted YFP-Spinach, the Spinach was flanked with 6-nucleotide sequences, i.e., six nucleotides after the translation stop codon and six nucleotides before the transcription termination sequence (Fig. 3). In the new construct, named mYFP-LL-Spinach, the 6-nucleotide linker upstream of Spinach was extended to 36 nucleotides with the intention to minimize possible interference between aptamer folding and ribosome activity. Three cloning sites were introduced into this linker, of which the *Eco*RV and the *Sma*XI are also present in the 18-nucleotide linker downstream of Spinach (Fig. 3). The palindromic nature of the *Eco*RV restriction site enables a partial hybridization of the new flanking regions as shown using

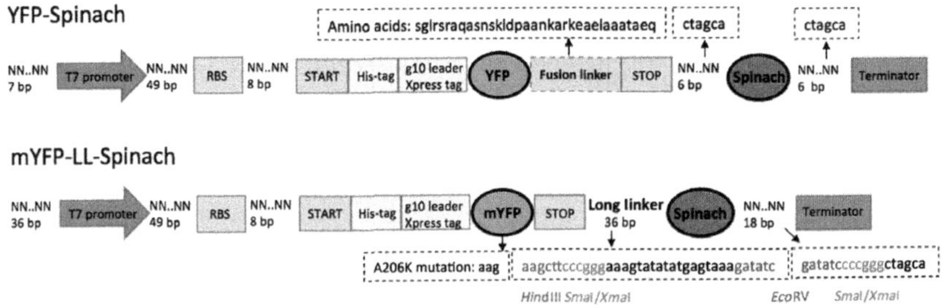

Figure 3 Schematic representation of the two DNA constructs primarily used in this study. Their main features and regulatory elements are depicted. The new construct mYFP-LL-Spinach has been designed based on our previously described YFP-Spinach gene (van Nies et al., 2013). The abbreviation RBS stands for ribosome binding site. (See the color plate.)

the Mfold Web Server (Zuker, 2003), thus extending and stabilizing the stem-I structure of the Spinach aptamer. The mYFP sequence was derived from the pRSETB-YFP plasmid with two modifications: the A206K mutation was introduced to generate a monomeric variant of YFP and the fusion linker was deleted (Fig. 3). Both YFP-Spinach and mYFP-LL-Spinach genes were cloned into the pmK_RQ vector using *Sfi*I sites by Life Technologies. The linear DNA templates used for gene expression were prepared by polymerase chain reaction (PCR) (Phusion polymerase, Finnzymes) with primers 5′-GCGAAATTAATACGACTCACTATAGGGAGACC-3′ (forward primer YFP-Spinach) or 5′-GAATTGAAGGAAGGCCGTCAAG-3′ (forward primer mYFP-LL-Spinach), and 5′-AAAAAACCCCTCAAGACCCGTTT AGAGG-3′ (reverse primer for both constructs).

The construct *mYFP* lacking the LL-Spinach domain was produced by removing the Spinach sequence with *Sma*I. The corresponding linear DNA template was generated by PCR using the same forward and reverse primers used for the mYFP-LL-Spinach.

2.4 *In vitro* transcription–translation with PURE*frex*

The PURE (protein synthesis using recombinant elements) system, a minimal gene expression system reconstituted from purified *E. coli* proteins (with the exception of the T7 bacteriophage RNA polymerase), was employed (Shimizu et al., 2001; Shimizu, Kanamori, & Ueda, 2005). Upon addition of a DNA template containing the relevant regulatory sequences (Fig. 3), the main reactions occurring in the PURE system are transcription, translation, tRNA aminoacylation, nucleotide triphosphate regeneration, and pyrophosphate hydrolysis, the fuel molecule being creatine phosphate. Since our ultimate goal is to construct an elementary—irreducible—cell, we consider the PURE system a preferred biosynthesis platform compared to cellular extracts.

Unless otherwise indicated, gene expression was conducted in the PURE*frex* (GeneFrontier, Japan). The constituting enzymes in the PURE*frex* are similar to those in the original expression system (Shimizu et al., 2001), with the exception that they are devoid of a histidine tag (Shimizu, Kuruma, Kanamori, & Ueda, 2014). The PURE*frex* is composed of three different solutions: the enzyme mixture (T7 RNA polymerase, translation factors, energy recycling system, etc.), the *E. coli* ribosomes, and the feeding mixture (amino acids, NTPs, tRNAs, creatine phosphate). All solutions were aliquoted in small volumes and stored at −80 °C.

For bulk experiments (in test tubes or cuvettes), the PURE*frex* reaction solution was prepared by mixing on ice 1 μL of enzyme mix, 1 μL of ribosome, 10 μL of feeding mix, 0.5 μL of Superase inhibitor (10 units final; SUPERase.In, Ambion), the DNA template (final concentration typically between 0.74 and 11.7 n*M*), 1 μL of DFHBI (20 μ*M* final concentration), and the volume was adjusted to 20 μL with nuclease-free water.

When indicated, the PURExpress system (New England Biolabs) was also used. The kit comes in two vials: the solution B containing the protein mix and the solution A containing the tRNAs and low-molecular weight compounds. Bulk reactions were carried out by mixing 7.5 μL of solution B with 10 μL of solution A along with the DNA template and 0.5 μL of Superase inhibitor, adjusted to a final volume of 25 μL with nuclease-free water.

2.5 Kinetics measurements by spectrofluorometry

Real-time monitoring of mRNA and protein synthesis was carried out on a fluorescence spectrophotometer (Cary Eclipse, Varian). The PURE*frex* reaction solution was transferred to a 15-μL cuvette (Hellma) positioned in a Peltier thermostated four-cell holder maintained at 37 °C. Spinach and YFP fluorescence signals were detected every 30 s in the high-voltage mode at (excitation/emission wavelengths) 460/502 and 515/528 nm, respectively, to minimize fluorescence overlap.

For each conditions tested, at least three independent (no pooling of the PURE system reaction solution) fluorescence curves have been generated and representative kinetics are displayed (Fig. 4).

The cuvettes need to be thoroughly cleaned immediately after each experiment to maximize data reproducibility. Washing was done by successively filling the cuvette with Hellmanex 2%, KOH 1 *M*, nuclease-free water, EtOH 100% for 1 min in a bath sonicator, applying three washes with nuclease-free water in between the different reagent treatments.

2.6 Characterization of improved Spinach fluorescence in the PURE system

Before performing kinetics measurements, the fluorescence spectra of Spinach should be recorded to setup the excitation and emission wavelengths optimally under the given experimental conditions. The fluorescence excitation and emission spectra of Spinach were similar for the YFP-Spinach and mYFP-LL-Spinach transcripts (Fig. 4A; van Nies et al., 2013). Next, one

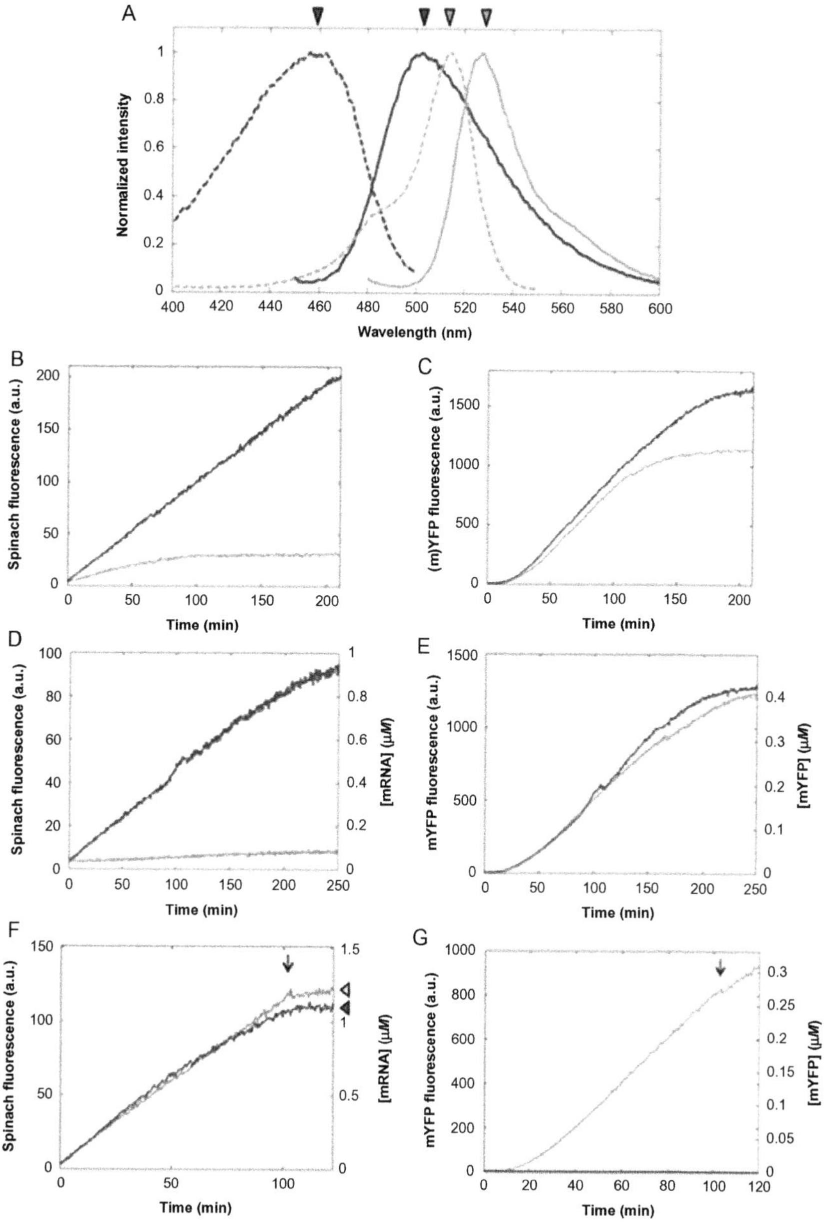

Figure 4 (A) Fluorescence excitation (dashed lines) and emission (solid lines) spectra of LL-Spinach (blue) and mYFP (green) measured in the PURE*frex* expressing the mYFP-LL-Spinach gene. The LL-Spinach spectra were measured in the PURE*frex* ΔR, that is devoid of ribosome, in the presence of 20 μ*M* DFHBI. The mYFP spectra were collected in the PURE*frex* without DFHBI. The arrowheads depict the excitation and emission

(Continued)

can examine the time courses of Spinach fluorescence arising from transcription of the YFP-Spinach and mYFP-LL-Spinach constructs (Fig. 4B). Strikingly, transcription of the mYFP-LL-Spinach DNA leads to eightfold higher Spinach fluorescence intensity (at 7.4 n*M* DNA after 3 h) compared to that with the YFP-Spinach construct. Moreover, the signal from the mYFP-LL-Spinach template increases for 7 h (not shown). In contrast, the fluorescence arising from transcription of the YFP-Spinach gene levels off after about 90 min (Fig. 4B), which is premature given the linear increase of mRNA amount observed on gel (van Nies et al., 2013). Compared to tRNA scaffold-stabilized Spinach (Chizzolini et al., 2014), our LL-Spinach is shorter, reducing consumption of NTPs.

Factors that could affect the Spinach fluorescence intensity and dynamics in transcription–translation reactions include aptamer misfolding due to thermal instability at 37 °C in the PURE system and interference of the translation process with folding. We consider a change in the photochemical properties of Spinach unlikely (molar extinction coefficient or quantum yield) (Strack et al., 2013). Note that the excitation and emission spectra of Spinach and LL-Spinach coincide (Fig. 4A). The possible influence of translating ribosomes on Spinach fluorescence can be examined by monitoring transcription activity in the presence (+R) or in the absence (ΔR) of ribosomes. The apparent kinetics of mRNA synthesis is the same whether the transcript is translated or not (Fig. 4F), indicating that translation does not affect the folding propensity of LL-Spinach. Furthermore, a change

Figure 4—Cont'd wavelengths used for kinetics measurements. (B) Fluorescence intensity profiles of Spinach produced from the mYFP-LL-Spinach (dark blue) or YFP-Spinach (light blue) construct. (C) Apparent kinetics of mYFP (dark green) and YFP (light green) synthesis monitored simultaneously as in (B). (B and C) DNA concentration for both genes was 7.4 n*M*. (D) Plots of LL-Spinach fluorescence versus time using the mYFP-LL-Spinach (dark blue) or mYFP (light blue) construct. (E) Apparent kinetics of mYFP produced from the mYFP-LL-Spinach (dark green) or mYFP (light green) construct and monitored simultaneously as in (D). (D and E) DNA concentration for both genes was 0.74 n*M*. (F) Progression of Spinach fluorescence versus time in a PURE*frex* ΔR (dark blue) or PURE*frex* (light blue) reaction starting from 11.7 n*M* of the mYFP-LL-Spinach DNA. The arrowheads on the right axis point to the final intensity values used for calculating the conversion factor between fluorescence a.u. and mRNA concentration. (G) Apparent kinetics of mYFP synthesis in a PURE*frex* ΔR (dark green) or PURE*frex* (light green) reaction monitored simultaneously as in (F). (F and G) The arrow at around 100 min indicates the addition of DNaseI to stop transcription. (D and F) Concentrations of mRNA were calculated using a conversion factor of 10 n*M*/a.u. (E and G) Concentrations of mYFP were calculated using a conversion factor of 0.33 n*M*/a.u. (See the color plate.)

of DNA concentration from 0.74 to 7.4 n*M* was accompanied by a 2.5-fold increase of the transcription rate (compare Fig. 4B with D).

We suspected that changing the nucleotide sequences upstream and downstream the Spinach could influence the folding time and thus the time delay between messenger production and Spinach fluorescence detection. This time delay can be decomposed into three consecutive events: the aptamer folding, the binding of DFHBI, and the fluorescence emission, the latter occurring at a shorter time scale than the earlier two. To investigate the lag time, DNase can be added to a running PURE*frex* reaction to immediately stop further mRNA production and the residual increase of the LL-Spinach fluorescence is monitored. It is important to inject the DNase while the Spinach intensity is linearly increasing and has reached sufficiently high signal-to-noise ratio for accurate measurement. The characteristic time of the increase in the remaining signal corresponds to the time delay defined above. We found that upon DNase injection the signal levels off instantaneously given our 30 s temporal resolution, resulting to a time delay estimate <1 min. This value is significantly lower than the 2.6 min delay found using the YFP-Spinach construct (van Nies et al., 2013). This result suggests that Spinach folding is the rate-limiting step for fluorescence detection, not the aptamer–DFHBI binding, which was anticipated since DFHBI is present in large excess.

2.7 Orthogonal detection of synthesized mRNA-Spinach and protein

As shown in Fig. 4A, the spectrally different Spinach and mYFP protein enable simultaneous detection when the appropriate excitation and emission wavelengths are selected. As a control, it is usually valuable to quantify the fluorescence crosstalk in Spinach and mYFP detection. This can be performed using the *mYFP* gene, where the LL-Spinach sequence is omitted. With the settings used, only 0.3% of the mYFP signal contributes to Spinach fluorescence (Fig. 4D and E). This residual component in Spinach fluorescence is negligible compared to the signal typically measured upon transcription of even subnanomolar concentration of mYFP-LL-Spinach DNA (Fig. 4D).

Next, it is necessary to study the influence of the insertion of the Spinach tag downstream the YFP coding sequence of the mRNA on the translation activity. Two interfering mechanisms could be envisaged: first, the larger depletion of ATP and GTP upon transcription of the longer mYFP-LL-Spinach gene compared to that with the mYFP template may lead to a shortening of the protein synthesis duration since ATP and GTP are also consumed during tRNA aminoacylation and translation, respectively. Second, the global folding of mRNA can differ in the presence of LL-Spinach, which may alter the rate of translation despite the helicase activity of the ribosome. The results indicate that the apparent kinetics of protein synthesis and the final concentrations are nearly identical for the mYFP-LL-Spinach and mYFP genes (Fig. 4E), consistent with a neutral effect of the Spinach reporter on the yield and cessation time of protein production. Furthermore, one should verify that the DFHBI chromophore does not interfere with the transcription and translation activities. This could be demonstrated by quantifying the amounts of produced mRNA and protein in the presence or absence of DFHBI (van Nies et al., 2013). Together, we found that Spinach labeling of mRNA is orthogonal to the essential reactions occurring in the PURE system and can thus be used in conjunction to fluorescent protein reporters for unbiased detection of the gene expression dynamics.

3. QUANTIFYING THE LEVELS OF mRNA AND PROTEIN SYNTHESIZED IN PURE SYSTEM BULK REACTIONS

3.1 Overall workflow

Quantitation of gene expression requires to convert arbitrary fluorescence intensity values into absolute mRNA and protein concentrations. The following complementary calibration experiments should therefore be performed and their results compared. The overall workflow is illustrated in Fig. 5. Gene expression reactions were performed in cuvettes and the fluorescence intensity of Spinach and YFP was monitored in real time (Fig. 4F and G). The mYFP-LL-Spinach construct was used as DNA template.

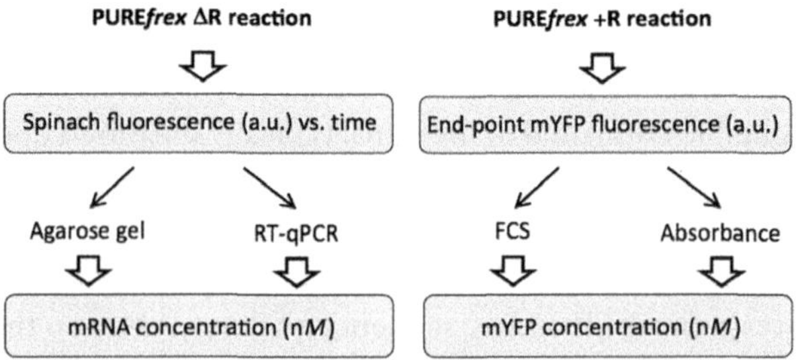

Figure 5 Flowchart of the protocol to quantify the amount of synthesized mRNA and protein, and to extract the corresponding conversion factors between a.u. fluorescence intensity measured on the spectrofluorometer and concentration.

Transcription only or coupled transcription–translation experiments were conducted in the absence (ΔR) or in the presence (+R) of ribosomes, respectively (Fig. 4F). After about 100 min, mRNA production should be stopped by adding DNase while the fluorescence signal is continuously monitored. The end-point fluorescence intensity is determined as the mean value of the Spinach signal during the last 10 min. Then, the reaction solution is harvested for further quantification using RNA gel and real-time quantitative PCR (RT-qPCR) analysis.

3.2 Preparation of reference RNA and purification

The reference RNA used for further quantification was produced from 14 n*M* of the mYFP-LL-Spinach DNA in the RiboMAX™ Large Scale RNA production kit (Promega) according to the recommended protocol. The RNA was then purified with the RNeasy MinElute Cleanup kit (Qiagen) following the manufacturer's protocol. The RNA concentration was determined using a Nanodrop (Thermo Scientific) with absorbance measurements performed at 280 nm and its purity was analyzed using the 260/280 nm absorbance ratio. The RNA content of PURE*frex* reactions treated with DNaseI was purified similarly.

3.3 Gel analysis of mRNA concentration

Two to six microliters of purified RNA samples were loaded on a 1.2% agarose gel containing EtBr and a voltage of 90 V was applied for 1.5 h. The gel was then imaged and the band intensities were analyzed using the ImageLab software (Fig. 6A). A calibration curve should be generated by plotting the measured band intensity values of purified RNA (reference RNA) against their predetermined amounts. The standard curve is then used to calculate the concentration of mRNA produced in the PURE system (Fig. 6B).

Comparing the mRNA band intensity in +R and ΔR expression conditions, we noticed that a lower amount is detected in the presence of ribosomal RNA (rRNA) (Fig. 6A and B), suggesting an rRNA-dependent loss of messenger during purification (though we paid attention that the purification column was not overloaded with total RNA) or less efficient migration through the gel. Therefore, PURE*frex* ΔR-produced mRNA samples only were used for quantification. A conversion factor corresponding to 10.0 ± 2.1 n*M* mRNA per fluorescence a.u. was obtained by plotting the mRNA band intensity onto the calibration curve and normalizing to the end-point fluorescence intensity measured with the spectrofluorometer.

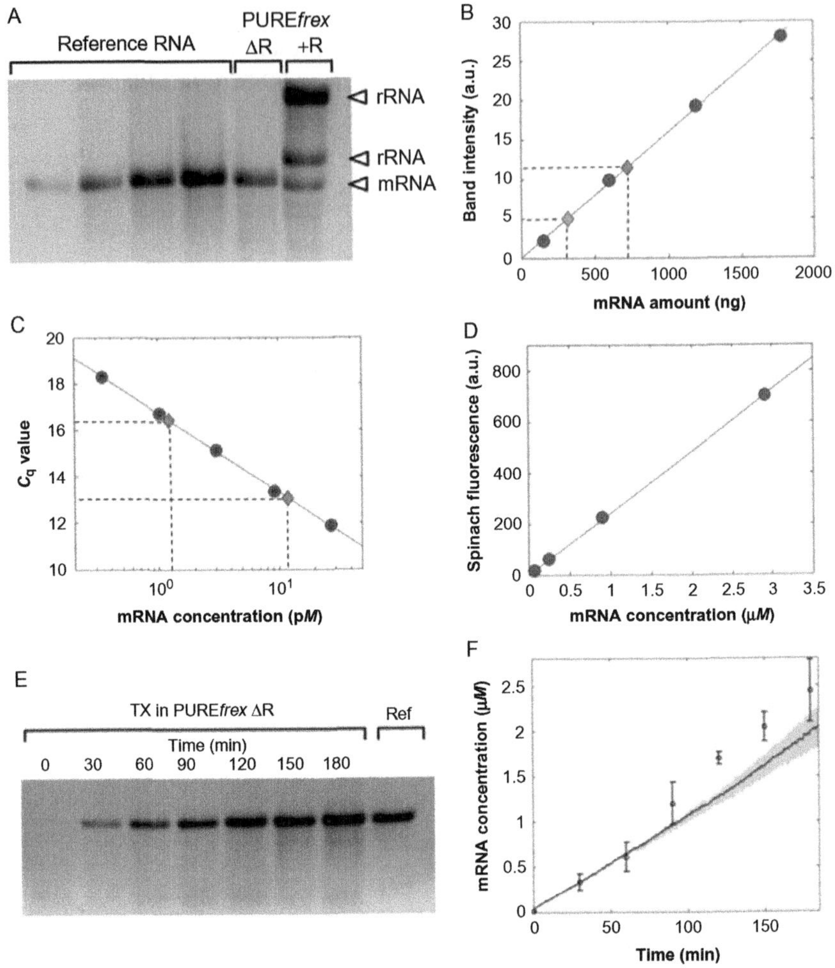

Figure 6 Quantification of mRNA synthesis in PURE system reactions starting from 7.4 n*M* of the mYFP-LL-Spinach DNA template. (A) RNA samples loaded on an agarose gel. The band intensities of mRNA produced in a PURE*frex* reaction (Fig. 4F) with (+R) or without ribosome (ΔR) were compared to reference RNA samples of known concentrations. The 1.5 and 2.9-kb ribosomal RNA bands are visible in the +R reaction condition. (B) Calibration curve plotted as the measured band intensities of reference RNA samples versus their predetermined concentrations. The mRNA band intensities of the PURE*frex* samples shown in (A) were appended on the calibration curve, after which the amount of synthesized transcript can be determined. (C) Reference RNA samples were analyzed by RT-qPCR and their C_q values plotted as a function of concentration. The obtained standard curve has a typical equation of $y = -1.479\ \ln(x) + 16.723$; $R^2 = 0.998$. The measured C_q values of diluted samples from PURE*frex* ΔR reactions were appended on the calibration curve and their concentrations were determined. Two samples of 10-fold different dilution factors are displayed. (D) Calibration curve consisting of Spinach fluorescence intensities measured for different concentrations of reference

3.4 Real-time quantitative PCR analysis

Absolute quantification of RNA samples can also be performed by RT-qPCR. We used the Eco Real-Time PCR System from Illumina. Three microliters of PURE*frex* ΔR-produced mRNA were harvested from a PURE system reaction treated with DNaseI, diluted 100-fold in RNase-free water and stored at −80 °C until used. No purification is needed. The samples were further diluted and 1 μL was added in 10 μL of RT-qPCR reactions corresponding to a final dilution factor of 10^5 or 10^6. The Power SYBR Green RNA-to-CT 1-Step kit (Applied Biosystems) was used according to the supplier's recommended protocol. Primers were designed to amplify a 267-bp long region of the mYFP gene. The forward (5′-CACCTACGGCAAGCTGACC-3′) and reverse (5′-TTCAGCTCGATGCGGTTC-3′) primers (Biolego) were used at 100 n*M* each. Prior mixing with the Power SYBR Green RNA-to-CT 1-Step kit, the reverse primer was incubated with the target RNA for 5 min at 65 °C and left for 10 min at room temperature. Reaction samples of 10 μL were loaded on a microplate (Eco Sample Dock, Illumina) and spun down for about 10 s at 3000–4000 rpm (Eppendorf 5810R centrifuge). Each sample was analyzed in triplicate.

Concentrations of mRNA were determined using a standard curve generated by serial dilution in autoclaved milliQ water of reference RNA at final concentrations ranging from 0.33 to 27.5 pM (five points, each in triplicate) (Fig. 6C). A total number of seven independent samples (gene expression reactions performed separately) were analyzed leading to a conversion factor of 9.0 ± 5.2 n*M* (mean ± standard deviation) of mRNA per fluorescence a.u., in close agreement with the value extracted from gel analysis.

3.5 mRNA quantification from Spinach fluorescence of reference RNA

Reference RNA samples were diluted at different factors in the PURE system buffer (50 m*M* HEPES, 100 m*M* potassium glutamate, 13 m*M* magnesium acetate, pH 7.6), heated for 5 min at 65 °C, and the solutions were

RNA solutions. The slope gives a conversion factor of 4.1 n*M*/a.u. (E) Time series gel analysis of mRNA produced in a PURE*frex* ΔR reaction. The band intensities were compared with that of a reference RNA (right-most lane). (F) The dynamics of transcription was reconstructed by plotting the concentrations of mRNA as determined in (E) at different time points. Error bars indicate SEM, $n = 3$. For comparison, the apparent kinetics obtained by monitoring the Spinach fluorescence in real time is overlaid. The blue curve is the mean of three independent measurements and the gray-shaded area denotes the min and max deviation. (See the color plate.)

allowed to cool down to room temperature in the presence of DFHBI. The Spinach fluorescence intensity of each sample was measured at 37 °C with the spectrofluorometer and its value plotted as a function of the corresponding RNA concentration (Fig. 6D). The obtained calibration curve served to determine the concentration of mRNA synthesized in the PURE*frex*, for which the end-point Spinach fluorescence signal was measured (Fig. 4F). A value of 4.1 n*M* mRNA per fluorescence a.u. was found, which is markedly lower than the conversion coefficient obtained with the gel and RT-qPCR methods. Though the exact reason of this discrepancy remains to be clarified, it is likely that the folding propensity of Spinach, and thus the fluorescence intensity, differs whether the mRNA is gradually produced *in situ*, i.e., in the PURE*frex* reaction, or is presynthesized and subsequently exposed to DFHBI. Therefore, we recommend to opt for the RNA gel and RT-qPCR as more reliable methods since the Spinach signal is also measured in the PURE*frex*.

3.6 Calculating mYFP concentration by fluorescence correlation spectroscopy and absorbance measurements

3.6.1 Fluorescence correlation spectroscopy setup

Fluorescence correlation spectroscopy (FCS) can be applied to determine the concentration of mYFP produced in the PURE system. The amount of synthesized mYFP is then measured *in situ* without protein purification. FCS measurements were performed using the MicroTime200 laser scanning confocal microscope (PicoQuant GmbH, Germany) equipped with a ×60 water-immersion objective (numerical aperture 1.2, Olympus UPlanSAPO). The 485-nm laser line (beam width 4.2 mm), a 50-μm pinhole, and a band-pass emission filter 520 ± 35 nm (Semrock) were used.

The calibration fluorophore or the *in vitro*-synthesized fluorescent protein was diluted in the PURE buffer devoid of magnesium (50 m*M* HEPES, 100 m*M* potassium glutamate, pH 7.6), supplemented with 0.05% Tween-20 to minimize depletion effects. A 50 μL droplet was deposited onto a clean glass coverslip (#1, thickness 0.13–0.16 mm) and the laser was focused 20 μm into the solution for measurements. All measurements were performed at room temperature (20 °C).

3.6.2 Calibration of the detection volume

The first step consists of calibrating the detection volume using a solution of Alexa-488 (Invitrogen) of 10 n*M*. The measured autocorrelation function $G(\tau)$ is fitted to the equation for free 3D diffusion:

$$G(\tau)=\frac{1}{N}\left(1+\frac{p}{1-p}\exp\left(-\frac{\tau}{\tau_t}\right)\right)\left(1+\frac{4D\tau}{w_{xy}^2}\right)^{-1}\left(1+\frac{4D\tau}{k^2w_{xy}^2}\right)^{-1/2}, \quad (1)$$

where τ is the autocorrelation delay, N is the average number of fluorescent molecules in the detection volume, D is the fluorophore diffusion coefficient, p is the fraction of molecules that occupy the triplet state, w_{xy} is the detection volume waist, k is the structure parameter, and τ_t is the triplet lifetime. The effective volume, V_{eff}, was determined at $G(\tau=0)$ as:

$$V_{\text{eff}}=\frac{N}{N_A C_{\text{Alexa}-488}}, \quad (2)$$

with N_A is the Avogadro's number and $C_{\text{Alexa-488}}$ is the concentration of Alexa-488. A volume of 1.1 fL is typically obtained. After pulsing effects were filtered out prior to calculating $G(0)$ through fluorescence lifetime correlation spectroscopy (FLCS) analysis. The FLCS algorithm was systematically applied on all collected FCS data. Fitting the autocorrelation curve of Alexa-488 to Eq. (1) gives typical values for $w_{xy}=300$ nm, $k=7.4$, $\tau_t=6$ μs, and $D=500$ μm^2/s, consistent with previously reported values (Petrásek & Schwille, 2008).

3.6.3 Measuring the concentration of synthesized mYFP

The mYFP protein was produced as described above and diluted in the PURE buffer supplemented with 0.05% Tween-20. Two different dilutions were prepared to confirm that the obtained concentrations fall into the linear sensitivity regime of the detection method. The autocorrelation curves were fitted to Eq. (1) with the parameters w_{xy} and k fixed to their values obtained by calibration performed on the same day (Fig. 7A). Typically, extracted values of the diffusion coefficient, triplet time, and triplet fraction of the mYFP are 95 ± 13 μm^2/s, 35 ± 2 μs, and 0.29 ± 0.01, respectively. The concentration of mYFP was determined as:

$$C_{\text{YFP}}=\frac{N}{N_A\pi^{3/2}w_{xy}^3k}. \quad (3)$$

The actual concentration of synthesized mYFP can then be obtained after correcting for the dilution factor.

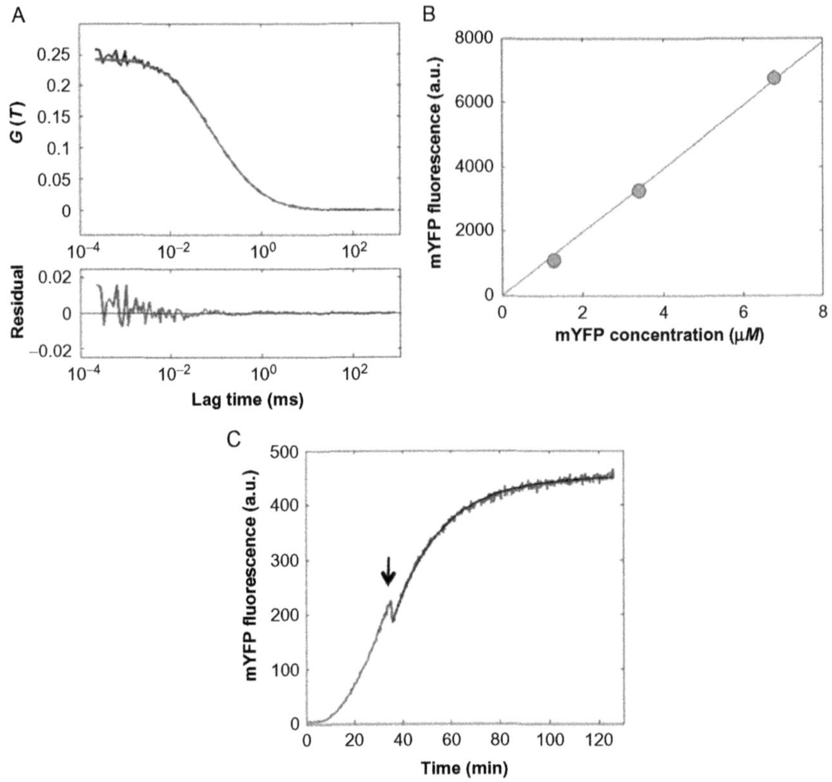

Figure 7 Quantification of YFP concentration and maturation time. (A) Fluorescence autocorrelation curve of a 50-fold diluted solution of mYFP synthesized in the PURE*frex* and analyzed by FCS. The fit to Eq. (1) (red (light gray in the print version) line) and the fit residual (bottom graph) are shown. (B) Quantification of PURExpress-synthesized mYFP concentration by absorbance measurements. The mYFP fluorescence intensity measured on the spectrofluorometer was plotted against the concentration determined by absorbance. The slope gives a conversion factor of 1.01 n*M*/a.u. (C) The maturation time of mYFP was experimentally measured by adding 3 μL of the translation inhibitor chloramphenicol (arrow). The residual increase in fluorescence was fitted with a monoexponential function (black curve), giving a maturation time of 20 min.

3.6.4 Converting mYFP fluorescence intensity into concentration

We found that 1 a.u. (fluorescence) measured on the Cary Eclipse spectrofluorometer corresponds to 0.33 n*M*.

Though we have not tried to apply FCS to quantify Spinach concentration, the technique is potentially of interest. However, the additional binding dynamics of DFHBI within the detection volume and the yet unknown triplet lifetime of Spinach will complicate the interpretation of the autocorrelation curve.

3.6.5 Absorbance measurements

Alternatively, the absorbance of mYFP generated in the PURE system can be measured (NanoDrop 2000, Thermo Scientific) and converted into a concentration using the Beer–Lambert equation. An excitation wavelength of 515 nm and a molar extinction coefficient of 83,400 *M*/cm (Shaner, Steinbach, & Tsien, 2005) were employed. The PURE system devoid of DNA (background) was used as the blank. Because the yield of protein production is too low using the PURE*frex*, the sample absorbance cannot reliably be measured. Thus, the most efficient PURExpress system was used to synthesize mYFP, leading to 8- to 10-fold larger amount of synthesized protein (according to fluorescence intensity) as compared to PURE*frex* expression. Then, the fluorescence intensity of the solution was measured on the Cary Eclipse spectrofluorometer and its absorbance was determined as described above (Fig. 7B). We found that 1.01 n*M* of mYFP corresponds to 1 a.u. fluorescence intensity. This value is threefold higher than that obtained by FCS.

3.6.6 FCS versus absorbance measurements

In this study, we decided to use the conversion factor calculated from FCS measurements for further quantitative analysis of mYFP kinetics. This choice is motivated by two facts: first, in contrast to our previous work (van Nies et al., 2013), the mYFP contains the mutation that makes the protein stable in the monomeric form, which is more suitable for FCS quantification. Second, FCS measurements can directly be performed in diluted PURE*frex*, whereas absorbance measurements require the substitution of the expression system for the PURExpress.

3.7 Quantitative analysis of mRNA and protein concentration versus time

Having determined the conversion factors between fluorescence intensity values and absolute concentrations of synthesized mRNA and YFP, one can quantitatively analyze the transcription and translation reactions. A complete quantitative understanding of the gene expression dynamics would require further mathematical modeling and computer simulations, but these studies are not within the scope of the work presented here.

The sensitivity of the two-reporter system can be assessed by calculating the lowest concentration of LL-Spinach and mYFP measured by spectrofluorometry. The lowest detected concentration of mRNA was calculated as two times the standard deviation of the LL-Spinach fluorescence signal collected within the first 20 min of expression after applying a baseline correction to eliminate the contribution of the rising average component of the signal. A value of 10 n*M* mRNA was obtained. The lowest detected concentration of mYFP was estimated as 2.3 n*M* using the same approach, except that a 20-min window taken in the linearly rising phase of the kinetics was used to compute the standard deviation.

To verify that the apparent kinetics reported by Spinach fluorescence reflects the actual profile of mRNA synthesis, it is essential to quantify on RNA gel the amount of transcript produced at different time points in PURE*frex* ΔR reactions. The mYFP-LL-Spinach DNA template was used. Two microliter samples were collected at 30 min intervals, diluted 50-fold in RNase-free water, purified, and gel analyzed according to the protocol described in Sections 3.2 and 3.3. The fluorescence and gel-based kinetics are nearly identical over the full time window covered (Fig. 6F), thus validating the use of the LL-Spinach tag as a reliable reporter for mRNA synthesis dynamics.

Given that the fluorescence emission signal of mYFP is not instantaneous upon synthesis owing to folding and chromophore maturation, experiments should be performed to determine the time delay between mYFP production and its fluorescence detection. We suggest to supplement the PURE*frex* reaction operating in the linear protein production regime with chloramphenicol (260 μg/mL final concentration), a translation inhibitor, and to measure the residual increase of mYFP fluorescence (Fig. 7C). Fitting the fluorescence signal to a first-order (monoexponential) kinetic equation led to a time delay of 20 min, corresponding to a maturation time of the mYFP chromophore of 0.05/min (Fig. 7C). This value is similar to that previously determined in the PURExpress with a different method (Iizuka, Yamagishi-Shirasaki, & Funatsu, 2011). Importantly, the actual kinetics of protein synthesis can now be reconstructed from the measured apparent kinetics accounting for 20 min time delay (van Nies et al., 2013).

The final amount of produced YFP and mYFP is between 0.4 and 0.5 μ*M* (Fig. 4C and E), assuming all the synthesized proteins emit fluorescence. This concentration is about 20-fold lower than the theoretically achievable production of 10.7 μ*M* calculated using an initial concentration of 0.3 m*M* for each amino acid (Shimizu et al., 2001). The low yield of protein synthesis contrasts with the more efficient production of mRNA that reaches the maximum theoretical concentration of 3.3 μ*M* after approximately 7 h reaction when starting with 7.4 n*M* of DNA template. In this calculation, the initial concentration of CTP and UTP (1 m*M*) was used, since they are not involved in translation-associated processes. This good agreement indicates that CTP and UTP depletion by reacting with the nucleoside diphosphate kinase for regenerating ATP and GTP is negligible. Together, these results indicate that only a small fraction of transcript is translated, which should definitely be taken into account for further improvement of the PURE system efficiency.

4. DETECTING GENE EXPRESSION INSIDE SEMIPERMEABLE LIPOSOMES

Cell-free gene expression inside liposomes can be seen as a platform for the construction of a genetically controlled minimal cell (Luisi, 2002; Noireaux, Maeda, & Libchaber, 2011). We recently established a protocol to trigger the biosynthesis of proteins inside surface-tethered lipid vesicles (Nourian et al., 2012; Fig. 1). The complete protocol is illustrated in Fig. 8. As the method is based on gentle rehydration of a lipid film, it is

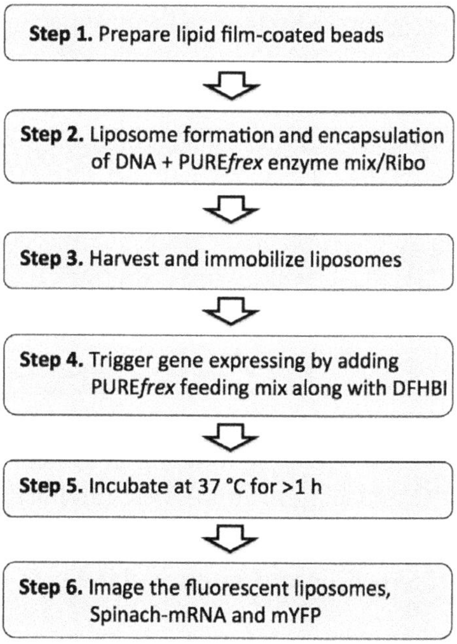

Figure 8 Flowchart of the protocol for liposome formation and confined gene expression.

compatible with a diversity of natural and synthetic lipids, and it is oil free. The vesicle membrane can be equipped with a number of functionalities, such as biotin–PEG (PEG = poly(ethylene glycol)) lipids for liposome immobilization on neutravidin-coated surfaces and TRITC (TRITC = *N*-(6-tetramethylrhodaminethiocarbamoyl))-conjugated lipids for membrane localization using fluorescence imaging. Moreover, the lipid composition can be tailored to tune the bilayer phase transition temperature or to regulate membrane permeability as a response to osmotic stress. We exploited these capacities to trigger gene expression inside semipermeable vesicles fed by external supply of the nutrients.

4.1 Preparation of lipid film-coated beads

In this study, we used short dimyristoylated, 14 carbon-acyl chain phospholipids, or long dioleoyl unsaturated phospholipids as they have been shown to be particularly suited for compartmentalized gene expression (Nourian & Danelon, 2013; Nourian et al., 2012). The following lipid compositions (approximate mol%) can be used: 1,2-dimyristoyl-*sn*-glycero-3-phosphocholine (DMPC, 80%), 1,2-dimyristoyl-*sn*-glycero-3-phospho-(1′-rac-glycerol) (DMPG, 20%), *N*-(6-tetramethylrhodaminethiocarbamoyl)-1,2-dihexadecanoyl-*sn*-glycero-3-phosphoethanolamine (TRITC-DHPE, 0.5%), and 1,2-distearoyl-*sn*-glycero-3-phosphoethanolamine-*N*-[biotinyl (polyethylene glycol)-2000] (DSPE-PEG-biotin, 0.5%). Alternatively, a different lipid mixture was used containing 1,2-dioleoyl-*sn*-glycero-3-phosphocholine (DOPC, 80%), 1,2-dioleoyl-*sn*-glycero-3-phospho-(1′-rac-glycerol) (sodium salt) (DOPG, 20%), TRITC-DHPE (0.5%), and DSPE-PEG-biotin (0.5%). All lipids were purchased from Avanti Polar Lipids except for TRITC-DHPE, which was from Invitrogen.

The lipids dissolved in chloroform were mixed at the desired molar ratio in a round-bottom glass flask in the presence of 212–300 μm glass beads (Sigma-Aldrich) to obtain a ratio of approximately 2 mg lipids per gram of beads. The organic solvent was evaporated in a rotavapor (Heidolph) at about 400 mbar overnight. The resulting lipid film-coated beads could be stored at −20 °C under argon atmosphere until used.

4.2 Liposome formation and encapsulation of the biosynthesis machinery

Around 10 μL of lipid film-coated beads were transferred in a 500-μL reaction tube and immersed in 11.5 μL swelling solution consisting of 1 μL of

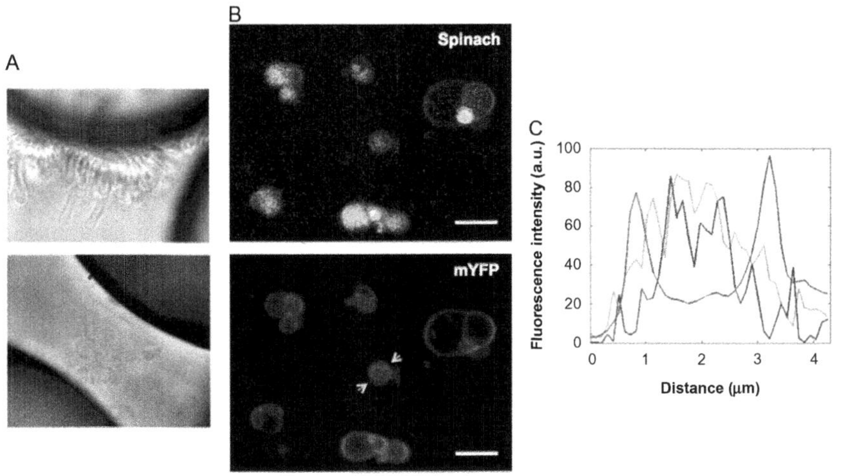

Figure 9 Fluorescence imaging of liposomes. (A) Phase contrast micrographs of lipid film swelling from glass bead surfaces (top). The tethered tubular liposomes eventually detach from the glass surface and remain trapped within the bead cavities (bottom) until the bead stack is gently disassembled for liposome harvesting. (B) Fluorescence confocal images of surface-tethered liposomes (dioleoyl phospholipids, composition 2) postexpression of the mYFP-LL-Spinach gene. The vesicle membrane (red) is localized using TRITC-labeled phospholipids. Scale bar is 5 μm. (C) Fluorescence intensity profiles of the TRITC, Spinach and mYFP signals measured along the line defined between the two arrows in (B). Color coding is the same as in (B). (See the color plate.)

PURE*frex* enzyme mix, 1 μL of PURE*frex* ribosome, 0.5 μL of DNA template (250 ng), 0.5 μL of Superase inhibitor (10 units final), and 8.5 μL of nuclease-free water. Liposomes were formed by swelling the lipid film at 30 °C (above the bilayer phase transition temperature) for 2 h and were then subjected to four freeze–thaw cycles to increase the yield of unilamellar vesicles (Fig. 9A). To collect intact liposomes efficiently, we advise to flip the tube and to rotate it gently to unpack the beads. This way, 1 μL of the liposome solution can carefully be harvested for immobilization. Using micrometer glass beads as a support of the lipid film offers a large surface area during swelling and enables microliter volume handling, which is difficult to achieve with conventional swelling methods.

4.3 Surface functionalization, liposome immobilization, and triggering of gene expression

Microscope coverslips were sonicated for 10 min in ethanol, rinsed with fresh ethanol, and dried under a stream of nitrogen. A 3-mm hole was punched into a 1–2-mm thick poly(dimethylsiloxane) (PDMS) piece that was then bounded onto a clean coverslip. The glass surface at the bottom of the PDMS chamber was incubated with 1 mg/mL BSA:BSA-biotin (1:1) (BSA was from Sigma-Aldrich, BSA-biotin was from Thermo Fisher Scientific) for 10 min, washed twice with reaction buffer (50 m*M* HEPES, 100 m*M* potassium glutamate), incubated for another 10 min with 1 mg/mL neutravidin (Sigma-Aldrich), and washed twice with reaction buffer.

Next, 1 μL of the liposome solution was added to the functionalized PDMS chamber and incubated at 37 °C for 3 min. At this stage, protein synthesis has not started, since the tRNAs and nutrients (amino acids, nucleotide triphosphates, and creatine phosphate) have not been supplied yet. To trigger gene expression, the surface-positioned vesicles were diluted with 5 μL of the PURE*frex* feeding mix along with DFHBI (20 μ*M* final concentration) and incubated at 37 °C for about 3 h after sealing the chamber with a clean glass coverslip.

4.4 Visualizing liposomes with fluorescence microscopy

A laser scanning confocal microscope (LSM710, Zeiss) equipped with a ×40 oil immersion objective was used to image the liposomes. The following laser lines and fluorescence emission windows were used: LL-Spinach 458/461–520 nm, mYFP 514/520–531 nm, and TRITC 543/557–797 nm. All measurements were performed at room temperature (20 °C). The fluorescence images were analyzed with the software ImageJ (Schneider, Rasband, & Eliceiri, 2012).

4.5 Factors influencing the levels of mRNA and protein produced in liposomes

Compared with the YFP-Spinach construct, imaging of synthesized mRNA inside liposomes was greatly enhanced using the LL-Spinach as shown by the high fluorescence intensity measured in transcriptionally active vesicles (Fig. 9B and C). Similar improvement was observed with both dimyristoyl- or dioleoyl-containing lipid compositions. Gene-expressing liposomes exhibit large signal disparities for both LL-Spinach and mYFP reporters, reflecting intrinsic heterogeneity in the levels of synthesized mRNA and protein. We showed previously that the vesicle-to-vesicle variability in YFP fluorescence intensity is not correlated neither with the amount of encapsulated DNA templates (Nourian & Danelon, 2013) nor with the mRNA level (van Nies et al., 2013). We attribute this heterogeneity (or stochasticity) primarily to the low-copy number of some constituents of the biosynthesis machinery confined within (sub-) micrometer-sized liposomes, which leads to a large compositional diversity of vesicles and, thus, to a great disparity in the transcription and translation rates between liposomes. As the resources have to translocate from the environment, another factor influencing the yield of internal production is the lamellarity of the liposome membrane, with lower exchange efficiency for vesicles with more bilayers.

Although the precise mechanism of molecular diffusion across the lipid bilayer remains to be explored, it is very likely that the osmolarity mismatch between the inside and outside liposome solutions generated when supplementing the feeding mixture leads to transient defects in the liposome membrane. Remarkably, these defects enable the uptake of tRNAs and nutrients, whereas the different PURE system constituents, engaged into functional macromolecular complexes, remain trapped inside the vesicles. When the 16 carbon-acyl chain saturated phospholipid DPPC (bilayer in the liquid-ordered phase at 37 °C) are used, liposomes fail to express fluorescent proteins (Nourian et al., 2012), indicating that the physicochemical properties of the lipid bilayer, in particular its mechanical response to osmotic pressure, govern the molecular exchange with the environment and thus the allocation of resources. Importantly, the efflux of toxic reactional products, such as pyrophosphate, across the vesicle membrane could contribute to the prolonged expression duration compared to batch mode reactions.

5. CONCLUSION AND OUTLOOK

The two-reporter assay described here, which uses a Spinach aptamer with improved folding stability to concurrently measure the levels of mRNA and protein synthesized from one gene, can be extended to measuring the expression dynamics of multiple genes. The development of multicolor reporters made of orthogonal pairs of RNA aptamers and GFP-like chromophores has unique potential to dissect the time-dependent levels of multiple mRNA species simultaneously (Paige et al., 2011), providing an additional level of control to tailor the temporal behavior of genetic circuits.

Further quantitative analysis is required to understand the expression dynamics of single and multiple genes inside lipid vesicles. However, it is clear that by providing cell-sized volumes, molecular crowding, a lipidic surface, and enabled molecular exchange with the environment, liposomes are unique reaction vessels to mimic the cellular milieu, offering great promise to characterize gene activity events as they take place inside the cell. Therefore, the "in vesiculo" gene expression platform that we developed fills the gap between the traditional oversimplified *in vitro* experiments and the complexity of live cells.

ACKNOWLEDGMENTS

We thank Jan van Esch, Rienk Eelkema, and Jos Poolman from the department of Chemical Engineering at the Delft University of Technology for synthesizing DFHBI. We are grateful to Roeland van Wijk for preliminary characterization of the mYFP-LL-Spinach construct, to Sabine van Schie for performing the chloramphenicol experiments, and to Ilja Westerlaken for assistance in designing and preparing the constructs. This work was supported by the Netherlands Organization for Scientific Research (NWO) through a VIDI grant and an ALW Open Programma grant to C. D.

REFERENCES

Babendure, J. R., Adams, S. R., & Tsien, R. Y. (2003). Aptamers switch on fluorescence of triphenylmethane dyes. *Journal of the American Chemical Society*, *125*(48), 14716–14717.

Bertrand, E., Chartrand, P., Schaefer, M., Shenoy, S. M., Singer, R. H., & Long, R. M. (1998). Localization of ASH1 mRNA particles in living yeast. *Molecular Cell*, *2*(4), 437–445.

Chizzolini, F., Forlin, M., Cecchi, D., & Mansy, S. S. (2014). Gene position more strongly influences cell-free protein expression from operons than T7 transcriptional promoter strength. *ACS Synthetic Biology*, *3*(6), 363–371 [Epub ahead of print].

Golding, I., Paulsson, J., Zawilski, S. M., & Cox, E. C. (2005). Real-time kinetics of gene activity in individual bacteria. *Cell*, *123*(6), 1025–1036.

Iizuka, R., Yamagishi-Shirasaki, M., & Funatsu, T. (2011). Kinetic study of de novo chromophore maturation of fluorescent proteins. *Analytical Biochemistry*, *414*(2), 173–178.

Kaern, M., Elston, T. C., Blake, W. J., & Collins, J. J. (2005). Stochasticity in gene expression: From theories to phenotypes. *Nature Reviews. Genetics*, *6*(6), 451–464.

Karzbrun, E., Shin, J., Bar-Ziv, R. H., & Noireaux, V. (2011). Coarse-grained dynamics of protein synthesis in a cell-free system. *Physical Review Letters*, *106*, 048104.

Kaufmann, B. B., & van Oudenaarden, A. (2007). Stochastic gene expression: From single molecules to the proteome. *Current Opinion in Genetics & Development*, *17*(2), 107–112.

Luisi, P. L. (2002). Toward the engineering of minimal living cells. *The Anatomical Record*, *268*(3), 208–214.

Maamar, H., Raj, A., & Dubnau, D. (2007). Noise in gene expression determines cell fate in *Bacillus subtilis*. *Science*, *317*(5837), 526–529.

Niederholtmeyer, H., Xu, L., & Maerkl, S. J. (2013). Real-time mRNA measurement during an in vitro transcription and translation reaction using binary probes. *ACS Synthetic Biology*, *2*(8), 411–417.

Noireaux, V., Bar-Ziv, R., & Libchaber, A. (2003). Principles of cell-free genetic circuit assembly. *Proceedings of the National Academy of Sciences of the United States of America*, *100*(22), 12672–12677.

Noireaux, V., & Libchaber, A. (2004). A vesicle bioreactor as a step toward an artificial cell assembly. *Proceedings of the National Academy of Sciences of the United States of America*, *101*, 17669–17674.

Noireaux, V., Maeda, Y. T., & Libchaber, A. (2011). Development of an artificial cell, from self-organization to computation and self-reproduction. *Proceedings of the National Academy of Sciences of the United States of America, 108*(9), 3473–3480.

Nomura, S.-I. M., Tsumoto, K., Hamada, T., Akiyoshi, K., Nakatani, Y., & Yoshikawa, K. (2003). Gene expression within cell-sized lipid vesicles. *ChemBioChem, 4*, 1172–1175.

Nourian, Z., & Danelon, C. (2013). Linking genotype and phenotype in protein synthesizing liposomes with external supply of resources. *ACS Synthetic Biology, 2*(4), 186–193.

Nourian, Z., Roelofsen, W., & Danelon, C. (2012). Triggered gene expression in fed-vesicle microreactors with a multifunctional membrane. *Angewandte Chemie, International Edition, 51*(13), 3114–3118.

Ozbudak, E. M., Thattai, M., Kurtser, I., Grossman, A. D., & van Oudenaarden, A. (2002). Regulation of noise in the expression of a single gene. *Nature Genetics, 31*, 69–73.

Paige, J. S., Nguyen-Duc, T., Song, W., & Jaffrey, S. R. (2012). Fluorescence imaging of cellular metabolites with RNA. *Science, 335*(6073), 1194.

Paige, J. S., Wu, K. Y., & Jaffrey, S. R. (2011). RNA mimics of green fluorescent protein. *Science, 333*(6042), 642–646.

Pedraza, J. M., & Paulsson, J. (2008). Effects of molecular memory and bursting on fluctuations in gene expression. *Science, 319*(5861), 339–343.

Petrásek, Z., & Schwille, P. (2008). Precise measurement of diffusion coefficients using scanning fluorescence correlation spectroscopy. *Biophysical Journal, 94*(4), 1437–1448.

Pothoulakis, G., Ceroni, F., Reeve, B., & Ellis, T. (2014). The Spinach RNA aptamer as a characterization tool for synthetic biology. *ACS Synthetic Biology, 3*(3), 182–187 [Epub ahead of print].

Rosenblum, G., Chen, C., Kaur, J., Cui, X., Goldman, Y. E., & Cooperman, B. S. (2012). Real-time assay for testing components of protein synthesis. *Nucleic Acids Research, 40*(12), e88.

Schneider, C. A., Rasband, W. S., & Eliceiri, K. W. (2012). NIH image to imageJ: 25 years of image analysis. *Nature Methods, 9*(7), 671–675.

Sei-Iida, Y., Koshimoto, H., Kondo, S., & Tsuji, A. (2000). Real-time monitoring of in vitro transcriptional RNA synthesis using fluorescence resonance energy transfer. *Nucleic Acids Research, 28*(12), E59.

Shaner, N. C., Steinbach, P. A., & Tsien, R. Y. (2005). A guide to choosing fluorescent proteins. *Nature Methods, 2*(12), 905–909.

Shimizu, Y., Inoue, A., Tomari, Y., Suzuki, T., Yokogawa, T., Nishikawa, K., et al. (2001). Cell-free translation reconstituted with purified components. *Nature Biotechnology, 19*, 751–755.

Shimizu, Y., Kanamori, T., & Ueda, T. (2005). Protein synthesis by pure translation systems. *Methods, 36*(3), 299–304.

Shimizu, Y., Kuruma, Y., Kanamori, T., & Ueda, T. (2014). The PURE system for protein production. *Methods in Molecular Biology, 1118*, 275–284.

Shin, J., & Noireaux, V. (2012). An *E. coli* cell-free expression toolbox: Application to synthetic gene circuits and artificial cells. *ACS Synthetic Biology, 1*(1), 29–41.

Siegal-Gaskins, D., Tuza, Z. A., Kim, J., Noireaux, V., & Murray, R. M. (2014). Gene circuit performance characterization and resource usage in a cell-free 'breadboard'. *ACS Synthetic Biology, 3*(6), 416–425 [Epub ahead of print].

Song, W., Strack, R. L., Svensen, N., & Jaffrey, S. R. (2014). Plug-and-play fluorophores extend the spectral properties of Spinach. *Journal of the American Chemical Society, 136*(4), 1198–1201.

Stögbauer, T., Windhager, L., Zimmer, R., & Rädler, J. O. (2012). Experiment and mathematical modeling of gene expression dynamics in a cell-free system. *Integrative Biology: Quantitative Biosciences from Nano to Macro, 4*(5), 494–501.

Strack, R. L., Disney, M. D., & Jaffrey, S. R. (2013). A superfolding Spinach2 reveals the dynamic nature of trinucleotide repeat-containing RNA. *Nature Methods*, *10*(12), 1219–1224.

Strack, R. L., Song, W., & Jaffrey, S. R. (2014). Using Spinach-based sensors for fluorescence imaging of intracellular metabolites and proteins in living bacteria. *Nature Protocols*, *9*(1), 146–155.

Tyagi, S. (2009). Imaging intracellular RNA distribution and dynamics in living cells. *Nature Methods*, *6*(5), 331–338.

Tyagi, S., & Kramer, F. R. (1996). Molecular beacons: Probes that fluoresce upon hybridization. *Nature Biotechnology*, *14*(3), 303–308.

van Nies, P., Nourian, Z., Kok, M., van Wijk, R., Moeskops, J., Westerlaken, I., et al. (2013). Unbiased tracking of the progression of mRNA and protein synthesis in bulk and inside lipid vesicles. *ChemBioChem*, *14*(15), 1963–1966.

Zuker, M. (2003). Mfold web server for nucleic acid folding and hybridization prediction. *Nucleic Acids Research*, *31*(13), 3406–3415.

CHAPTER ELEVEN

Design, Synthesis, and Application of Spinach Molecular Beacons Triggered by Strand Displacement

Sanchita Bhadra, Andrew D. Ellington[1]
Department of Chemistry and Biochemistry, Institute for Cellular and Molecular Biology, Center for Systems and Synthetic Biology, University of Texas at Austin, Austin, Texas, USA
[1]Corresponding author: e-mail address: ellingtonlab@gmail.com

Contents

Abstract

We describe design parameters for the synthesis and analytical application of a label-free RNA molecular beacon, termed Spinach.ST. The RNA aptamer Spinach fluoresces upon binding the small-molecule fluorophore DFHBI ((*Z*)-4-(3,5-difluoro-4-hydroxybenzylidene)-1,2-dimethyl-1*H*-imidazol-5(4*H*)-one). Spinach has been reengineered by extending its 5′- and 3′-ends to create Spinach.ST, which is

Methods in Enzymology, Volume 550
ISSN 0076-6879
http://dx.doi.org/10.1016/bs.mie.2014.10.049

predicted to fold into an inactive conformation that fails to bind DHFBI. Hybridization of a trigger oligonucleotide to a designed toehold on Spinach.ST initiates toehold-mediated strand displacement and restores the DFHBI-binding, fluorescence-enhancing conformation of Spinach. The versatile Spinach.ST sensor can detect DNA or RNA trigger sequences and can readily distinguish single-nucleotide mismatches in the trigger toehold. Primer design techniques are described that augment amplicons produced by enzymatic amplification with Spinach.ST triggers. Interaction between these triggers and Spinach.ST molecular beacons leads to the real-time, sequence-specific quantitation of these amplicons. The use of Spinach.ST with isothermal amplification reactions such as nucleic acid sequence-based amplification (NASBA) may enable point-of-care applications. The same design principles could also be used to adapt Spinach reporters to the assay of nonnucleic acid analytes in *trans*.

1. INTRODUCTION

Nucleic acid probes generated through the covalent conjugation of fluorophores and quenchers have become indispensable for sequence-specific detection and quantitation of nucleic acid molecules and amplicons. Molecular beacons are widely used for real-time, sequence-specific quantitation of nucleic acids (Tyagi & Kramer, 1996) and consist of a central, target-specific, single-stranded loop flanked by a stretch of five to seven complementary nucleotides that can base-pair to form a stem that terminates in a paired fluorophore and quencher. In the absence of a specific sequence target, fluorescence remains quenched due to the proximity of the fluorophore and quencher at the 5′- and 3′-ends of the molecular beacon. Binding of a complementary oligonucleotide to the loop leads to a conformational change that forces the beacon stem to unpair. The ensuing separation of the fluorophore from the quencher results in sequence-specific fluorescence enhancement.

While the widespread use of labeled nucleic acid probes *in vitro* is well documented, their application as *in vivo* reporters remains less than ideal due to the expense of chemical synthesis as well as the inefficiency of intracellular delivery. As one possible alternative, several RNA aptamers (Ellington & Szostak, 1990; Tuerk & Gold, 1990) have been described that bind and enhance the fluorescence of small-molecule dyes such as the Hoechst 33258 derivative 2,6-di-*tert*-butyl-4-[5-(4-methylpiperazin-1-yl)-1*H*,1′*H*-2,5′-bibenzo[*d*]imidazol-2′-yl]phenol (Sando, Narita, Hayami, & Aoyama, 2008); triphenylmethane dyes including malachite green, patent blue VF, and patent blue violet (Babendure, Adams, & Tsien, 2003; Grate & Wilson, 1999; Holeman, Robinson, Szostak, & Wilson, 1998);

cyanine dyes such as dimethylindole red (Constantin et al., 2008) and thiazole orange conjugates (Pei, Rothman, Xie, & Stojanovic, 2009), in addition to photoinduced *N*-(*p*-methoxyphenyl)piperazine derivatives of 2′,7′-dichlorofluorescein (Sparano & Koide, 2005, 2007). Some of these aptamers have been adapted to function as biosensors whose fluorescence enhancement is triggered by analytes such as nucleotides, theophylline (Furutani, Shinomiya, Aoyama, Yamada, & Sando, 2010; Stojanovic & Kolpashchikov, 2004), and nucleic acid sequences (Afonin, Danilov, Novikova, & Leontis, 2008; Kolpashchikov, 2005).

Recently, another RNA aptamer, termed Spinach, was selected to bind DFHBI ((*Z*)-4-(3,5-difluoro-4-hydroxybenzylidene)-1,2-dimethyl-1*H*-imidazol-5(4*H*)-one), a fluorophore that resembles those found in fluorescent proteins such as green fluorescent protein. A large increase in the green fluorescence emission of DFHBI occurs upon its binding to the aptamer (Paige, Wu, & Jaffrey, 2011; Strack, Disney, & Jaffrey, 2013).

Theoretically, any of these fluorescent RNA aptamers could be adapted to function as *in vivo* reporters via transcription and dye uptake. In particular, Spinach has been adapted to act as a genetically encoded sensor for RNA transcription in cells (Pothoulakis, Ceroni, Reeve, & Ellis, 2013), as an imaging agent, as a sensor for the intracellular detection of small-molecule analytes such as adenosine 5′-diphosphate (ADP) and *S*-adenosylmethionine (SAM) (Paige, Nguyen-Duc, Song, & Jaffrey, 2012), and as a sensor for proteins such as streptavidin, thrombin, and the MS2 coat protein (Song, Strack, & Jaffrey, 2013).

To create a more versatile, *trans*-acting reporter module, we reengineered Spinach to act as a conformational switch that can be triggered by toehold-mediated nucleic acid strand displacement (Bhadra & Ellington, 2014a; Zhang & Seelig, 2011). Similar programmable nucleic acid strand displacement reactions underlie multiple nucleic acid assays and devices, including molecular logic circuits and motors, catalytic amplifiers, and reconfigurable self-assembled nanostructures (Andersen et al., 2009; Chen & Ellington, 2010; Han, Pal, Liu, & Yan, 2010; Li, Ellington, & Chen, 2011; Qian & Winfree, 2011; Qian, Winfree, & Bruck, 2011; Seelig, Soloveichik, Zhang, & Winfree, 2006; Yin, Choi, Calvert, & Pierce, 2008; Yurke, Turberfield, Mills, Simmel, & Neumann, 2000; Zhang & Seelig, 2011). The resultant sequence-dependent Spinach (Spinach.ST) is the aptamer-based equivalent of a molecular beacon that fluoresces only upon hybridization with specific target sequences. However, this reagent no longer relies on the chemical conjugation of fluorophore and quencher moieties.

The sequence specificity of Spinach.ST can be readily altered by *in silico* design and cotranscriptional folding algorithms, and thus new sensors can be quickly built and then directly transcribed either *in vitro* or potentially *in vivo*. Spinach.ST can be readily inserted into layered nucleic acid circuits to integrate and report function. As an example, we have used Spinach.ST for the real-time quantitation of the output from a catalyzed hairpin assembly (CHA) circuit in which the circuit input, processors, and output are all composed of transcribed RNA. We have also developed novel *in vitro* one-pot enzymatic isothermal nucleic acid assay systems in which Spinach.ST molecular beacons are transcriptionally generated *in situ* and then report the accumulation of target amplicons in real-time. Single-nucleotide mismatches can also be distinguished within the target amplicons.

In the following sections, we elaborate the design principles for engineering Spinach.ST molecular beacons and their triggers. We also describe techniques to synthesize and measure the activity of Spinach.ST beacons *in vitro*. Finally, we detail the application of Spinach.ST molecular beacons for sequence-specific signal transduction of isothermal nucleic acid sequence-based amplification (NASBA).

2. HOW TO ENGINEER SPINACH MOLECULAR BEACONS TRIGGERED BY TOEHOLD-MEDIATED STRAND DISPLACEMENT

The 80-nucleotide minimized Spinach aptamer (24-2-min) (Paige et al., 2011) can be reengineered into a molecular beacon (Spinach.ST) by placing oligonucleotide extensions at its 5′- and 3′-ends, forming a nonbinding conformation. In the presence of a trigger nucleic acid, toehold-mediated strand displacement will lead to the formation of a binding conformation and aptamer fluorescence (Figs. 1 and 2).

2.1 Engineering conformations and sequence modules for Spinach beacons

In its predicted minimum free energy (MFE) conformation at 37 °C, the minimized Spinach aptamer has a predicted folding free energy of −30.4 kcal/mol and bears three internal stem loops and a basal stem that is formed by base pairing between the nine terminal nucleotides at its 5′- and 3′-ends (Fig. 1A). While stem loops 1 and 3 have sequence-specific roles in inducing DFHBI fluorescence, stem 2 appears to play a structural role in aptamer function (Paige et al., 2012).

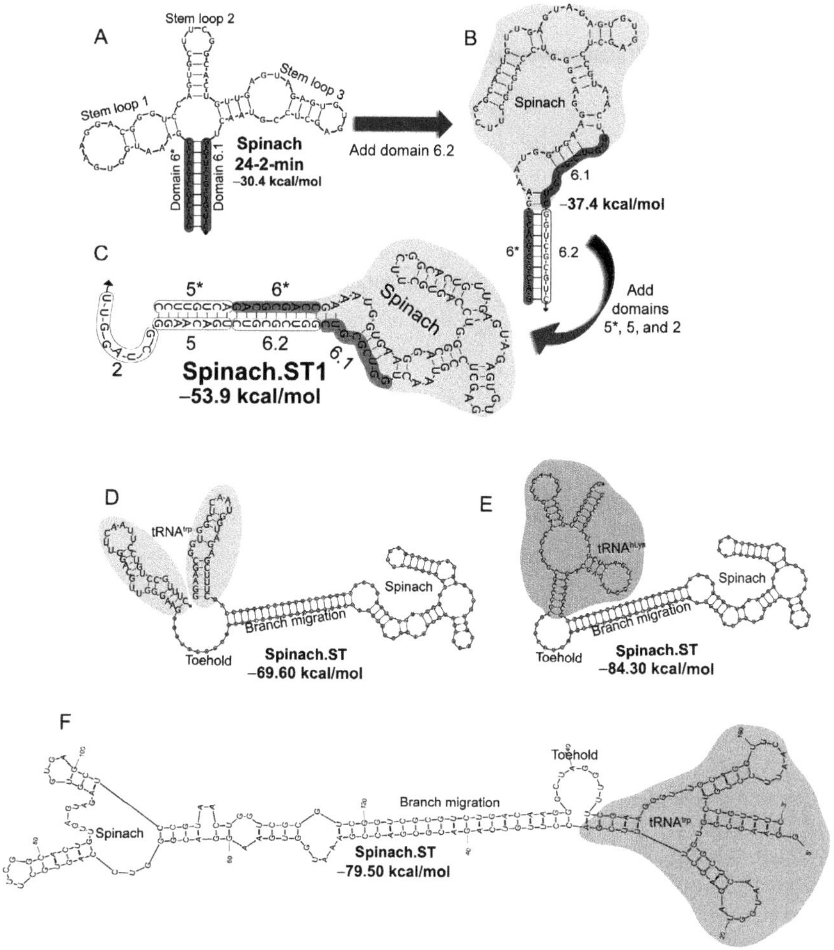

Figure 1 Engineering the Spinach.ST molecular beacon. (A) The minimum free energy (MFE) structure of the minimized 80-mer RNA aptamer Spinach 24-2-min as predicted by NUPACK. Nine nucleotides at the 5′- and the 3′-ends of the aptamer (designated as domains 6* and 6.1, respectively) hybridize to form the basal stem (highlighted in gray). (B) The minimized Spinach aptamer was extended at its 3′-end with a duplicate domain 6.1 (designated as domain 6.2) and is predicted to fold into an alternate MFE conformation by NUPACK. Domain 6.2 hybridizes with domain 6* and disrupts the basal stem of the Spinach aptamer. (C) The NUPACK-predicted MFE conformation of Spinach when extended with an additional domain 5* at its 5′-end and with domains 2 and 5 at its 3′-end. The Spinach.ST1 sequence is depicted. The aptamer domains 6* and 6.1 that form the Spinach 24-2-min basal stem are unable to interact. The extended domains 5 and 6.2 (outlined) hybridize with complementary domains 5* and 6* to form a long stem that leaves the unpaired domain 2 free to act as a toehold. (D–F) The Spinach.ST molecular beacon is stabilized by flanking it with tRNA sequences that fold into hairpins. Structures of *Saccharomyces cerevisiae* tRNA^{trp} flanks (D) and human tRNA^{Lys} flanks (E) as predicted by NUPACK are depicted. Structures of *Saccharomyces cerevisiae* tRNA^{trp} flanks generated by mFold are depicted in (E).

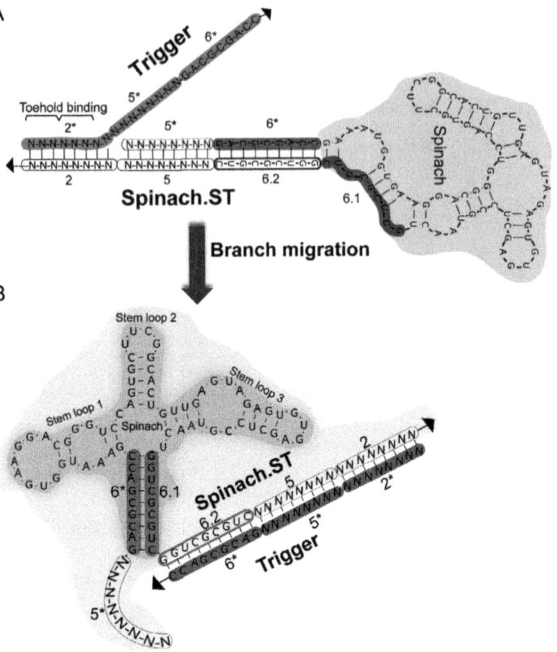

Figure 2 Sequence-specific activation of Spinach.ST molecular beacon by toehold-mediated strand displacement. Sequences of the invariant (aptamer-derived) domains 6*, 6.1, and 6.2 are shown, while the variable (target-derived) domains 2 and 5 (and their complementary domains 2* and 5*) are depicted as strings of "N." (A) Trigger toehold domain 2* initiates strand displacement upon binding to its complementary domain 2 in Spinach.ST. (B) Toehold hybridization initiates branch migration leading to contiguous hybridization of the trigger domains 2*, 5*, and 6* to the Spinach.ST 3′-end extended domains 2, 5, and 6.2, respectively. The ensuing conformational rearrangement restores the Spinach 24-2-min aptamer structure, including the basal stem formed by base pairing between domain 6* and 6.1. All structures were generated using NUPACK.

For simplicity, we designate the terminal nine-nucleotide-long complementary sections that form the basal stem of the aptamer as domains 6* (at the 5′-end) and 6.1 (at the 3′-end) (Fig. 1A). In nucleic acid devices and circuits driven by toehold-mediated strand displacement, the term "domain" is typically used for referring to short stretches of sequences that act as a unit in hybridization, branch migration, dissociation, structure, or function (Zhang, 2010). To create Spinach.ST, we appended a duplicate domain 6.1 (now designated as domain 6.2) to the 3′-end of the aptamer (Fig. 1B). This molecule is predicted to assume a conformation (with a free energy of −37.4 kcal/mol) in which the 5′-domain 6* hybridizes with the duplicate domain 6.2 and forces the aptamer to misfold such that stem loops 1 and 3 are disrupted and stem loop 2 is altered. To favor this misfolding pathway, we also appended eight-nucleotide-long complementary domains 5* and 5 to the 5′- and 3′-ends of the aptamer, respectively. The resulting continuously base-paired stretch of domains 5* and 6* lends additional thermodynamic stability to the misfolded state of Spinach.ST (Fig. 1C).

Disruption of the stem loop organization in Spinach.ST should render it unable to enhance the fluorescence of DFHBI. To allow sequence-specific activation of Spinach.ST, we further extended the molecular beacon at its 3′-end with a unique eight-nucleotide-long sequence termed domain 2 (Fig. 1C). This domain should remain unpaired in the predicted secondary structures of Spinach.ST (Fig. 1C). This single-stranded domain 2 can thus act as a toehold for binding complementary domain 2* in trigger oligonucleotides. A contiguous arrangement of domains 2*-5*-6* should, in turn, lead to the initiation and propagation of branch migration (Fig. 2A), ultimately releasing domains 5* and 6* at the 5′-end of Spinach.ST while sequestering domains 2, 5, and 6.2 at the 3′-end. The newly exposed 6* domain can potentially pair with domain 6.1, leading to conformational rearrangement of Spinach into an active structure that enhances DFHBI fluorescence (Fig. 2B).

It should be noted that the Spinach.ST domain numbering is a user-defined appellation. We have numbered the Spinach.ST domains discontinuously starting with domain 2 instead of 1 to maintain constancy with our previous publications. The prototype Spinach.ST was designed as a signal transducer for an upstream RNA circuit that included sequence domains 1 through 6 (Bhadra & Ellington, 2014a). The Spinach.ST domains that allowed it to interact with the circuit were accordingly designated as domains 2, 5, and 6.2.

As detailed above, there are two critical structural constraints for engineering Spinach.ST molecular beacons: (i) the toehold domain should remain fully exposed and (ii) the branch migration domains should sequester aptamer domain 6* by hybridization with the duplicate domain 6.2 (Fig. 1C). However, the design principles are broadly applicable and the sequence space available for designing the aptamer extension domains is expected to be large. We have designed three versions of Spinach.ST that are exclusively activated by different trigger sequences (Bhadra & Ellington, 2014b). The lengths and sequences of the toehold (domain 2) and branch migration (domains 5 and 6.2) domains appended to the minimal Spinach sequence were chosen based on previously described kinetic and thermodynamic considerations (Bhadra & Ellington, 2014a; Li et al., 2011). The effective rate constant of strand displacement is known to rise exponentially with increasing toehold length, reaching a plateau somewhere between 5- and 10-nucleotide toeholds (Yurke & Mills, 2003; Zhang & Winfree, 2009). In general, toehold binding should be strong enough to initiate strand displacement effectively. We commonly use eight-nucleotide-long toeholds (with 37–50% GC content and −10 to −13 kcal/mol free energy of binding) at 37–52 °C for operation of toehold-mediated strand displacement circuits composed of RNA under physiological as well as other salt concentrations.

The duplex structures of the branch migration domains of Spinach.ST should be strong enough to prevent unwanted intra- or intermolecular hybridizations. Furthermore, the total number of base pairs that can form between the cognate trigger and Spinach.ST (a combination of the toehold domain 2 and branch migration domains 5 and 6.2) should be long enough to discriminate effectively against nontarget sequences. Unnecessarily long extensions would likely slow down the conformational switching and/or frustrate the folding pathway for the Spinach beacon. The sequence and length (nine nucleotides) of the appended domain 6.2 is fixed in advance because it is designed to be complementary to the Spinach aptamer domain 6*. The other appended branch migration domains (5 and its complement 5*) are typically designed to be eight-nucleotides long and have a near 50% GC ratio. Thus, the effective trigger length is 25 nucleotides (eight-nucleotide-long toehold domain 2* combined with eight-nucleotide-long branch migration domain 5* and the nine-nucleotide-long branch migration domain 6*), which should ensure a very high degree of sequence specificity in Spinach.ST activation. We have also successfully created Spinach.ST designs for probing longer trigger sequences by including an additional eight- nucleotide-long branch migration domain between domains 5 and 6.2 (Bhadra & Ellington, 2014b).

The branch migration domains of Spinach.ST may be designed to include mismatches that create bubbles within the stem; such mismatches can impact both the thermodynamics and kinetics of reporter function. We have tested molecules containing two mismatches, one located at position 6 of domain 6.2 and the other at position 3 of domain 5 (Fig. 3A). Such designs might be more suitable for favoring strand displacement of longer stem domains by perfectly matched trigger oligonucleotides (especially DNA triggers). Also, periodic interruption of long RNA duplex regions with unpaired, bulged out nucleotides provides protection from RNase III-mediated cleavage in bacteria and disfavors the activation of

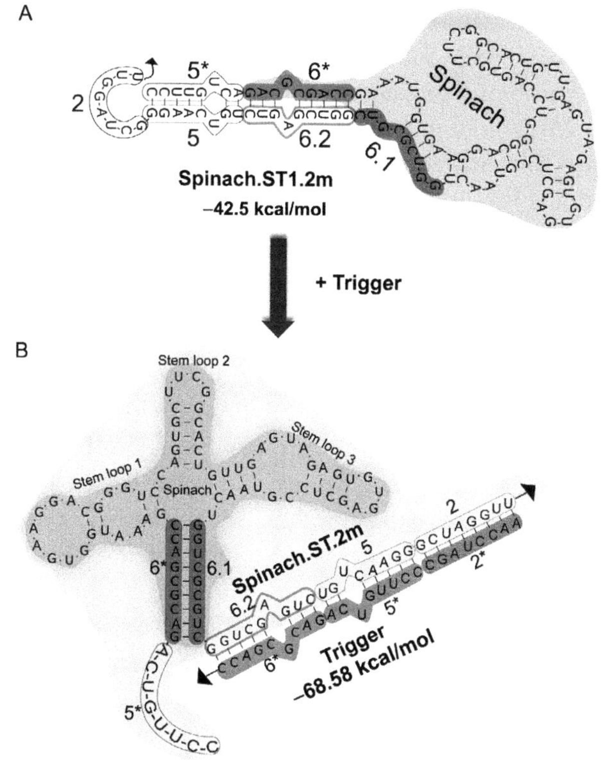

Figure 3 Engineering Spinach.ST with mismatched nucleotides. (A) Single mismatched bases introduced into the Spinach.ST1 domains 5 and 6.2 are depicted. The resulting molecule is still predicted to fold into the characteristic Spinach.ST conformation, but with two bulges within the extended duplex. (B) Spinach.ST trigger oligonucleotides harboring single mismatched bases in domains 5* and 6* are predicted to bind Spinach.ST and trigger its conformational rearrangement. All structures were generated using NUPACK.

RNA-activated protein kinase (PKR), a vital component of the mammalian innate immune response (Court, 1993; Heinicke, Nallagatla, Hull, & Bevilacqua, 2011; Hjalt & Wagner, 1995). We have found that trigger sequences that preserve these mismatches in the branch migration domains are still capable of activating cognate Spinach.ST molecules (Figs. 3B and 6A).

2.2 Using NUPACK and KineFold for design

The sequences of the Spinach extension domains were determined by *in silico* conformation modeling. We have designed Spinach.ST molecular beacons by an iterative process of manual sequence design guided by *in silico* thermodynamic modeling of RNA MFE structures and probabilities of intermolecular hybridization, and by stochastic folding simulations of cotranscriptional RNA folding kinetics. The goal of these design and modeling iterations was to generate sequences that should display the desired Spinach.ST MFE nonbinding structure (depicted in Fig. 1) and also to ensure that this structure will be kinetically favored along the cotranscriptional RNA folding pathway.

We typically use the NUPACK Web server (accessible at http://www.nupack.org/) for modeling MFE conformations of individual RNA sequences and for determining the structure and probability of complex formation with an interacting RNA (Dirks, Bois, Schaeffer, Winfree, & Pierce, 2007; Dirks & Pierce, 2003, 2004; Zadeh et al., 2011). All RNA analyses are performed in the "Analysis" mode of NUPACK with the following parameters. RNA energy parameters can currently be toggled between "Serra and Turner (1995)" and "Mathews, Sabina, Zuker, & Turner (1999)," and we typically perform Spinach and Spinach.ST folding and interaction analyses at 37 °C using the Serra and Turner RNA folding energy parameters (Serra & Turner, 1995). Three options (None, Some, All) are available on the NUPACK Web server for factoring in the effect of unpaired (dangling) bases on the folding landscape. We use the default "Some" dangle treatment option, in which dangle energy is incorporated for each unpaired base flanking a duplex. In addition, the NUPACK Web server does not allow the incorporation of pseudoknots into RNA structures of sequences longer than 100 nucleotides. Since most of the engineered Spinach sequences are greater than 100 nucleotides, we do not allow pseudoknots during the NUPACK analysis.

Both kinetic and thermodynamic folding considerations must be taken into account during design. Spinach.ST and its RNA trigger sequences will eventually be synthesized by enzymatic transcription of duplex DNA templates. RNA is known to undergo sequential or cotranscriptional folding that often involves the formation of transient structures that guide the folding pathway toward kinetically trapped secondary structures (Kramer & Mills, 1981). It is possible that the thermodynamically predicted, nonbinding conformations of the Spinach.ST molecular beacons will be disfavored during cotranscriptional folding of RNA. Therefore, it is critical to guide the iterative design process by modeling the cotranscriptional folding of the engineered RNA sequences. To this end, we typically use the KineFold Web server (accessible at http://kinefold.curie.fr/) to model cotranscriptional folding (Xayaphoummine, Bucher, & Isambert, 2005). Since we use T7 RNA polymerase for *in vitro* transcription of Spinach.

ST, the cotranscriptional folding of the molecular beacons was simulated at the transcription speed of T7 RNA polymerase.

The end product of the design process is RNA sequences that are predicted (by both NUPACK and KineFold) to fold into the Spinach.ST molecular beacon conformation as depicted in Fig. 1. The interaction of these newly designed Spinach.ST molecular beacons with their complementary trigger sequences (with the domain organization of 2*-5*-6*) can then be modeled using NUPACK to ensure that the Spinach molecular beacon will bind its trigger sequence and result in structural reorganization of the Spinach.ST beacon into a binding, fluorescent conformation (Fig. 2). For analyzing intermolecular interactions using NUPACK, set the "Number of strand species to 2": Spinach.ST and its trigger. Also set the maximum number of interacting species in NUPACK to 2. It is best to model the interaction between Spinach.ST and its trigger at various stoichiometries, starting with equimolar conditions and potentially at different concentrations. Under all conditions tested, the trigger should form a complex with its cognate Spinach.ST molecular beacon with near 100% probability (as predicted by NUPACK), indicating activation of the molecular beacon. However, since NUPACK does not take kinetic traps into account when predicting intermolecular complexes, it is critical to examine the predicted structures visually. For example, if the toeholds in the trigger sequence or Spinach.ST molecular beacon were occluded, NUPACK would still predict complete hybridization between the two molecules, although of course, in reality, the absence of an exposed toehold will prevent any such productive interactions. To evaluate whether and how the Spinach molecular beacon might be nonspecifically triggered, scrambled or mismatched trigger sequences can be substituted for the cognate trigger during modeling.

2.3 Promoters for Spinach.ST expression

The speed of transcription is known to influence RNA folding pathways (Meyer & Miklos, 2004; Xayaphoummine et al., 2005). Slowing down or speeding up transcription might yield inactive transcripts. Thus far, we have expressed Spinach.ST using T7 RNA polymerase from the canonical 17-mer T7 promoter "TAATACGACTCACTATA" (Dunn & Studier, 1983; Martin & Coleman, 1987; Milligan, Groebe, Witherell, & Uhlenbeck, 1987). We have most commonly employed "GGAAGC" and to a lesser extent "GGGTGG" as the +1 to +6 initiation sequences (part of the 5′-end tRNA scaffold, see Section 2.4). Because of the potential impact of transcription kinetics on folding, users should always examine the effects of altering the promoter and/or initiation sequences on Spinach.ST function.

2.4 Stabilization of Spinach.ST within a tRNA scaffold

We have embedded the designed Spinach.ST molecular beacon within a tRNA scaffold to stabilize the RNA (Iioka, Loiselle, Haystead, & Macara, 2011). The tRNA flanks are predicted to form stable hairpins at the 5′- and 3′-ends of Spinach.ST (Fig. 1D–F). Such scaffolds may improve the structural stability of aptamers *in vitro* as well as improve the life span of RNA *in vivo* (Paige et al., 2011; Ponchon & Dardel, 2007). We have used sequences derived from *Saccharomyces cerevisiae* $tRNA^{trp}$ (Fig. 1D and F) as well as human $tRNA^{Lys}$ (Fig. 1E) to flank various Spinach.ST molecules. It should be noted that while the human $tRNA^{Lys}$ is predicted by NUPACK to fold into the expected tRNA conformation (Fig. 1E), the MFE structure of the *S. cerevisiae* $tRNA^{trp}$ flanks that is predicted by NUPACK does not conform to the tRNA fold (Fig. 1D). In contrast, mFold depicts a normal tRNA conformation for the *S. cerevisiae* $tRNA^{trp}$ flanks of Spinach.ST1 (Fig. 1F) and predicts an overall free energy that is about 10 kcal/mol lower than that predicted by NUPACK. This discrepancy is atypical and does not significantly impact the design of modular folding units. NUPACK remains the most versatile tool available for modeling (and designing) base pairing probabilities in ordered complexes.

2.5 Programming triggers

The choice of the trigger will also impact Spinach.ST signaling. In general, trigger conformation should also be analyzed using NUPACK to ensure that it is not prohibitively structured and that its toehold is readily accessible for binding to Spinach.ST and for initiating branch migration. Beyond that, triggers may be configured for various applications. For instance, single-nucleotide mismatches in target sequences may be distinguished by positioning the mismatched nucleotide within the toehold domains of trigger oligonucleotides, thus increasing the ΔG of toehold binding and slowing the kinetics of Spinach.ST activation (detailed in Section 2.5.1). Since the magnitude of Spinach.ST activation is dependent on the trigger concentration, trigger sequences can also potentially be quantitated, allowing the measurement of nucleic acid circuits (see Section 2.5.2) and amplicons (see Section 5) with Spinach.ST. The simultaneous presence of two different target sequences, including device outputs, may be logically processed using Spinach.ST by restricting the generation of a functional trigger in association with two targets (detailed in Section 2.5.3).

2.5.1 Designing triggers to discriminate single-nucleotide mismatches

It has previously been shown that toehold design can significantly modulate the kinetics of strand displacement (Zhang, Chen, & Yin, 2012; Zhang & Winfree, 2009). We have demonstrated that mismatches at the first or second nucleotide position of the toehold (trigger domain 2*) do not significantly disrupt Spinach.ST strand displacement (Bhadra & Ellington, 2014b). However, toehold mismatches occurring at nucleotide positions 3, 4, or 5 almost completely destroy the ability of a trigger to activate Spinach.ST1, resulting in near-zero initial rates similar to those observed with an unrelated trigger.

In general, to achieve maximal distinction among single-nucleotide polymorphisms (SNPs), mismatches are localized to trigger toehold domain 2* at positions that result in maximal increase in the ΔG of toehold binding (Fig. 4B). However, SNPs located deep in the branch migration domain are not expected to be able to distinguish triggers; previous work with RNA strand displacement circuits has demonstrated that branch migration can readily jump over two contiguous mismatches (Bhadra & Ellington, 2014a). Also, as depicted in Fig. 6A, trigger sequences with mismatches designed in the branch migration domains are capable of activating Spinach.ST.

2.5.2 Designing triggers to function in the context of nucleic acid circuits

Nucleic acid devices, circuits, and sensors can be engineered to undergo programmed intra- or intermolecular conformational rearrangements that lead to the exposure of previously sequestered trigger sequences. Spinach.ST activation can potentially be used as a reporter for such conformational changes. Trigger sequestration may be achieved by ensconcing trigger sequences in base pairing interactions with the functional nucleic acid such that the single-stranded toehold is unavailable for initiating nucleic acid strand displacement and thus activation of Spinach.ST. Triggers may be either covalently appended to (design detailed below in the context of a CHA circuit) or intermolecularly annealed to the functional nucleic acid. Release of the single-stranded trigger (or its toehold domain) occurs upon

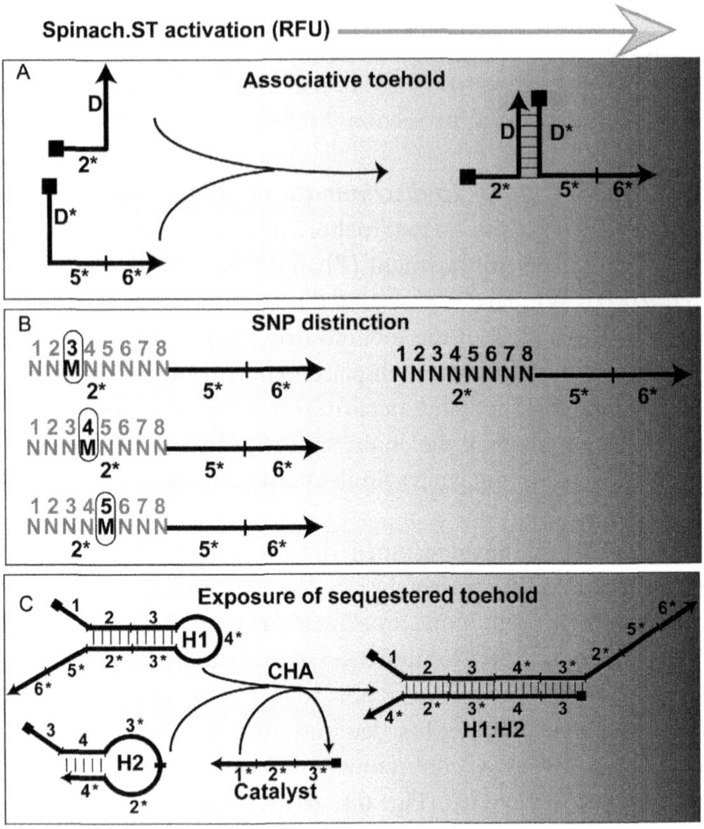

Figure 4 Design options for Spinach.ST trigger sequences. (A) Generation of Spinach.ST trigger sequences by associative toehold activation. The trigger toehold domain 2* is placed at the 5′-end of one oligonucleotide, while the trigger branch migration domains 5* and 6* are placed at the 3′-end of a second oligonucleotide. Neither oligonucleotide alone is capable of activating Spinach.ST. However, when brought together by hybridization between the complementary domains D and D*, the toehold and branch migration domains activate Spinach.ST. (B) Presence of a single mismatched nucleotide (M) at position 3, 4, or 5 within the trigger toehold domain 2* significantly decreases the efficiency of Spinach.ST activation. The fully complementary toehold is denoted as a string of "N" with nucleotide positions numbered above. (C) Detection of RNA CHA circuit output using Spinach.ST. Toehold-mediated strand displacement by a catalyst oligonucleotide leads to the formation of H1:H2 from the kinetically trapped CHA hairpins H1 and H2. The Spinach.ST trigger is appended to H1 such that the toehold domain 2* remains sequestered within the H1 stem, and hence is unable to initiate activation of the Spinach molecular beacon. Upon formation of the H1:H2 circuit output complex, the trigger toehold 2* is exposed and can be quantitated by measuring the fluorescence accumulation of Spinach.ST molecules that undergo conformational activation.

input-dependent operation of the functional nucleic acid. For instance, we have used Spinach.ST for measuring the kinetic development of a CHA circuit (Bhadra & Ellington, 2014a). In CHA, two partially complementary nucleic acid oligonucleotides are ensconced within hairpin structures (H1 and H2) leading to kinetic trapping that prevents them from interacting with one another (Fig. 4C). The Spinach.ST trigger sequence appended to one of the hairpins cannot activate Spinach.ST, since the toehold domain 2* is sequestered within the double-stranded hairpin stem. A short, single-stranded oligonucleotide "catalyst" that initiates toehold-mediated strand displacement of one of the hairpins in turn leads to revelation of sequences that can interact with the other hairpin, the formation of a double-stranded product, and the recycling of the catalyst. The trigger toehold 2* that is exposed in the double-stranded circuit output (H1:H2) can be quantitated by measuring fluorescence accumulation of trigger-activated Spinach.ST molecules. Using this approach, we have created completely RNA-based circuits, wherein the circuit input, processor, output, and sensor molecules are all composed of transcribed RNA (Bhadra & Ellington, 2014a).

Similar design strategies can be employed to adapt Spinach.ST to the detection of nonnucleic acid analytes. Structure-switching aptamers can be engineered to undergo analyte-induced intra- or intermolecular rearrangements that lead to the display of sequestered Spinach.ST trigger sequences or toeholds. These exposed sequences would in turn activate standalone Spinach.ST (Cho, Lee, & Ellington, 2009; Nutiu & Li, 2003). Structure-switching aptamers have already been widely used in developing aptamer-based optical (Li & Ho, 2008) and electrochemical sensors (Lubin & Plaxco, 2010) for analytes such as nucleotides, cocaine, and thrombin (Liu & Lu, 2006; Tang et al., 2008), and their adaptation to Spinach.ST signaling should be straightforward.

Another strategy that has been previously described for Spinach-based ligand-sensing relies on interruption of the Spinach sequence with ligand-specific aptamer domains (Paige et al., 2012; Strack, Song, & Jaffrey, 2014). In this approach, unique Spinach biosensors must be designed for each analyte, wherein (a) the stem loop 2 of the Spinach aptamer is replaced with the ligand-binding aptamer sequence and (b) Spinach is forced to misfold into a nonfluorescent conformation. Upon binding of the cognate analyte, a "communication module" (Koizumi, Soukup, Kerr, & Breaker, 1999) between the aptamer and Spinach promotes a conformational change of the Spinach aptamer into its fluorescent state.

2.5.3 Associative toehold triggers

The trigger need not be a contiguous sequence. Indeed, in some instances it is advantageous to disrupt the trigger or spread it out over multiple nucleic acids. Spinach.ST activation should then occur only upon the programmed assembly of the separate trigger units into one complex that would be jointly capable of toehold-mediated strand displacement. Such discontinuous triggers should prove useful for diverse applications such as nucleic acid amplicon validation (Section 5), determining the coincident presence or expression of multiple nucleic acid sequences and logical processing of nucleic acid circuit outputs.

Discontinuous trigger molecules may be built by "associative toehold activation" wherein the toehold and branch migration domains become connected via the hybridization of auxiliary domains, and then allow strand displacement across a three-way junction (Chen, 2012). In one instantiation, the trigger may be split between two separate nucleic acids such that the toehold domain 2* is presented by oligonucleotide 1, while the branch migration domains 5* and 6* are located on oligonucleotide 2 (Fig. 4A). Either oligonucleotide alone will fail to initiate Spinach.ST strand displacement. One of the ways to facilitate association between the two oligonucleotides would be to append domain D at the 3′-end of the toehold-bearing oligonucleotide 1 and to append the complementary domain D* at the 5′-end of oligonucleotide 2 that bears the branch migration domains. Hybridization of domains D and D* will juxtapose the toehold and branch migration domains and allow activation of Spinach.ST (Fig. 4A).

We have previously used random 20- to 28-nucleotide-long domains D and D* with approximately 50% GC contents and free energies (at 37 °C) ranging between −37 and −53 kcal/mol to create associative toeholds. However, the choice of length and sequence should be flexible as long as hybridization probability and thermodynamic stability under assay conditions remains high. The presence of two bulged thymidine bases at the three-way junction of associative DNA toeholds is known to accelerate strand displacement substantially, especially in reactions mediated by reversible toehold binding (Chen, 2012). Although the eight-nucleotide-trigger toehold is expected to bind reversibly (as is evident from the operation of multiple turnover RNA CHA circuits using eight-nucleotide toeholds; Bhadra & Ellington, 2014a), we have found it unnecessary to include bulged thymidines for stabilization of the RNA three-way junctions. All domain structures and interactions should be verified using NUPACK prior to experimentation.

2.6 Spinach.ST function within larger RNA contexts

The conformational rearrangements required for Spinach.ST function can occur within the context of larger RNAs (Figs. 5 and 6). The ability of Spinach.ST to function when it and/or its trigger is embedded within a larger RNA is not only important for *in vitro* applications such as signal transduction from nucleic acid amplicons (see Section 5) but also might be of particular importance for *in vivo* applications. Potential examples include detection of RNA localization and/or interaction using

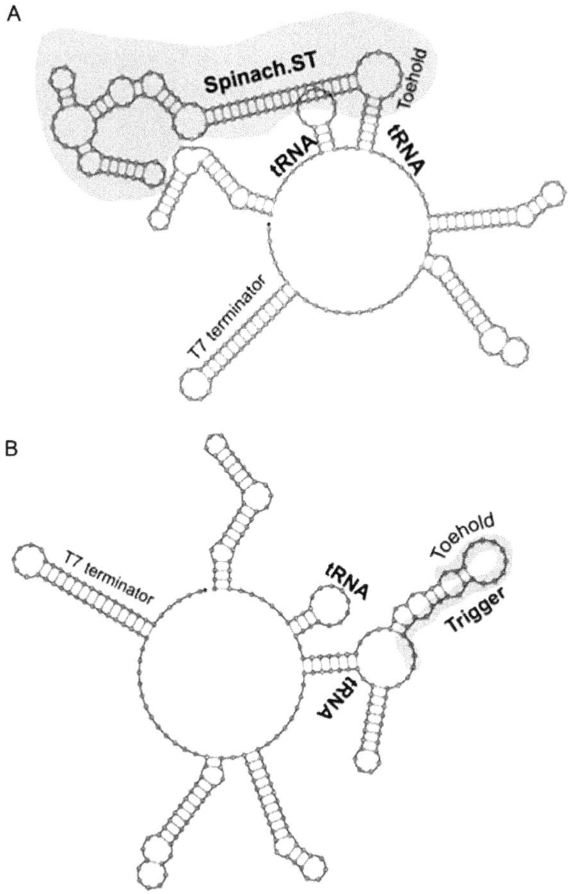

Figure 5 Spinach.ST and trigger modules embedded in larger sequence contexts. Spinach.ST (A) and its trigger oligonucleotide (B) (both highlighted in gray) flanked with tRNA sequences and expressed as part of a larger transcript. All structures were generated using NUPACK.

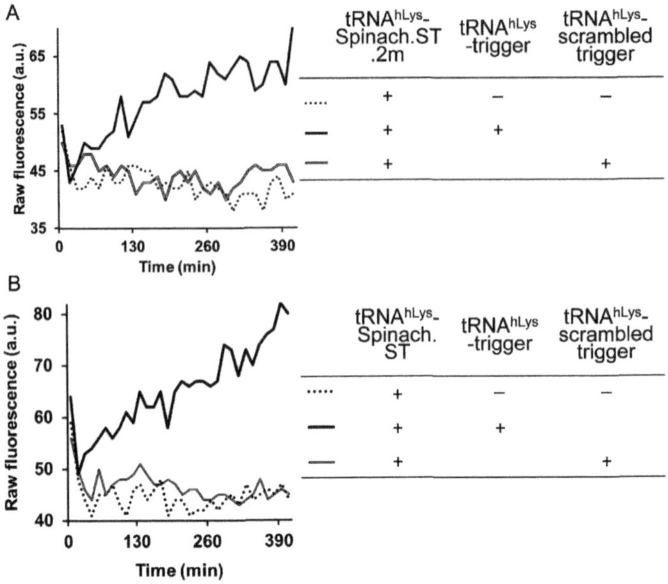

Figure 6 Sequence-specific activation of Spinach.ST molecular beacons. Spinach.ST.2m (A) (for structure, see Fig. 3) and Spinach.ST (B) (for structure, see Fig. 1) flanked by tRNA and embedded within larger transcripts (for structure, see Fig. 4) were incubated with cognate or scrambled trigger oligonucleotides that were similarly flanked by tRNA and expressed as a portions of larger transcripts. Spinach molecular beacons were activated only by cognate trigger-containing transcripts and led to the accumulation of DFHBI fluorescence over time.

programmed Spinach.ST molecules as probes and monitoring RNA splicing by detecting associative trigger sequences whose toehold and branch migration domains become juxtaposed only upon splicing. We embedded the tRNA–Spinach.ST–tRNA cassette within a larger transcript such that it was flanked by 35 nucleotides on the left and 55 nucleotides on the right followed by a 42-nucleotide-long T7 terminator sequence (Fig. 5A). The Spinach.ST beacon in this transcript fluoresced only in the presence of cognate trigger oligonucleotides (Fig. 6B). The trigger sequence could also be successfully embedded in a larger transcript (Fig. 5B). Embedded scrambled sequences failed to activate the Spinach.ST molecular beacon. Intracellularly generated Spinach.ST transcripts may contain upstream or downstream sequences (for example, promoter or terminator regions) that might prevent folding and function. Such flanking sequences can potentially be removed by appending ribozyme cleavases at each end (Bhadra & Ellington, 2014a).

3. HOW TO SYNTHESIZE SPINACH.ST MOLECULAR BEACONS ENZYMATICALLY

3.1 DNA oligonucleotides

We have generally purchased all oligonucleotides from Integrated DNA Technologies (IDT, Coralville, IA, USA). Resuspend oligonucleotides at a final concentration of 100 μ*M* in TE (10:0.1) buffer (10 m*M* Tris–HCl, pH 7.5, 0.1 m*M* EDTA, pH 8.0), and measure the concentration of all nucleic acid solutions by UV spectrophotometry. We typically use a Nanodrop 1000 spectrophotometer (Thermo Scientific, Wilmington, DE, USA) for this purpose. Store dissolved oligonucleotides at −20 °C.

3.2 How to generate transcription templates

We typically use fully double-stranded DNA as templates for T7 RNA polymerase-driven transcription. Short transcription templates ($\leq$60 bp) can be prepared by annealing two complementary synthetic oligonucleotides that have been mixed at an equimolar concentration in TE (10:0.1) buffer containing 50 m*M* NaCl. Mixtures of oligonucleotides are denatured for 5 min in a thermocycler at 95 °C and then annealed by slow cooling (0.1 °C/s) to 25 °C. Annealed oligonucleotide templates should be quantitated by UV spectrophotometry and directly used for *in vitro* transcription reactions.

Longer transcription templates may be purchased from IDT as double-stranded gBlocks or single-stranded Ultramers and then amplified by the polymerase chain reaction (PCR). Longer templates may also be sequentially assembled from sets of shorter, overlapping oligonucleotides by oligonucleotide annealing, primer extension, and PCR. The design of overlapping oligonucleotides can be automated using tools such as DNAWorks (http://helixweb.nih.gov/dnaworks/). However, since Spinach.ST contains numerous secondary structures, it is important that the 3′-ends of the overlapping oligonucleotides are sufficiently single-stranded to allow base pairing and polymerase-mediated extension. We usually perform all enzymatic amplification reactions using high fidelity DNA polymerases such as the Phusion DNA polymerase (New England Biolabs (NEB), Ipswich, MA, USA).

While longer transcription templates can be used directly for transcription, we typically prefer to first clone them into vectors. Cloning creates a convenient repository and a homogenous, error-free source of synthetic transcription templates for continued use. Enzymatically synthesized transcription templates (as above) should first be purified by agarose gel electrophoresis, and then extracted from the gel using Wizard SV clean-up columns (Promega, Madison, WI, USA). Purified transcription templates may then be inserted into vectors such as pCR2.1 (Life Technologies) by TOPO-TA cloning (Life Technologies, Grand Island, NY, USA) or by Gibson cloning (NEB).

Cloned transcription templates should be amplified from previously sequence-verified plasmids by PCR using Phusion DNA polymerase. We typically amplify a Spinach.ST expression cassette using primers that reach into the flanking plasmid sequence at the 5′-end and primers specific to the 3′-end sequence of Spinach.ST, creating a template for runoff transcription. A small aliquot of the amplification reaction should be analyzed by agarose gel electrophoresis prior to transcription, to ensure that PCR has yielded a product of the appropriate size and purity. The remainder of the amplification reaction should be purified using the Wizard SV Clean-up system (Promega) and quantitated by UV spectrophotometry prior to setting up the *in vitro* transcription reaction.

4. HOW TO PERFORM FUNCTIONAL ASSAYS OF SPINACH.ST MOLECULAR BEACONS

In our experience, Spinach.ST molecules are readily activated upon posttranscriptional incubation with cognate RNA or DNA trigger oligonucleotides (Bhadra & Ellington, 2014b) (Section 4.2). Spinach.ST molecules can also be activated by triggers that are cotranscribed along with them (Section 4.3). However, the performance of each new construct should of course be evaluated.

4.1 *In vitro* transcription reactions

We typically transcribe some 100–500 ng of double-stranded DNA template using 100 units of T7 RNA polymerase (NEB) in a 50 μl reaction containing 40 m*M* Tris–HCl, pH 7.9, 30 m*M* $MgCl_2$, 10 m*M* DTT, 2 m*M* spermidine, 4 m*M* of each ribonucleotide (rNTPs), and 20 units of the recombinant ribonuclease inhibitor RNaseOUT (Life Technologies). Transcription reactions are incubated at 42 °C for 120 min.

KineFold predicts that initial base pairing during transcription would occur between domain 6* and the first domain 6.1, but as transcription proceeds the molecule should rearrange into the nonfluorescent, trapped conformation. Our experimental observations have also verified that the Spinach.ST molecular beacons cotranscriptionally fold into their inactive conformation. Thus, it is unnecessary to perform posttranscriptional thermal equilibration of Spinach.ST prior to use. Upon completion of transcription, the Spinach.ST RNA molecules may be used directly or they can be purified by filtration through Sephadex G25 prior to use. We commonly employ the Illustra MicroSpin G-25 columns (GE Healthcare, Piscataway, NJ, USA) for crude transcript purification. For applications requiring even higher purity, Spinach.ST can be isolated by denaturing polyacrylamide gel electrophoresis followed by gel extraction. Unused portions of Spinach.ST transcripts may be stored frozen at −20 or −80 °C and reused immediately upon thawing.

4.2 Reactions with added trigger and control sequences

Activation assays can be set up as 15 μl reactions in 1 × TNaK buffer (20 m*M* Tris–HCl, pH 7.5, 140 m*M* NaCl, 5 m*M* KCl). The Spinach.ST system is compatible with buffers such as those used for transcription and enzymatic nucleic acid amplification reactions (see Sections 4.3 and 5.2), with physiological buffers such as phosphate buffered saline (137 m*M* NaCl, 2.7 m*M* KCl, 10 m*M* Na_2HPO_4, 1.8 m*M* KH_2PO_4, pH 7.2) with or without 3 m*M* $MgCl_2$, and in buffers that mimic intracellular salt concentrations such as the IC buffer (20 m*M* HEPES, 140 m*M* KCl, 10 m*M* NaCl, 2 m*M* $MgCl_2$, 5 m*M* KH_2PO_4, pH 7.2) (Lodish & Darnell, 1995). However, we do not recommend using the Spinach aptamer selection buffer (40 m*M* HEPES, pH 7.4, 125 m*M* KCl, 5 m*M* $MgCl_2$, and 5% DMSO), since the Spinach molecular beacon shows significant background with this buffer. This is likely due to increased folding of Spinach.ST into the native Spinach conformation even in the absence of the trigger oligonucleotide.

The reaction also contains 70 μ*M* DFHBI, 20 units of RNaseOUT, and some 3–6 μl aliquots of the 50 μl G25-filtered Spinach.ST transcripts. DFHBI may be synthesized using the published pathway (Paige et al., 2011) or it may be purchased from Lucerna, Inc. (New York, NY, USA). Prepare a 10 m*M* solution of DFHBI in dimethyl sulfoxide and store in the dark at −20 °C.

Trial activation reactions should include 0.5 μ*M* single-stranded trigger DNA oligonucleotides and 3–6 μl of G25 Sephadex-purified Spinach.ST transcripts, prepared as described above. The Spinach.ST molecular beacons can be assayed by incubation with equimolar trigger oligonucleotides. Increasing trigger concentrations should increase the rate and amplitude of Spinach.ST activation. Reactions containing no and nonspecific trigger oligonucleotides should also be included as controls. Reactions without Spinach.ST transcripts or trigger oligonucleotides will provide values for subtracting the baseline fluorescence of DFHBI.

Reactions can be assembled in 0.2-ml PCR tubes and then transferred to alternate wells of a Nunc™ 384-well flat-bottom black plate (Thermo Scientific) covered with a MicroAmp® optical adhesive film (Life Technologies). Spinach.ST activation is then monitored by measuring fluorescence accumulation as a function of time at 37 °C. The Spinach.ST molecular beacons at micromolar concentrations typically require approximately 30–60 min to develop maximal fluorescence in response to cognate trigger oligonucleotides. We perform fluorimetry as a function of time using the TECAN Safire, a monochromator-based microplate reader, and set to 469 nm excitation and 501 nm emission wavelengths.

While the fluorescence quantum yield of unmodified Spinach–DFHBI complex is reported to be over a 1000-fold greater than DFHBI alone, activated Spinach biosensors are typically not as bright (Kellenberger, Wilson, Sales-Lee, & Hammond, 2013). The fluorescence intensity of Spinach.ST rises with increasing trigger concentration, typically demonstrating a 4- to 70-fold increment over the fluorescence of DFHBI alone. This level of activation is similar to the approximately 4- to 32-fold fluorescence reported for aptamer-based Spinach-based biosensors when incubated with 10^3- to 10^4-fold excess of small-molecule ligands such as adenosine, ADP, SAM, guanine, guanosine 5′-triphosphate (GTP), cyclic di-GMP, and cyclic AMP–GMP (Kellenberger et al., 2013; Paige et al., 2012).

4.3 Real-time cotranscriptional functional assays with Spinach.ST

Sequence-dependent Spinach.ST activation can also be measured by transcribing the beacon in parallel with trigger sequences. Cotranscriptions can be performed *in vitro* in 25 μl reaction volumes in Nunc™ 384-well flat-bottom black plates. As an example, some 500 ng of double-stranded Spinach.ST and trigger transcription templates can be transcribed using 100 units of T7 RNA polymerase in a buffer composed of 40 m*M* Tris–HCl, pH 7.9, 6 m*M* $MgCl_2$, 10 m*M* DTT, 2 m*M* spermidine, 50 m*M* NaCl, 4.8 m*M* rNTPs, 40 μ*M* DFHBI, and 20 units of RNaseOUT. The plate is then covered with an optical film and incubated at 37–39 °C in a TECAN Safire plate reader, and real-time fluorescence measurements are recorded at regular intervals. Specificity of the cotranscriptional activation can be determined by substitution of the cognate trigger transcription templates with templates that generate scrambled trigger RNAs. Control reactions with only trigger or Spinach.ST transcription templates should be included to determine background and nonspecific activation. Significant accumulation of Spinach.ST fluorescence should only be observed when Spinach.ST is cotranscribed with its cognate trigger.

5. APPLICATION: REAL-TIME SPINACH.ST-BASED DETECTION OF NASBA

We have previously used nucleic acid circuits for transducing signals from the amplification reactions that underlie molecular diagnostics (Jiang, Li, Milligan, Bhadra, & Ellington, 2013; Li, Chen, & Ellington, 2012). In one such amplification reaction, NASBA (Compton, 1991), RNA templates are reverse transcribed into double-stranded DNA using forward and reverse primers, one of which carries the T7 RNA polymerase promoter sequence. The resulting double-stranded DNA is then transcribed by T7 RNA polymerase to generate additional RNA amplicons, which can in turn continuously be converted into additional double-stranded DNA templates. The application of Spinach.ST for real-time, sequence-specific signal transduction of NASBA amplicons is especially appealing, because the Spinach. ST molecular beacons can be enzymatically synthesized *in situ* by simply including their double-stranded transcription templates in one-pot NASBA reactions.

5.1 Including Spinach.ST trigger components in NASBA primers

Choose a relatively structure-free region of the target for amplification and design a Spinach.ST molecular beacon with domains 2 and 5 derived from a continuous stretch of target sequence. We routinely model the RNA amplicon structure at the amplification reaction temperature using NUPACK. The amplicon should not present major structural challenges to primer binding, polymerase progression, primer-driven amplicon structural organization (see below), or toehold accessibility. Conformations and probabilities of complex formation between NASBA amplicons, primers, and Spinach.ST should be verified using NUPACK.

Guidelines commonly used for designing PCR primers should also be used to design the target-binding domains (PBS-F and PBS-R*, respectively) of the NASBA forward (NF) and reverse (NR) primers (Fig. 7).

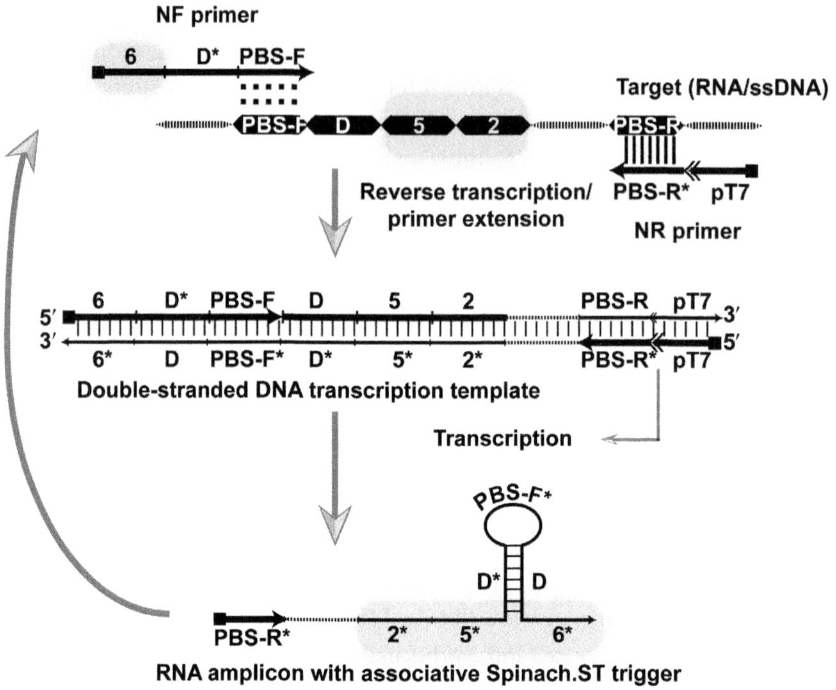

Figure 7 Schematic of trigger-generating primers for real-time signal transduction with NASBA. Regions on the amplicon that are involved in primer binding (PBS-F and PBS-R), associative toehold formation (domain D), and partial Spinach.ST trigger generation (domains 2 and 5) are depicted as dark boxes. Hybridization is depicted using vertical slashes, while the potential hybridization of the PBS-F domain of the NF primer to the complementary strand is indicated with colons. Dashed regions in the target, double-stranded DNA transcription template and RNA transcript denote target sequences that are not involved in priming or in Spinach.ST activation.

For clarity, we provide example sequences of a model NASBA template, primers, transcription templates, and the resulting RNA amplicons in Fig. 8. The NR primer-binding site (PBS-R*) should be downstream of amplicon domain 2. The NF binding site (PBS-F) should reside at a position at least 20-nucleotide upstream of amplicon domain 5. Extend the NR primer by adding the 17-mer T7 promoter sequence "TAATACGACTCACTATA" to its 5′-end. If the adjoining PBS-R* domain does not begin with two guanine nucleotides, then these should be appended immediately following the T7 promoter sequence; for example, the PBS-R domain of the model NASBA template sequence depicted in Fig. 8 does not end with two cytosine (CC) groups at its 3′-end. Therefore, to allow more efficient transcription, two guanine residues were inserted

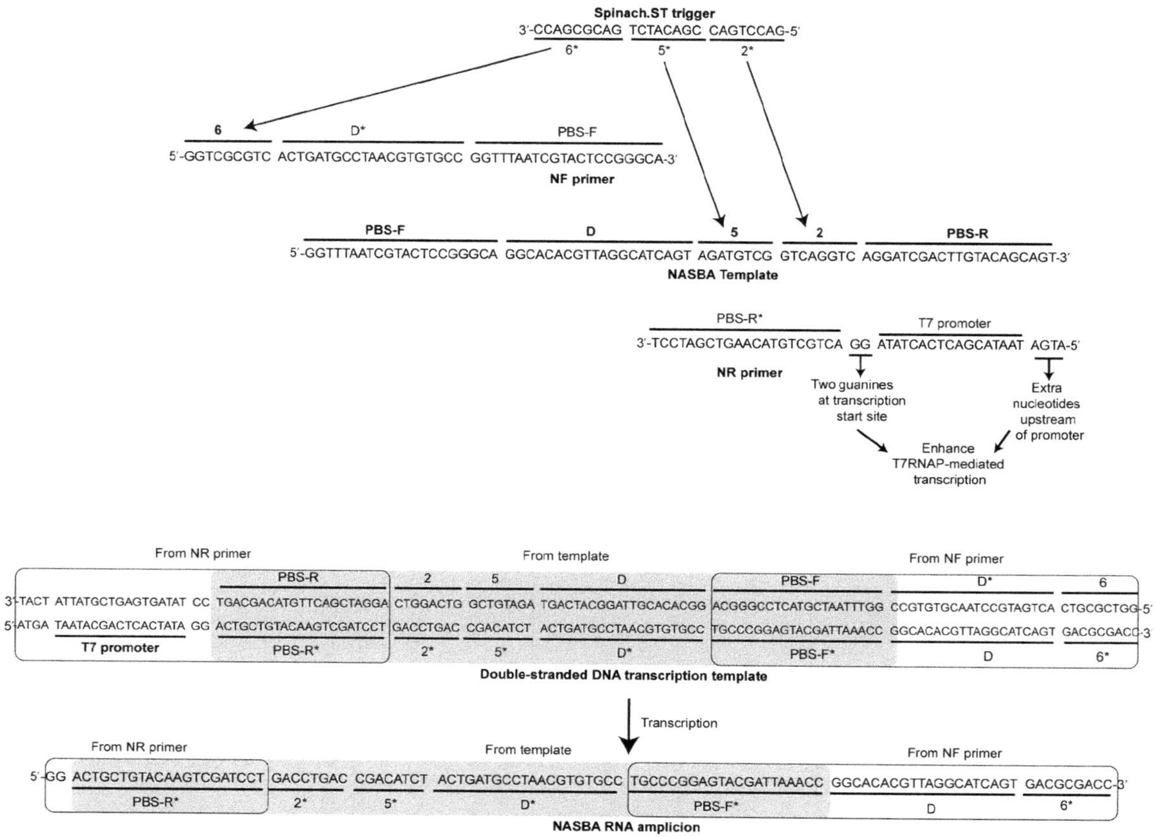

Figure 8 See legend on next page.

in the NR primer immediately prior to the complementary PBS-R* domain. We also typically include a sequence of four additional nucleotides (such as "ATGA") upstream of the promoter to enhance the transcription efficiency (Fig. 8).

To allow hybridization of Spinach.ST to the NASBA amplicon the NF primer should be extended according to the following guidelines. Although Spinach.ST can be designed to accept many different sequences via trigger domains 2* and 5*, domain 6* should remain invariant as it is a part of the aptamer basal stem (Fig. 1A). Only domains 2* and 5* of the trigger should originate from target sequences (Fig. 7). The nine-nucleotide invariant complementary domain 6 should be appended to the 5′-end of the NF primer such that all resulting RNA products will contain the requisite trigger domain 6*. To facilitate understanding the design process, an example of template sequence and the corresponding NASBA primers that facilitate Spinach.ST-based detection is depicted in Fig. 8.

As few as eight intervening, nonpaired nucleotides in trigger sequences render them incapable of activating Spinach.ST (Bhadra & Ellington, 2014b). Therefore, the gap between the RNA amplicon trigger domains 6* and 5* (composed of PBS-F* and the adjoining 20 nucleotides) must

Figure 8 Examples of nucleic acid sequence-based amplification (NASBA) primer and template design. Sequences complementary to the desired trigger domains 2* and 5* are designated in the templates. Sequences complementary to the trigger domain 6* are appended at the 5′-end of the NF primer. The resulting NASBA RNA amplicons contain the trigger domains 2* and 5* (transcribed from the template) and domain 6* (transcribed from the NF primer). The domain D* is complementary to a stretch of template sequence, termed D, that is inserted in the NF primer. Hybridization of D and D* juxtaposes the otherwise distributed trigger domains via hairpin formation (see Fig. 7). Sequences for domains 2*, 5*, and 6* are derived from the trigger sequence for Spinach.ST2 molecular beacon with the sequence GGA AGC GGT GGC TCA ATG GTA GAG CTT TCG ACG ACA TCT GAC GCG ACC GAA ATG GTG AAG GAC GGG TCC AGT GCT TCG GCA CTG TTG AGT AGA GTG TGA GCT CCG TAA CTG GTC GCG TCG GTC GCG TCA GAT GTC GGT CAG GTC TCG AAG GGT TGC AGG TTC AAT TCC TGT CCG TTT C (Bhadra & Ellington, 2014b). For the synthetic target primer-binding sites (PBS-F and PBS-R) with approximately 50% GC content were used; this was also the case for the hybridization domains (D and its complement D*). The strong 17-mer T7 RNA polymerase promoter sequence (including the two guanine nucleotides at the transcription initiation site) was obtained from literature (Dunn & Studier, 1983; Martin & Coleman, 1987; Milligan et al., 1987). An additional four nucleotide sequence was included at the 5′-end of the promoter based on reports that suggested that T7 polymerase contacts DNA through the −21 position (Baklanov, Golikova, & Malygin, 1996; Ikeda & Richardson, 1986).

be bridged by what we term "associative toehold activation" (see Section 2.5.3). The associative toeholds will be juxtaposed by hairpin formation during transcription (Fig. 7). In the example NASBA system, we show in Fig. 8, hairpin formation is achieved by inserting domain D* in the NF primer between domains 6 and PBS-F. Domain D* has been designed to be complementary to a template sequence that lies immediately upstream of template domain 5 (Fig. 8). The resultant NASBA-generated RNA amplicons should therefore have the following domain organization: 5′-PBS-R*—2*—5*—D*—PBS-F*—D—6*-3′ (Figs. 7 and 8). Hybridization during transcription between the complementary domains D* and D will lead to sequestration of intervening sequences into a stem–loop structure and will also bring domains 5* and 6* into apposition, thereby generating the associative toehold that should lead to the Spinach.ST activation. In our designs, we typically use a 20-nucleotide-long domain D* for intramolecular base pairing with a 20-nucleotide-long domain D leading to formation of a double-stranded stem whose loop is composed of a 21-nucleotide-long PBS-F domain. Since the PBS-F domain acts as the primer, its length should at least be 18 nucleotides and have a predicted melting temperature that is greater than the isothermal amplification reaction temperature to maintain target specificity. The lengths of domains D and D* may also be varied as long as the probability of their intramolecular hybridization remains close to 100%. Unnecessarily long domain lengths should be avoided as these might frustrate the folding pathways.

Strand displacement across the three-way RNA junction formed by a transcribed NASBA amplicon not only activates the Spinach.ST reporter but can also allow SNPs located in the middle of the toehold domain to be readily distinguished in real-time. Similar to the substantial disruption in Spinach.ST activation that is observed with standalone trigger oligonucleotides harboring mismatches within the toehold (see Section 2.5.1 and Fig. 6B), we have achieved significant discrimination between wild-type and SNP-containing amplicons by placing the mismatched nucleotide at position 4 of amplicon toehold domain 2* (Bhadra & Ellington, 2014b). Spinach.ST molecules that perfectly matched the wild-type or the SNP amplicons could detect their targets with a seven- to ninefold signal-to-noise ratio. In contrast, the signal-to-noise ratio was less than 2 for mismatched amplicons.

A single Spinach.ST molecular beacon can potentially be used as a universal reporter for a wide variety of amplicons (Fig. 9). In one approach the constant Spinach.ST trigger is split between the two NASBA primers such

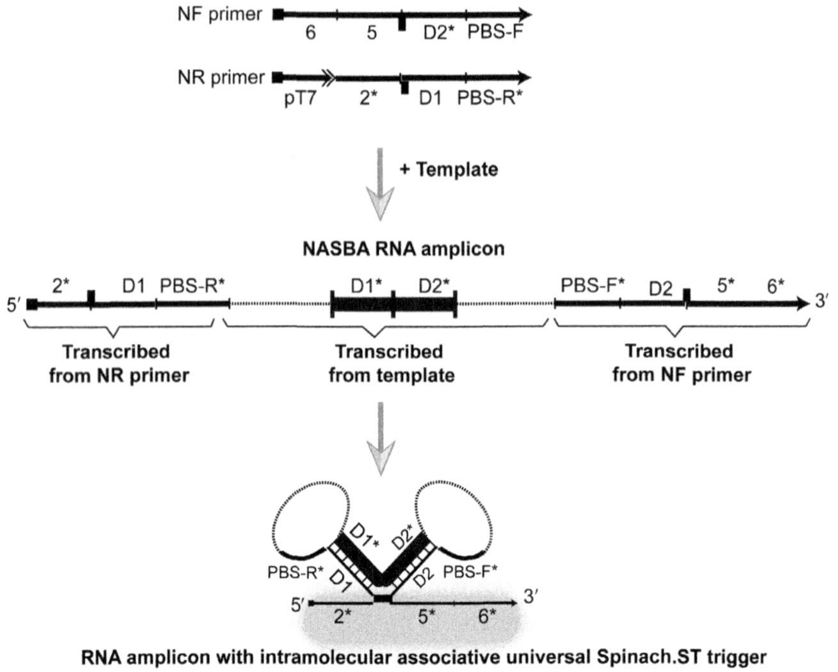

Figure 9 Potential real-time nucleic acid sequence-based amplification (NASBA) amplicon detection using universal Spinach.ST molecular beacons. A scheme for generating NASBA primers that upon transcription creates an intramolecular RNA fold that can activate a universal Spinach.ST trigger is depicted. Transcribed, primer-derived sequences are shown as black lines. Transcribed, target-derived sequences are shown as dashed lines. Regions D1 and D2 that are involved in intramolecular RNA stem–loop formation are highlighted as thick gray blocks. Hybridization is depicted using vertical slashes. The dark bars at the junctions of domains 2* and D1 and domains D2 and 5* designate possible complementary base pairs that might stabilize the associative trigger.

that the complementary branch migration domains 6 and 5 are appended at the 5′-end of the NF primer, while the toehold domain 2* is inserted at the 5′-end of the NR primer, immediately downstream of the T7 RNA polymerase promoter (Fig. 9). Associative toehold activation via template sequence-dependent apposition of the toehold and branch migration domains of the trigger can then be facilitated by engineering the formation of two adjacent intramolecular stem–loop structures within each RNA amplicon. Two stretches of RNA sequence that are flanked by the PBS-F and PBS-R primer-binding domains are designated as hybridization domains D1 and D2. As D1 and D2 are transcribed, they can form duplexes with transcribed D1* and D2* that are inserted into the NF and NR primers, respectively (Fig. 9). The resulting NASBA RNA amplicons will have the domain organization: 5′—2*—D1—PBS-R*—D1*—D2*—PBS-F*—D2—5*—6*—3′, and hybridization of D1:D1* and D2:D2* should lead to formation of two adjoining stem loops that juxtapose the trigger domains 2*, 5*, and 6* that would otherwise be separated on either end of the RNA amplicons. The conformation of the junction between 2* and 5* can potentially be adjusted with single-nucleotide bulges or base pairs to optimize signaling. Of course, the entire trigger sequence could be appended to one primer; however, given the propensity for spurious amplification by isothermal processes, we prefer the target sequence-dependent assembly of the Spinach.ST triggers.

5.2 Detecting NASBA amplification with Spinach.ST

NASBA reactions can be set up starting from either ssDNA or RNA templates that have been freshly diluted into TE (10:0.1) buffer containing 1 μ*M* oligo dT_{17}. We typically set up the NASBA reactions in two steps. First, increasing concentrations of template molecules are mixed with 24 m*M* $MgCl_2$, 20 m*M* KCl, 50 m*M* NaCl, 100 n*M* each of forward and reverse primers, and 500 μ*M* dNTP mix in a total volume of 15 μl. This solution is heated to 65 °C for 5 min followed by incubation on ice for 2 min to allow primer annealing. Subsequently, a 10 μl aliquot containing the remaining reaction components including 100–138 ng of Spinach.ST dsDNA transcription templates, 1× RNAPol reaction buffer (NEB; 40 m*M* Tris–HCl, 6 m*M* $MgCl_2$, 10 m*M* DTT, 2 m*M* spermidine), 6 m*M* of each rNTP (NEB), 100 units of MMLV reverse transcriptase (NEB), 50 units of T7 RNA polymerase, 20 units of RNaseOUT, and 40 μ*M* DFHBI are added (final volume = 25 μl). The reactions are immediately transferred to NUNC™ 384-well flat-bottom black plates and covered with an optical film. The plate is incubated for 6–8 h in a TECAN Safire microplate reader at 37 °C, and real-time Spinach–DFHBI fluorescence measurements are recorded at regular intervals. Allelic discrimination may be improved by incubation at 42 °C instead of 37 °C.

To determine the specificity of NASBA RNA amplicon-mediated activation of Spinach.ST molecular beacons, control reactions that generate nonspecific NASBA amplicons should be also be carried out. The most convenient way to create nonspecific amplicons is to place the T7 promoter at the 5′-end of the NF primer instead of the NR primer. RNA amplicons generated using such a primer pair would harbor domains 6, 5, and 2 instead of the complementary domains 2*, 5*, and 6*. As an additional control, remove 1–2 μl aliquots at the completion of NASBA reactions and analyze these via denaturing polyacrylamide gel electrophoresis. This will ensure that amplification has actually occurred, although all NASBA reactions (containing either specific or nonspecific RNA amplicons) will likely display similar amounts of Spinach.ST transcripts. Visualization by gel electrophoresis is also a very intuitive way of observing how Spinach.ST is useful for discriminating true from artifactual amplicons.

6. CONCLUSIONS

We have described the design principles for engineering the fluorescent RNA aptamer Spinach to be a sequence-dependent molecular beacon that is trapped during transcription into a nonfluorescent conformation and is then switched to an active, fluorescent state upon sequence-specific, toehold-mediated strand displacement. Spinach.ST reporters and their triggers are very amenable to *in silico* engineering since their experimentally determined behavior has proven to be consistent with the predictions of design and cotranscriptional modeling algorithms.

Overall, the potential utility of Spinach.ST as a signal transducer for detecting DNA and RNA sequences *in vitro* is based on its simplicity relative to competing methods. A variety of other fluorescent probes have been described for detection of specific nucleic acid sequences *in vitro*. These include single as well as dual-labeled probes such as the molecular beacons (Tyagi & Kramer, 1996), assimilation probes (Kubota et al., 2013), molecular zippers (Yi, Zhang, & Zhang, 2006), TaqMan® probes (Holland, Abramson, Watson, & Gelfand, 1991), cycling probes (Duck, Alvarado-Urbina, Burdick, & Collier, 1990), adjacent hybridization probes (Didenko, 2001), scorpion primers (Whitcombe, Theaker, Guy, Brown, & Little, 1999), double-stranded hybridization probes (Didenko, 2001), eclipse probes (Afonina, Reed, Lusby, Shishkina, & Belousov, 2002), LUX primers (Kusser, 2006), and the Qzyme probes (Mokany, Todd, Fuery, & Applegate, 2006). Many of these detection methods require prior synthesis, and new probes must be resynthesized for every new target. Cotranscriptional generation of Spinach.ST should be comparatively cost-effective, including in terms of sequence redesign. This is especially true because Spinach.ST can fold and function correctly without requiring purification of transcripts, unlike most synthetic probes that require postsynthetic purification. This capability not only reduces the cost of probe generation but also potentially enables novel applications involving *in situ* probe synthesis.

We have demonstrated several permutations for how different Spinach. ST trigger sequences can enable Spinach.ST molecular beacons to act as reporters in varied applications such as sequence validation of enzymatic isothermal nucleic acid amplification reactions, SNP distinction, and nucleic acid circuit output measurement. Even in the complex *milieu* of NASBA or other isothermal amplification reactions, Spinach.ST beacons fold cotranscriptionally into stable, inactive conformations that are only activated in the presence of cognate trigger RNA sequences. The sensor is able to readily distinguish single-nucleotide mismatches within NASBA-generated sequence targets.

Into the future, since Spinach.ST molecular beacons operate well in physiological buffers, we anticipate that they should be useful tools for intracellular sensing as well. It is conceivable that Spinach.ST might be used as an information processor in a more complex nucleic acid strand displacement circuit. Spinach.ST could both detect a target sequence and also transmit this information to other portions of a circuit via the newly exposed domain 5* (Fig. 2).

ACKNOWLEDGMENTS

This work was supported by the Welch Foundation (F1654), by the National Institutes of Health in conjunction with the Boston University (1U54EB015403), and by the Gates Foundation (OPP1028808).

REFERENCES

Afonin, K. A., Danilov, E. O., Novikova, I. V., & Leontis, N. B. (2008). TokenRNA: A new type of sequence-specific, label-free fluorescent biosensor for folded RNA molecules. *Chembiochem*, *9*(12), 1902–1905. http://dx.doi.org/10.1002/cbic.200800183.

Afonina, I. A., Reed, M. W., Lusby, E., Shishkina, I. G., & Belousov, Y. S. (2002). Minor groove binder-conjugated DNA probes for quantitative DNA detection by hybridization-triggered fluorescence. *Biotechniques*, *32*(4), 940–944, 946–949.

Andersen, E. S., Dong, M., Nielsen, M. M., Jahn, K., Subramani, R., Mamdouh, W., et al. (2009). Self-assembly of a nanoscale DNA box with a controllable lid. *Nature*, *459*(7243), 73–76. http://dx.doi.org/10.1038/Nature07971.

Babendure, J. R., Adams, S. R., & Tsien, R. Y. (2003). Aptamers switch on fluorescence of triphenylmethane dyes. *Journal of the American Chemical Society*, *125*(48), 14716–14717. http://dx.doi.org/10.1021/Ja037994o.

Baklanov, M. M., Golikova, L. N., & Malygin, E. G. (1996). Effect on DNA transcription of nucleotide sequences upstream to T7 promoter. *Nucleic Acids Research*, *24*(18), 3659–3660.

Bhadra, S., & Ellington, A. D. (2014a). Design and application of cotranscriptional non-enzymatic RNA circuits and signal transducers. *Nucleic Acids Research*, *42*(7), e58. http://dx.doi.org/10.1093/nar/gku074.

Bhadra, S., & Ellington, A. D. (2014b). A Spinach molecular beacon triggered by strand displacement. *RNA*, *20*(8), 1183–1194. http://dx.doi.org/10.1261/rna.045047.114.

Chen, X. (2012). Expanding the rule set of DNA circuitry with associative toehold activation. *Journal of the American Chemical Society*, *134*(1), 263–271. http://dx.doi.org/10.1021/ja206690a.

Chen, X., & Ellington, A. D. (2010). Shaping up nucleic acid computation. *Current Opinion in Biotechnology*, *21*(4), 392–400. http://dx.doi.org/10.1016/j.copbio.2010.05.003.

Cho, E. J., Lee, J. W., & Ellington, A. D. (2009). Applications of aptamers as sensors. *Annual Review of Analytical Chemistry*, *2*, 241–264. http://dx.doi.org/10.1146/annurev.anchem.1.031207.112851.

Compton, J. (1991). Nucleic acid sequence-based amplification. *Nature*, *350*(6313), 91–92. http://dx.doi.org/10.1038/350091a0.

Constantin, T. P., Silva, G. L., Robertson, K. L., Hamilton, T. P., Fague, K., Waggoner, A. S., et al. (2008). Synthesis of new fluorogenic cyanine dyes and incorporation into RNA fluoromodules. *Organic Letters*, *10*(8), 1561–1564. http://dx.doi.org/10.1021/Ol702920e.

Court, D. (1993). *RNA processing and degradation by RNase III.* New York: Academic Press.

Didenko, V. V. (2001). DNA probes using fluorescence resonance energy transfer (FRET): Designs and applications. *Biotechniques*, *31*(5), 1106–1116, 1118, 1120–1101.

Dirks, R. M., Bois, J. S., Schaeffer, J. M., Winfree, E., & Pierce, N. A. (2007). Thermodynamic analysis of interacting nucleic acid strands. *SIAM Review*, *49*(1), 65–88. http://dx.doi.org/10.1137/060651100.

Dirks, R. M., & Pierce, N. A. (2003). A partition function algorithm for nucleic acid secondary structure including pseudoknots. *Journal of Computational Chemistry*, *24*(13), 1664–1677. http://dx.doi.org/10.1002/jcc.10296.

Dirks, R. M., & Pierce, N. A. (2004). An algorithm for computing nucleic acid base-pairing probabilities including pseudoknots. *Journal of Computational Chemistry*, *25*(10), 1295–1304. http://dx.doi.org/10.1002/jcc.20057.

Duck, P., Alvarado-Urbina, G., Burdick, B., & Collier, B. (1990). Probe amplifier system based on chimeric cycling oligonucleotides. *Biotechniques*, *9*(2), 142–148.

Dunn, J. J., & Studier, F. W. (1983). Complete nucleotide-sequence of bacteriophage-T7 DNA and the locations of T7 genetic elements. *Journal of Molecular Biology*, *166*(4), 477–535. http://dx.doi.org/10.1016/S0022-2836(83)80282-4.

Ellington, A. D., & Szostak, J. W. (1990). In vitro selection of RNA molecules that bind specific ligands. *Nature*, *346*(6287), 818–822. http://dx.doi.org/10.1038/346818a0.

Furutani, C., Shinomiya, K., Aoyama, Y., Yamada, K., & Sando, S. (2010). Modular blue fluorescent RNA sensors for label-free detection of target molecules. *Molecular BioSystems*, *6*(9), 1569–1571. http://dx.doi.org/10.1039/c001230k.

Grate, D., & Wilson, C. (1999). Laser-mediated, site-specific inactivation of RNA transcripts. *Proceedings of the National Academy of Sciences of the United States of America*, *96*(11), 6131–6136. http://dx.doi.org/10.1073/pnas.96.11.6131.

Han, D. R., Pal, S., Liu, Y., & Yan, H. (2010). Folding and cutting DNA into reconfigurable topological nanostructures. *Nature Nanotechnology*, *5*(10), 712–717. http://dx.doi.org/10.1038/nnano.2010.193.

Heinicke, L. A., Nallagatla, S. R., Hull, C. M., & Bevilacqua, P. C. (2011). RNA helical imperfections regulate activation of the protein kinase PKR: Effects of bulge position, size, and geometry. *RNA*, *17*(5), 957–966. http://dx.doi.org/10.1261/rna.2636911.

Hjalt, T. A., & Wagner, E. G. (1995). Bulged-out nucleotides protect an antisense RNA from RNase III cleavage. *Nucleic Acids Research*, *23*(4), 571–579.

Holeman, L. A., Robinson, S. L., Szostak, J. W., & Wilson, C. (1998). Isolation and characterization of fluorophore-binding RNA aptamers. *Folding and Design*, *3*(6), 423–431. http://dx.doi.org/10.1016/S1359-0278(98)00059-5.

Holland, P. M., Abramson, R. D., Watson, R., & Gelfand, D. H. (1991). Detection of specific polymerase chain reaction product by utilizing the 5′—3′ exonuclease activity of *Thermus aquaticus* DNA polymerase. *Proceedings of the National Academy of Sciences of the United States of America*, *88*(16), 7276–7280.

Iioka, H., Loiselle, D., Haystead, T. A., & Macara, I. G. (2011). Efficient detection of RNA–protein interactions using tethered RNAs. *Nucleic Acids Research*, *39*(8), e53. http://dx.doi.org/10.1093/nar/gkq1316.

Ikeda, R. A., & Richardson, C. C. (1986). Interactions of the RNA polymerase of bacteriophage T7 with its promoter during binding and initiation of transcription. *Proceedings of the National Academy of Sciences of the United States of America*, *83*(11), 3614–3618.

Jiang, Y. S., Li, B., Milligan, J. N., Bhadra, S., & Ellington, A. D. (2013). Real-time detection of isothermal amplification reactions with thermostable catalytic hairpin assembly. *Journal of the American Chemical Society*, *135*(20), 7430–7433. http://dx.doi.org/10.1021/ja4023978.

Kellenberger, C. A., Wilson, S. C., Sales-Lee, J., & Hammond, M. C. (2013). RNA-based fluorescent biosensors for live cell imaging of second messengers cyclic di-GMP and cyclic AMP-GMP. *Journal of the American Chemical Society*, *135*(13), 4906–4909. http://dx.doi.org/10.1021/ja311960g.

Koizumi, M., Soukup, G. A., Kerr, J. N., & Breaker, R. R. (1999). Allosteric selection of ribozymes that respond to the second messengers cGMP and cAMP. *Nature Structural Biology*, *6*(11), 1062–1071. http://dx.doi.org/10.1038/14947.

Kolpashchikov, D. M. (2005). Binary malachite green aptamer for fluorescent detection of nucleic acids. *Journal of the American Chemical Society*, *127*(36), 12442–12443. http://dx.doi.org/10.1021/Ja0529788.

Kramer, F. R., & Mills, D. R. (1981). Secondary structure formation during RNA synthesis. *Nucleic Acids Research*, *9*(19), 5109–5124.

Kubota, R., Labarre, P., Weigl, B., Li, Y., Haydock, P., & Jenkins, D. (2013). Molecular diagnostics in a teacup: Non-instrumented nucleic acid amplification (NINA) for rapid, low cost detection of *Salmonella enterica*. *Chinese Science Bulletin*, *58*(10), 1162–1168. http://dx.doi.org/10.1007/s11434-012-5634-9.

Kusser, W. (2006). Use of self-quenched, fluorogenic LUX primers for gene expression profiling. *Methods in Molecular Biology*, *335*, 115–133. http://dx.doi.org/10.1385/1-59745-069-3:115.

Li, B., Chen, X., & Ellington, A. D. (2012). Adapting enzyme-free DNA circuits to the detection of loop-mediated isothermal amplification reactions. *Analytical Chemistry*, *84*(19), 8371–8377. http://dx.doi.org/10.1021/ac301944v.

Li, B., Ellington, A. D., & Chen, X. (2011). Rational, modular adaptation of enzyme-free DNA circuits to multiple detection methods. *Nucleic Acids Research*, *39*(16), e110. http://dx.doi.org/10.1093/nar/gkr504.

Li, N., & Ho, C. M. (2008). Aptamer-based optical probes with separated molecular recognition and signal transduction modules. *Journal of the American Chemical Society*, *130*(8), 2380–2381. http://dx.doi.org/10.1021/ja076787b.

Liu, J. W., & Lu, Y. (2006). Fast colorimetric sensing of adenosine and cocaine based on a general sensor design involving aptamers and nanoparticles. *Angewandte Chemie-International Edition*, *45*(1), 90–94. http://dx.doi.org/10.1002/anie.200502589.

Lodish, H. F., & Darnell, J. E. (1995). *Molecular cell biology* (3rd ed.). New York: Scientific American Books: Distributed by W.H. Freeman and Co.

Lubin, A. A., & Plaxco, K. W. (2010). Folding-based electrochemical biosensors: The case for responsive nucleic acid architectures. *Accounts of Chemical Research*, *43*(4), 496–505. http://dx.doi.org/10.1021/Ar900165x.

Martin, C. T., & Coleman, J. E. (1987). Kinetic analysis of T7 RNA polymerase–promoter interactions with small synthetic promoters. *Biochemistry*, *26*(10), 2690–2696.

Mathews, D. H., Sabina, J., Zuker, M., & Turner, D. H. (1999). Expanded sequence dependence of thermodynamic parameters improves prediction of RNA secondary structure. *Journal of Molecular Biology*, *288*(5), 911–940. http://dx.doi.org/10.1006/jmbi.1999.2700.

Meyer, I. M., & Miklos, I. (2004). Co-transcriptional folding is encoded within RNA genes. *BMC Molecular Biology*, *5*, 10. http://dx.doi.org/10.1186/1471-2199-5-10.

Milligan, J. F., Groebe, D. R., Witherell, G. W., & Uhlenbeck, O. C. (1987). Oligoribonucleotide synthesis using T7 RNA polymerase and synthetic DNA templates. *Nucleic Acids Research*, *15*(21), 8783–8798.

Mokany, E., Todd, A. V., Fuery, C. J., & Applegate, T. L. (2006). Diagnosis and monitoring of PML-RARalpha-positive acute promyelocytic leukemia by quantitative RT-PCR. *Methods in Molecular Medicine*, *125*, 127–147.

Nutiu, R., & Li, Y. (2003). Structure-switching signaling aptamers. *Journal of the American Chemical Society*, *125*(16), 4771–4778. http://dx.doi.org/10.1021/ja028962o.

Paige, J. S., Nguyen-Duc, T., Song, W., & Jaffrey, S. R. (2012). Fluorescence imaging of cellular metabolites with RNA. *Science*, *335*(6073), 1194. http://dx.doi.org/10.1126/science.1218298.

Paige, J. S., Wu, K. Y., & Jaffrey, S. R. (2011). RNA mimics of green fluorescent protein. *Science*, *333*(6042), 642–646. http://dx.doi.org/10.1126/science.1207339.

Pei, R., Rothman, J., Xie, Y. L., & Stojanovic, M. N. (2009). Light-up properties of complexes between thiazole orange-small molecule conjugates and aptamers. *Nucleic Acids Research*, *37*(8), e59. http://dx.doi.org/10.1093/nar/gkp154.

Ponchon, L., & Dardel, F. (2007). Recombinant RNA technology: The tRNA scaffold. *Nature Methods*, *4*(7), 571–576. http://dx.doi.org/10.1038/nmeth1058.

Pothoulakis, G., Ceroni, F., Reeve, B., & Ellis, T. (2013). The Spinach RNA aptamer as a characterization tool for synthetic biology. *ACS Synthetic Biology*, *3*(3), 182–187. http://dx.doi.org/10.1021/sb400089c.

Qian, L., & Winfree, E. (2011). Scaling up digital circuit computation with DNA strand displacement cascades. *Science*, *332*(6034), 1196–1201. http://dx.doi.org/10.1126/science.1200520.

Qian, L., Winfree, E., & Bruck, J. (2011). Neural network computation with DNA strand displacement cascades. *Nature*, *475*(7356), 368–372. http://dx.doi.org/10.1038/nature10262.

Sando, S., Narita, A., Hayami, M., & Aoyama, Y. (2008). Transcription monitoring using fused RNA with a dye-binding light-up aptamer as a tag: A blue fluorescent RNA. *Chemical Communications*, *September*(33), 3858–3860. http://dx.doi.org/10.1039/B808449a.

Seelig, G., Soloveichik, D., Zhang, D. Y., & Winfree, E. (2006). Enzyme-free nucleic acid logic circuits. *Science*, *314*(5805), 1585–1588. http://dx.doi.org/10.1126/science.1132493.

Serra, M. J., & Turner, D. H. (1995). Predicting thermodynamic properties of RNA. *Methods in Enzymology*, *259*, 242–261.

Song, W., Strack, R. L., & Jaffrey, S. R. (2013). Imaging bacterial protein expression using genetically encoded RNA sensors. *Nature Methods*, *10*(9), 873–875. http://dx.doi.org/10.1038/nmeth.2568.

Sparano, B. A., & Koide, K. (2005). A strategy for the development of small-molecule-based sensors that strongly fluoresce when bound to a specific RNA. *Journal of the American Chemical Society*, *127*(43), 14954–14955. http://dx.doi.org/10.1021/Ja0530319.

Sparano, B. A., & Koide, K. (2007). Fluorescent sensors for specific RNA: A general paradigm using chemistry and combinatorial biology. *Journal of the American Chemical Society*, *129*(15), 4785–4794. http://dx.doi.org/10.1021/Ja070111z.

Stojanovic, M. N., & Kolpashchikov, D. M. (2004). Modular aptameric sensors. *Journal of the American Chemical Society*, *126*(30), 9266–9270. http://dx.doi.org/10.1021/ja032013t.
Strack, R. L., Disney, M. D., & Jaffrey, S. R. (2013). A superfolding Spinach2 reveals the dynamic nature of trinucleotide repeat-containing RNA. *Nature Methods*, *10*(12), 1219–1224. http://dx.doi.org/10.1038/Nmeth.2701.
Strack, R. L., Song, W., & Jaffrey, S. R. (2014). Using Spinach-based sensors for fluorescence imaging of intracellular metabolites and proteins in living bacteria. *Nature Protocols*, *9*(1), 146–155. http://dx.doi.org/10.1038/nprot.2014.001.
Tang, Z. W., Mallikaratchy, P., Yang, R. H., Kim, Y. M., Zhu, Z., Wang, H., et al. (2008). Aptamer switch probe based on intramolecular displacement. *Journal of the American Chemical Society*, *130*(34), 11268–11269. http://dx.doi.org/10.1021/Ja804119s.
Tuerk, C., & Gold, L. (1990). Systematic evolution of ligands by exponential enrichment: RNA ligands to bacteriophage T4 DNA polymerase. *Science*, *249*(4968), 505–510. http://dx.doi.org/10.1126/science.2200121.
Tyagi, S., & Kramer, F. R. (1996). Molecular beacons: Probes that fluoresce upon hybridization. *Nature Biotechnology*, *14*(3), 303–308. http://dx.doi.org/10.1038/Nbt0396-303.
Whitcombe, D., Theaker, J., Guy, S. P., Brown, T., & Little, S. (1999). Detection of PCR products using self-probing amplicons and fluorescence. *Nature Biotechnology*, *17*(8), 804–807. http://dx.doi.org/10.1038/11751.
Xayaphoummine, A., Bucher, T., & Isambert, H. (2005). Kinefold web server for RNA/DNA folding path and structure prediction including pseudoknots and knots. *Nucleic Acids Research*, *33*(Web server issue), W605–W610. http://dx.doi.org/10.1093/nar/gki447.
Yi, J., Zhang, W., & Zhang, D. Y. (2006). Molecular zipper: A fluorescent probe for real-time isothermal DNA amplification. *Nucleic Acids Research*, *34*(11), e81. http://dx.doi.org/10.1093/nar/gkl261.
Yin, P., Choi, H. M., Calvert, C. R., & Pierce, N. A. (2008). Programming biomolecular self-assembly pathways. *Nature*, *451*(7176), 318–322. http://dx.doi.org/10.1038/nature06451.
Yurke, B., & Mills, A., Jr. (2003). Using DNA to power nanostructures. *Genetic Programming and Evolvable Machines*, *4*(2), 111–122. http://dx.doi.org/10.1023/a:1023928811651.
Yurke, B., Turberfield, A. J., Mills, A. P., Simmel, F. C., & Neumann, J. L. (2000). A DNA-fuelled molecular machine made of DNA. *Nature*, *406*(6796), 605–608.
Zadeh, J. N., Steenberg, C. D., Bois, J. S., Wolfe, B. R., Pierce, M. B., Khan, A. R., et al. (2011). NUPACK: Analysis and design of nucleic acid systems. *Journal of Computational Chemistry*, *32*(1), 170–173. http://dx.doi.org/10.1002/jcc.21596.
Zhang, D. Y. (2010). *Dynamic DNA strand displacement circuits*. Pasadena, CA: California Institute of Technology.
Zhang, D. Y., Chen, S. X., & Yin, P. (2012). Optimizing the specificity of nucleic acid hybridization. *Nature Chemistry*, *4*(3), 208–214. http://dx.doi.org/10.1038/Nchem.1246.
Zhang, D. Y., & Seelig, G. (2011). Dynamic DNA nanotechnology using strand-displacement reactions. *Nature Chemistry*, *3*(2), 103–113. http://dx.doi.org/10.1038/nchem.957.
Zhang, D. Y., & Winfree, E. (2009). Control of DNA strand displacement kinetics using toehold exchange. *Journal of the American Chemical Society*, *131*(47), 17303–17314. http://dx.doi.org/10.1021/ja906987s.

CHAPTER TWELVE

Using Riboswitches to Regulate Gene Expression and Define Gene Function in Mycobacteria

Erik R. Van Vlack*, Jessica C. Seeliger[†,1]
*Department of Chemistry, Stony Brook University, Stony Brook, New York, USA
[†]Department of Pharmacological Sciences, Stony Brook University, Stony Brook, New York, USA
[1]Corresponding author: e-mail address: jessica.seeliger@stonybrook.edu

Contents

Abstract

Mycobacteria include both environmental species and many pathogenic species such as *Mycobacterium tuberculosis*, an intracellular pathogen that is the causative agent of tuberculosis in humans. Inducible gene expression is a powerful tool for examining gene function and essentiality, both in *in vitro* culture and in host cell infections. The theophylline-inducible artificial riboswitch has recently emerged as an alternative to protein repressor-based systems. The riboswitch is translationally regulated and is combined with a mycobacterial promoter that provides transcriptional control. We here provide methods used by our laboratory to characterize the riboswitch response to theophylline in reporter strains, recombinant organisms containing riboswitch-regulated endogenous genes, and in host cell infections. These protocols should facilitate the application of both existing and novel artificial riboswitches to the exploration of gene function in mycobacteria.

Methods in Enzymology, Volume 550
ISSN 0076-6879
http://dx.doi.org/10.1016/bs.mie.2014.10.034

1. INTRODUCTION

Experimental control over gene expression is a critical tool in molecular genetics. Inducible gene regulation by small molecules offers time- and dose-dependent control. Inducible systems are commonly used to express proteins for purification and biochemical characterization or to silence genes by expressing antisense mRNA or directly controlling target gene transcription. For protein expression, the dosing control afforded by inducible regulation can help ameliorate the harmful effects of toxic proteins or overexpression. For gene silencing, time-dependent control can be used to test how gene function changes as experimental conditions vary. In pathogens, the essentiality of a gene may evolve over the course of infection depending on the needs and metabolic state of the pathogen. Inducible systems have been used to show, for example, that the proteasome and the gluconeogenic enzyme phosphoenolpyruvate carboxykinase are necessary for survival of the human pathogen *Mycobacterium tuberculosis* during both the acute and chronic stages of infection in mice (Gandotra, Schnappinger, Monteleone, Hillen, & Ehrt, 2007; Marrero, Rhee, Schnappinger, Pethe, & Ehrt, 2010).

In general, inducible gene regulation tools for Gram-positive bacteria are not as numerous or diverse as those available for their Gram-negative brethren (see also the chapter "Conditional control of gene expression by synthetic riboswitches in *Streptomyces coelicolor*" by Rudolph et al.). For *Mycobacteria*, which include numerous pathogenic species such as *M. tuberculosis*, gene regulatory systems have been validated for induction by acetamide, pristinamycin, thiostrepton, nitriles, and tetracycline derivatives (Carroll, Muttucumaru, & Parish, 2005; Ehrt et al., 2005; Forti, Crosta, & Ghisotti, 2009; Hernandez-Abanto, Woolwine, Jain, & Bishai, 2006; Klotzsche, Ehrt, & Schnappinger, 2009; Pandey et al., 2009; Parish, Mahenthiralingam, Draper, Davis, & Colston, 1997; Sassetti, 2008). The Tn10-based Tet repressor and its variants have become some of the most widely used and the only systems that have been validated in an animal model of *M. tuberculosis* infection (Gandotra et al., 2007). All of these systems rely on allosteric control of a protein repressor that must be expressed in tandem with a target gene controlled by the corresponding promoter. Not all genes respond identically to a given promoter/repressor pair, and empirical testing can be necessary to balance repressor and target expression and achieve the desired level of control.

As an alternative to these protein-based systems, we recently reported the application of a theophylline-inducible synthetic riboswitch to gene regulation in mycobacteria such as *M. tuberculosis* (Seeliger et al., 2012; Topp et al., 2010). This system combined transcription driven by the strong constitutive promoter (from the *hsp60* gene of the vaccine strain *Mycobacterium bovis* BCG) (Stover et al., 1991) with translational control by an optimized synthetic theophylline riboswitch (the E* variant) (Topp et al., 2010). All the regulatory elements are encoded within the 5′UTR of the mRNA and no additional factors are required. Using fluorescence, enzyme activity, immunoblot, and phenotypic assays, we demonstrated similar responses to millimolar theophylline for two reporter genes and one endogenous mycobacterial gene (Seeliger et al., 2012). In this last application, we confirmed that the catalase-peroxidase KatG confers sensitivity to the frontline tuberculosis prodrug isoniazid. In the absence of theophylline, KatG was repressed and the half-maximum effective concentration (EC_{50}) for isoniazid was 10-fold higher. Theophylline induction restored both protein expression and antibiotic sensitivity in a dose-dependent manner (Fig. 5). We have made similar observations of theophylline-riboswitch function for the overexpression of several mycobacterial proteins in both *Mycobacterium smegmatis* and *M. tuberculosis*, with dose-dependent responses and no expression detected by immunoblot in the absence of theophylline (unpublished results). Induction is reversible upon the removal of theophylline. The riboswitch also functions to control GFP expression in *M. tuberculosis* in an infected macrophage-like cell line. An activation ratio of ~70 was observed *in vitro*, affording a moderate dynamic range over 0–2 m*M* theophylline.

Here, we provide protocols for assaying riboswitch function and implementing riboswitch-controlled gene expression in mycobacteria. We also report results from a panel of mycobacterial promoters paired with the theophylline riboswitch. Guidelines and representative data follow below for (1) cloning promoters into a theophylline-riboswitch reporter system, (2) assaying promoter–riboswitch reporter constructs by fluorescence or enzyme activity assays, (3) engineering strains by single-crossover homologous recombination for riboswitch-regulated control of endogenous genes, and (4) inducing riboswitch-controlled bacterial genes in infected host cells.

2. RIBOSWITCH REPORTER ASSAYS

Because the theophylline riboswitch controls translation, it must be combined with a promoter that drives transcription. The choice of promoter can alter the response of the riboswitch by influencing the number of mRNA transcripts and introducing additional regulatory elements. Based on our goals of achieving generality and high dynamic range in mycobacteria, we paired the riboswitch with the strong constitutive promoter P*hsp60*, but users may want to use different promoters depending on their requirements.

Based on our experiences, we recommend that new promoter–riboswitch combinations be tested in one or more reporter gene systems to confirm and characterize the dose-dependent response to theophylline. Thus far, in addition to P*hsp60*, we have paired four constitutive mycobacterial promoters, both naturally occurring (smyc, imyc) (Ehrt et al., 2005) and synthetic (MOP, A37) (Agarwal & Tyagi, 2006; Hickey et al., 1996), with the E* riboswitch variant (Topp et al., 2010). We found that the strength of the promoter does not correlate with the behavior of the corresponding promoter–riboswitch combination (Fig. 1). When driven by the original promoters, GFP expression varied over a half an order of magnitude. When the promoters were combined with the riboswitch, however, the theophylline-dependent GFP response was nearly identical for all constructs, with the exception of P*smyc*-ribo. This construct showed higher maximum expression and higher sensitivity to theophylline, but also a small but reproducible signal above background that indicates incomplete repression in the absence of theophylline.

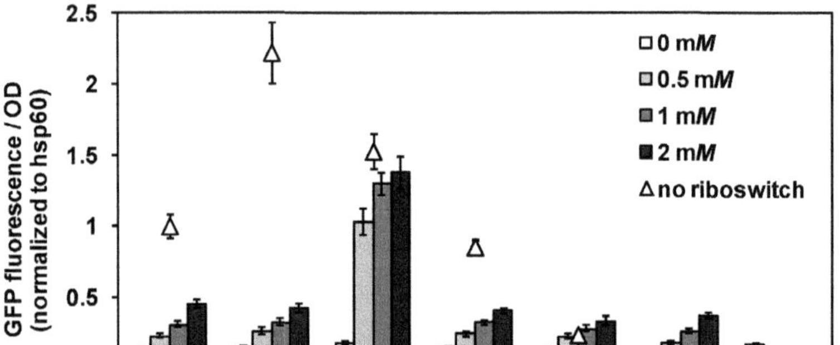

Figure 1 The P*smyc* promoter in combination with the riboswitch has the greatest dynamic range and sensitivity to theophylline. *M. smegmatis* transformed with promoter-gfp (triangles) and promoter-*ribo*-gfp (bars) reporter constructs were assayed for GFP fluorescence in response to theophylline over 6–8 h. For comparison all data were normalized to the GFP fluorescence from P*hsp60*-gfp.

Factors that could influence theophylline-induced behavior when compared to the native promoter include differences that affect translational strength (such as the Shine-Dalgarno (SD) sequence and the distance between the SD and the start codon) or riboswitch function (such the sequence of the 5′UTR upstream of the riboswitch). However, we did not find any apparent correlation between these factors and the behavior of the promoter–riboswitch constructs (Table 1). P*smyc*-ribo is intriguing for its greater dynamic range and as an alternative to P*hsp60*-ribo for applications in which stringent repression is not required. Efforts are also underway to generate variants with improved repression.

Of the two most commonly used reporters, β-galactosidase and GFP, we favor GFP as the primary reporter system. The β-galactosidase assay requires lysing cells and adding a chromogenic substrate. In contrast, GFP fluorescence can be monitored continuously in whole cells and is therefore conducive to screening in multi-well plate format, measuring induction kinetics, and analyzing by flow cytometry. The ability to monitor whole cells is especially advantageous when working with pathogenic strains, as it minimizes handling and exposure. For β-galactosidase a membrane-permeable fluorogenic substrate 5-acetylaminofluorescein di-β-D-galactopyranoside (C2FDG) has been used in whole *M. tuberculosis* cells (Ehrt et al., 2005). At the time of writing, only the more lipophilic alternative (C12FDG) is commercially available and is still significantly more costly than the chromogenic substrate, 2-nitrophenyl β-D-galactopyranoside. Independent of the method of detection, β-galactosidase is an important secondary reporter gene for confirming that the theophylline-induced response is not strongly dependent on the downstream coding sequence and for testing the stringency of repression, since the accumulated signal from substrate turnover makes enzyme activity the more sensitive assay.

The plasmids pST5552 (ribo-gfp) and pST5832 (ribo-lacZ) are reporter constructs for P*hsp60*-ribo (Seeliger et al., 2012). When testing novel promoter–riboswitch combinations, these plasmids can be used for comparison and as positive controls for theophylline-induced expression (Fig. 2). Below we provide guidance for creating promoter–riboswitch reporter plasmids and detailed protocols for the GFP and β-galactosidase assays. The responses of P*hsp60*-ribo in nonpathogenic *M. smegmatis* and pathogenic *M. tuberculosis* are similar (Seeliger et al., 2012). Therefore, unless the target promoter is known to be species-specific, we recommend using *M. smegmatis* as the faster, easier, and safer choice for reporter assays. Our protocols assume the use of *M. smegmatis* unless otherwise noted. Additional references should

Table 1 Comparison of promoter and promoter–riboswitch properties

	hsp60	**gs**	**smyc**	**imyc**	**MOP**	**A37**	**ribo**
Promoter strength	1	2.5	2	1	0.1		
Maximum response of promoter-ribo	1	1	4	1	0.6	1	
Shine-Dalgarno (deduced)	GGAGGAA	AAAGGAG	AAGGAGA	AAGGAG	AGGAG	AAGGAGG	AAGGAGG
# of bases between SD and ATG	10	6	6	7	9	7	7
# of bases between transcriptional start and SD (promoter)	166	32	91	–[a]	37	22	
# of bases between transcriptional start and riboswitch (promoter-ribo)	16	33	77	–[a]	59	22	
$\Delta\Delta G$ (kcal/mol)	−7.9	−13.8	−30.0	–	−22.2	−10.9	
Constant sequence	Y	Y	N	Y	Y	Y	

[a]The sequence of the 5′UTR is unknown because the transcriptional start site of P*imyc* has not been mapped.

For comparison, the approximate promoter strengths and promoter-ribo responses (as measured by GFP fluorescence) are normalized to P*hsp60* and P*hsp60*-ribo, respectively (summarized from Fig. 1). For the promoter-ribo constructs, the UNAFold server (https://www.idtdna.com/UNAFold) was used to calculate the predicted RNA secondary structures and calculated free energies for sequences from the transcriptional start through the start codon. The calculated free energy of the riboswitch alone was then subtracted to obtain $\Delta\Delta G$ values as a metric for the secondary structure introduced by the promoter-dependent 5′UTR.

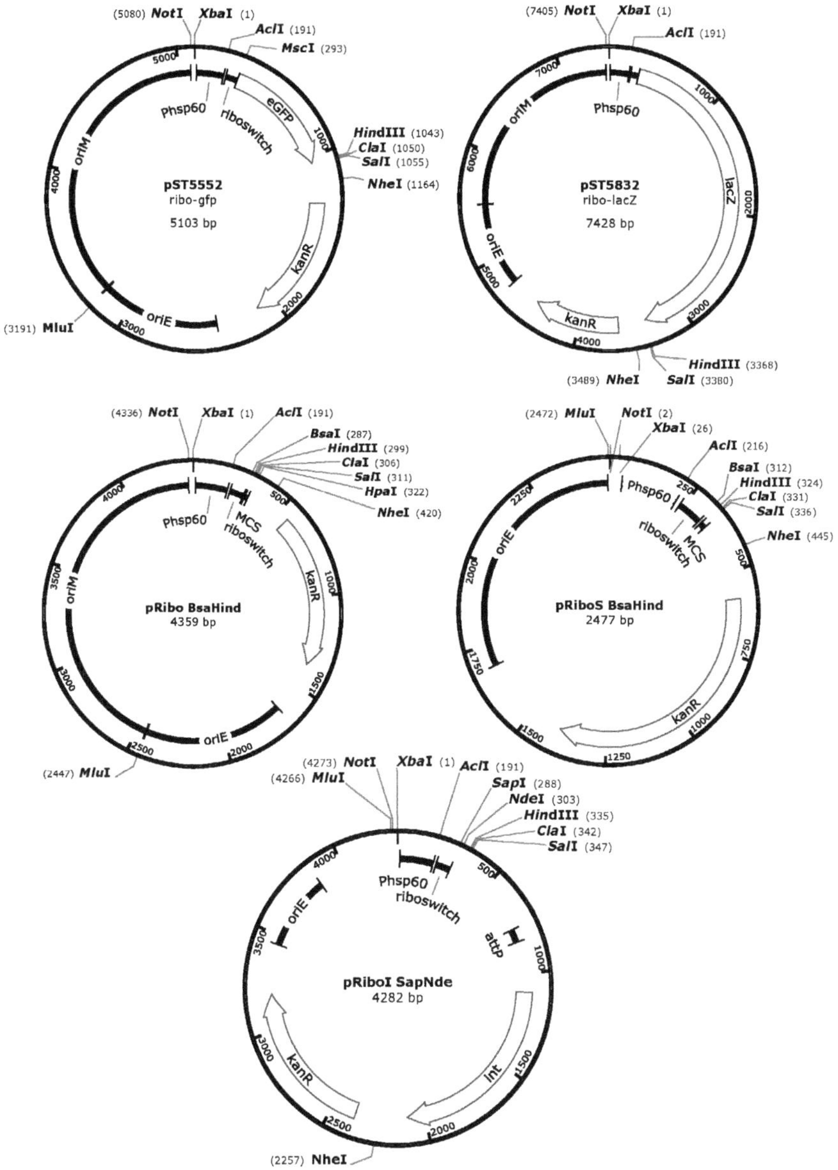

Figure 2 Maps for reporter and cloning plasmids containing the P*hsp60*-ribo regulatory element. The restriction sites most relevant to cloning are indicated. Plasmids and partial sequences are available from the AddGene depository. Full maps and sequences are available from the authors upon request.

be consulted for information on the handling and manipulation of *M. smegmatis* and *M. tuberculosis* (Larsen, Biermann, & Jacobs, 2007; Larsen, Biermann, Tandberg, Hsu, & Jacobs, 2007; Singh & Reyrat, 2009).

2.1. Construction of promoter–riboswitch GFP or β-galactosidase reporter plasmids

Because pST5552 and pST5832 are both based on pMV261, an episomal low-copy number shuttle vector, cells containing these constructs or derivatives thereof must be grown in the presence of kanamycin (25 μg/mL in mycobacteria; 50 μg/mL in *Escherichia coli*) to maintain the plasmid (Stover et al., 1991).

To insert a promoter 5′ of the E* riboswitch, digest the parent vector with *Xba*I and *Acl*I to excise the P*hsp60* promoter (~180 bp). Isolate the vector backbone by agarose gel purification and ligate with the appropriate PCR product containing the desired promoter sequence. This method preserves 14 bases of 5′UTR from the *hsp60* gene and 15 bases of a constant sequence that is recommended for poorly characterized promoters (Topp et al., 2010). If removal of the entire 5′UTR and constant sequence is desired, a single digest with *Kpn*I will remove a segment from P*hsp60* up to the aptamer (Fig. 3).

General cloning vectors are available to express other proteins as reporters for riboswitch-inducible expression. These include the episomal vector pRibo BsaHind (Seeliger et al., 2012) and the integrating vector

```
    KpnI                XbaI  Phsp60
  1 GGTACCAGATCTTTAAATCTAGAGGTGACCACAACGCGCCCGCTTTGATC 50

 51 GGGGACGTCTGCGGCCGACCATTTACGGGTCTTGTTGTCGTTGGCGGTCA 100

101 TGGGCCGAACATACTCACCCGGATCGGAGGGCCGAGGACAAGGTCGAACG 150

151 AGGGGCATGACCCGGTGCGGGGCTTCTTGCACTCGGCATAGGCGAGTGCT 200

           AclI  +1                 constant seq.  KpnI
201 AAGAATAACGTTGGCACTCGCGACCGGGATACGACTCACTATAGGTACCG 250

    aptamer                                      RBS
251 GTGATACCAGCATCGTCTTGATGCCCTTGGCAGCACCCTGCTAAGGAGGC 300

           start
301 AACAAGATG
```

Figure 3 Sequence of P*hsp60*-ribo regulatory element. The 5′UTR of P*hsp60* is truncated as described (Topp et al., 2010). Restriction sites *Kpn*I or *Xba*I/*Acl*I can be used to replace P*hsp60* with a different promoter.

pRiboI SapNde (unpublished work; Fig. 2). The Type II restriction endonucleases *Bsa*I and *Sap*I cut outside their recognition sites and allow seamless cloning of target genes. To clone using traditional Type II endonucleases, use pRibo EcoHind. All plasmids described above are available from the AddGene depository.

2.2. GFP fluorescence endpoint assay

For greatest accuracy, grow the bacteria in culture tubes followed by aliquoting into a 96-well plate to measure GFP fluorescence and cell density (OD_{600}). The protocol below is appropriate for all mycobacteria, except for any modifications to the culturing conditions necessary to account for differences in growth rate and pathogenicity. The doubling time for *M. smegmatis* in standard 7H9 medium is approximately 2.5–3 h. The time to maximum induction of GFP in fully aerated culture is approximately two doubling times after additional of theophylline.

Due to its low solubility in aqueous solution, theophylline should be dissolved directly in growth medium such as 7H9 and sterilized by 0.2 μm filtration. A 20 m*M* stock solution may be obtained with sonication and gentle warming at 30–50 °C. At this concentration, the stock is stable at room temperature; avoid storing at 4 °C, as the theophylline tends to crystallize out of solution over time.

1. Inoculate starter cultures of 3 mL 7H9 medium [4.7 g dehydrated 7H9 (BD and Company), 0.5% glycerol, 10% ADC supplement (BD and Company), and 0.05% Tween 80 per liter] per 14-mL culture tube so that cells will reach late log phase (OD_{600} of 0.8–1) by the time of subculturing and induction. Inoculate strains containing the appropriate positive control (promoter without riboswitch), negative control (cells carrying empty vector or vector with *gfp*, but no promoter), and riboswitch (promoter-ribo-gfp) vectors. Incubate with vigorous shaking (250 rpm) at 37 °C.
2. The next day (after ~12–16 h), measure the OD_{600}. Aliquot the appropriate volume of overnight culture to obtain OD_{600} of 0.2 in 2 mL per sample. Pellet cells for 5 min at 12,000 × *g* and aspirate the supernatant. Resuspend cells in fresh media containing 0, 0.5, 1, or 2 m*M* theophylline and incubate for 6 h at 37 °C to induce target gene expression.
3. Aliquot 200 μL of each culture into a black, clear-bottom 96-well plate. Also aliquot 200 μL of growth medium as a blank. The excitation and emission should be set by either monochromator or the choice of filters near 488 and 508 nm, respectively. For example, we have used

Molecular Devices SpectraMax Gemini XPS (450 nm excitation, 510 nm emission with 495 nm cutoff) and F5 (485/10 nm excitation filter, 535/13 nm emission filter) plate readers. Measure the OD_{600} for each sample and subtract the buffer blank. Calculate the normalized fluorescence F_{norm} for each sample by dividing by the OD_{600} and subtracting the negative control. Calculate the activation ratio as F_{norm}(2 m*M*)/ F_{norm}(0 m*M*). While the absolute fluorescence values will depend on the individual plate reader, relative trends should be comparable and reproducible.

2.2.1 Note on culturing bacteria in multi-well plates

M. smegmatis cultured in a 96-well plate grows ~40% more slowly (with a doubling time of ~3.5 vs. ~2.5 h) and the induction kinetics are correspondingly slower. Nevertheless, we have found that the relative behavior of riboswitches can be accurately compared in a multi-well plate, and this format is preferable when appropriate, as it greatly simplifies the procedure and reduces the volume of growth medium required.

If using a 96-well plate, inoculate strains at OD 0.2 in 200 μL 7H9 (with 0–2 m*M* theophylline) per well in a black, clear-bottom plate. Cover with a breathable membrane (e.g., Nunc 241205 sealing tape). Shake during incubation and do not cover with a lid, as this will introduce aeration-dependent artifacts in growth and GFP expression. Without humidification, approximately 10–15% of the total volume will evaporate over 6–8 h, so this format is recommended only for fast-growing nonpathogenic mycobacteria.

2.3. GFP flow cytometry

As shown by flow cytometry of *M. smegmatis*, the theophylline riboswitch is titratable in response to inducer (Fig. 4). Due to the tendency of mycobacteria to aggregate, especially in the absence of detergent, cell clumps must be dispersed to obtain single-cell suspensions prior to flow cytometry analysis. A single-setting, low-power sonicator was used for this protocol. The use of sonicators with variable power settings may require optimization to obtain the optimal balance between dispersing and lysing cells. In contrast to the nonpathogeneic *M. smegmatis*, flow cytometry of pathogenic strains should be performed only when an instrument dedicated to an appropriate Biosafety Level 3 facility is available, due to the generation of aerosols.

1. Grow overnight starter cultures and induce cells with theophylline as described in Section 2.2, Steps 1 and 2. To afford sufficient sample for analysis, a culture volume of 3 mL per sample is recommended for

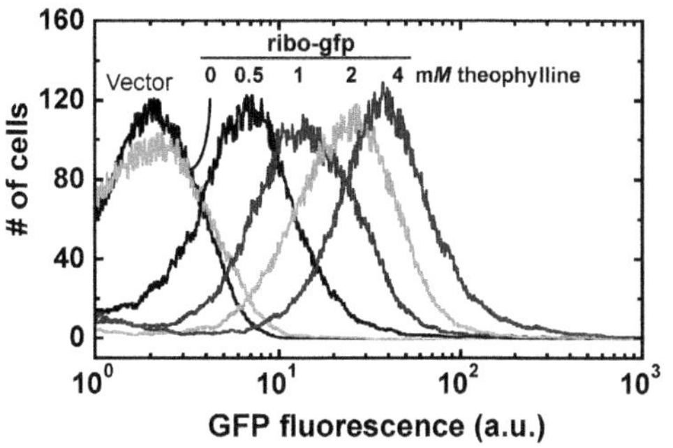

Figure 4 Flow cytometry of theophylline-dependent GFP expression shows that the E* riboswitch is titratable. *Adapted from Seeliger et al. (2012).* (See the color plate.)

theophylline induction. After the desired period of induction, pellet 1–3 × 10^8 cells (based on OD_{600} of 1 = 3 × 10^8 cells/mL) for each strain by centrifugation (5 min at 12,000 × *g*) and wash twice with 1 mL of PBS. Resuspend the final pellet in 1 mL 10% formalin.

2. After sonication in an ice water bath for 2 min, pellet cell clumps by centrifugation for 10 min at 200 × *g*. Reserve 900 μL supernatant for flow cytometry analysis.
3. Use the negative control sample (empty vector or no-GFP control strain) to set a gate based on forward and side scatter channels and select against cell debris and any remaining cell clumps. Record flow cytometry data for at least 1 × 10^4 cells per sample. Calculate the activation ratio as above using the mean fluorescence intensity (MFI) for each sample and first subtracting the negative control. Activation ratio = MFI(2 m*M*)/MFI (0 m*M*).

2.4. β-Galactosidase activity endpoint assay

For 1 mL of cells at OD_{600} of ~1, the cell debris after lysis does not scatter significantly at 420 nm, so in this protocol, lysates are not cleared after stopping the reaction. However, if lysates are turbid, debris should be removed by centrifugation and the supernatant transferred to a fresh plate for OD_{420} measurement.

1. Grow overnight starter culture and induce cells with theophylline as described above in Section 2.2. Measure the OD_{600} for each sample.
2. Pellet 1 mL (or ~1 × 10^8 cells) of cells for 5 min at 12,000 × *g* and aspirate supernatant. Resuspend cells in 1 mL Z buffer (60 m*M* Na_2HPO_4, 40 m*M* NaH_2PO_4, 10 m*M* KCl, 1 m*M* $MgSO_4$, 50 m*M* β-mercaptoethanol, pH 7.0). Chill cells on ice.

3. Lyse cell suspension with a tip sonicator with two pulses of 20 s with 20 s rest in between. Aliquot 200 μL of uncleared lysate per sample into a clear 96-well plate. Aliquot Z buffer as a negative control. Pre-warm samples at 30 ° C for 5 min.
4. At $t = 0$ min, add 50 μL 2-nitrophenyl β-D-galactopyranoside (4 mg/mL in Z buffer) to each well. Incubate at 30 ° C until yellow color is visible (~10 min). Record the time and stop the reaction with 125 μL 1 *M* sodium bicarbonate.
5. Measure the absorbance at 420 nm. Subtract the Z buffer control to account for nonspecific hydrolysis and calculate substrate turnover in Miller units: $(OD_{420} \times 1000)/(OD_{600} \times \text{reaction time in min})$. Calculate the activation ratio as above in the GFP endpoint assay.

3. CONSTRUCTION OF RECOMBINANT STRAINS WITH RIBOSWITCH-REGULATED GENES

The P*hsp60*-ribo element can be used to control the expression of endogenous genes if inserted directly upstream of the target gene in the chromosome. Below we provide a protocol for generating such a recombinant strain by a single-crossover event, as we have done previously for *katG* in *M. smegmatis* (Seeliger et al., 2012). This strategy relies on homologous recombination of a suicide plasmid (pRiboS; Fig. 2) to generate one functional copy of the gene controlled by P*hsp60*-ribo and one nonfunctional

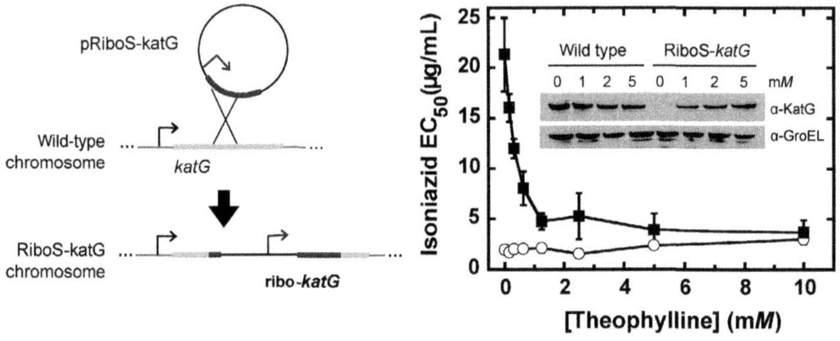

Figure 5 An endogenous riboswitch-controlled gene can be obtained by single crossover. Left: The suicide plasmid pRiboS-katG recombines homologously with the *M. smegmatis* genome to generate a strain with one truncated copy and one full-length riboswitch-controlled copy of *katG*. Right: Theophylline-dependent expression of KatG was confirmed by immunoblot (inset) and phenotypic resistance to isoniazid. *Adapted from Seeliger et al. (2012).*

copy controlled by the native promoter (Fig. 5). The major advantage of this method is its relative ease; the disadvantages are the low frequency of legitimate homologous recombination in mycobacteria and the retention of a residual, truncated gene copy.

Other methods that are possible, but have not yet been tested, are to introduce an antibiotic resistance marker and P*hsp60*-ribo upstream of the target gene by homologous recombination using (1) specialized phage transduction (Bardarov et al., 2002; Larsen, Biermann, Tandberg, et al., 2007), (2) recombineering (Van Kessel & Hatfull, 2007), or (3) selection and counter-selection to obtain double-crossover strains (Parish & Stoker, 2000).

1. Digest pRiboS BsaHind with *Bsa*I and ligate with appropriate digested PCR product encoding an N-terminal fragment of the target gene with an in-frame stop codon.
2. After sequence confirmation, aliquot 1–2 μg of plasmid DNA and treat with 100 mJ/cm^2 UV light to promote recombination in the next step (Hinds et al., 1999).
3. Electroporate DNA into electrocompetent cells (Goude & Parish, 2009). Spread on selective medium and incubate at 37 °C until colonies are visible (~2–3 days).
4. Select single clones and confirm recombination by PCR and sequencing.

4. INDUCTION OF MYCOBACTERIAL GENES IN INFECTED HOST CELLS

Mycobacteria residing in infected macrophage cells respond to theophylline in the culture medium (Seeliger et al., 2012). The suggested concentration and period of induction are detailed in the protocol below, although these parameters may need to be optimized depending on host cell type used and experimental needs. Theophylline is moderately toxic to mammalian cells at millimolar concentrations. We therefore recommend treating the selected host cells with theophylline and checking for cell death before performing the infection/induction experiment. The protocol below uses incubation times appropriate for infection with *M. tuberculosis*:

1. Seed mammalian cells (e.g., primary bone marrow-derived macrophages, or cell lines such as RAW264.7, J774.1, or PMA-differentiated THP-1) on 22 × 22 mm sterile glass coverslips in 6-well plates at 3 × 10^5 cells per well and culture for 1 day. Synchronize the growth of wild type

and riboswitch-containing mycobacteria to reach late log phase by the time the seeded cells have recovered.

2. Pellet a 10-mL aliquot of bacteria and spin at 500 × *g* for 5 min to remove cell clumps. Transfer the supernatant by pipette to a fresh tube and spin at 3500 × *g* for 5 min. Wash the cell pellet twice with an equal volume of PBS.
3. Measure OD_{600} of final cell suspension. Serial dilute in culture medium with 10% horse serum (to promote infectivity) to obtain the concentration necessary for a multiplicity of infection (MOI) of 5 bacteria per host cell at 1 mL/well.
4. Aspirate medium off mammalian cell monolayers and cover with bacterial cell suspension to initiate infection. Incubate at 37 °C for 4 h.
5. Aspirate medium and wash cell monolayers twice with an equal volume of PBS. Add 1 mL culture medium without theophylline and allow infected macrophages to recover for 24 h at 37 °C.
6. Aspirate medium and replace with fresh medium containing 0 or 0.5 m*M* theophylline. Return to incubator.
7. After an additional 24 h, wash monolayers with PBS and fix in phosphate-buffered 10% formalin for 1 h. Mount coverslips on glass slides with antifade reagent and stain nuclei with DAPI (e.g., Vectorlabs Vectashield Mounting Medium with DAPI). Image cells by phase contrast and by GFP and DAPI fluorescence.

ACKNOWLEDGMENTS

We thank Shana Topp, Mary Lou Previti, Larisa Kamga, and Lawton Chung for helpful discussions and experimental assistance. This work was supported by NIH 1R21AI103321.

REFERENCES

Agarwal, N., & Tyagi, A. K. (2006). Mycobacterial transcriptional signals: Requirements for recognition by RNA polymerase and optimal transcriptional activity. *Nucleic Acids Research, 34*, 4245–4257.

Bardarov, S., Bardarov, S., Jr., Pavelka, M. S., Jr., Sambandamurthy, V., Larsen, M., Tufariello, J., et al. (2002). Specialized transduction: An efficient method for generating marked and unmarked targeted gene disruptions in *Mycobacterium tuberculosis, M. bovis* BCG and *M. smegmatis*. *Microbiology, 148*, 3007–3017.

Carroll, P., Muttucumaru, D. G. N., & Parish, T. (2005). Use of a tetracycline-inducible system for conditional expression in *Mycobacterium tuberculosis* and *Mycobacterium smegmatis*. *Applied and Environmental Microbiology, 71*, 3077–3084.

Ehrt, S., Guo, X. V., Hickey, C. M., Ryou, M., Monteleone, M., Riley, L. W., et al. (2005). Controlling gene expression in mycobacteria with anhydrotetracycline and Tet repressor. *Nucleic Acids Research, 33*, e21.

Forti, F., Crosta, A., & Ghisotti, D. (2009). Pristinamycin-inducible gene regulation in mycobacteria. *Journal of Biotechnology, 140*, 270–277.

Gandotra, S., Schnappinger, D., Monteleone, M., Hillen, W., & Ehrt, S. (2007). In vivo gene silencing identifies the *Mycobacterium tuberculosis* proteasome as essential for the bacteria to persist in mice. *Nature Medicine, 13*, 1515–1520.

Goude, R., & Parish, T. (2009). Electroporation of mycobacteria. *Methods in Molecular Biology, 465*, 203–215.

Hernandez-Abanto, S., Woolwine, S., Jain, S., & Bishai, W. (2006). Tetracycline-inducible gene expression in mycobacteria within an animal host using modified *Streptomyces* tcp830 regulatory elements. *Archives of Microbiology, 186*, 459–464.

Hickey, M. J., Arain, T. M., Shawar, R. M., Humble, D. J., Langhorne, M. H., Morgenroth, J. N., et al. (1996). Luciferase in vivo expression technology: Use of recombinant mycobacterial reporter strains to evaluate antimycobacterial activity in mice. *Antimicrobial Agents and Chemotherapy, 40*, 400–407.

Hinds, J., Mahenthiralingam, E., Kempsell, K. E., Duncan, K., Stokes, R. W., Parish, T., et al. (1999). Enhanced gene replacement in mycobacteria. *Microbiology, 145*, 519–527.

Klotzsche, M., Ehrt, S., & Schnappinger, D. (2009). Improved tetracycline repressors for gene silencing in mycobacteria. *Nucleic Acids Research, 37*, 1778–1788.

Larsen, M. H., Biermann, K., & Jacobs, W. R., Jr. (2007). Laboratory maintenance of *Mycobacterium tuberculosis*. *Current Protocols in Microbiology, Chapter 10*, Unit 10A.1.

Larsen, M. H., Biermann, K., Tandberg, S., Hsu, T., & Jacobs, W. R., Jr. (2007). Genetic manipulation of *Mycobacterium tuberculosis*. *Current Protocols in Microbiology, Chapter 10*, Unit 10A.2.

Marrero, J., Rhee, K. Y., Schnappinger, D., Pethe, K., & Ehrt, S. (2010). Gluconeogenic carbon flow of tricarboxylic acid cycle intermediates is critical for *Mycobacterium tuberculosis* to establish and maintain infection. *Proceedings of the National Academy of Sciences of the United States of America, 107*, 9819–9824.

Pandey, A. K., Raman, S., Proff, R., Joshi, S., Kang, C. M., Rubin, E. J., et al. (2009). Nitrile-inducible gene expression in mycobacteria. *Tuberculosis, 89*, 12–16.

Parish, T., Mahenthiralingam, E., Draper, P., Davis, E. O., & Colston, E. O. (1997). Regulation of the inducible acetamidase gene of *Mycobacterium smegmatis*. *Microbiology, 143*, 2267–2276.

Parish, T., & Stoker, N. G. (2000). Use of a flexible cassette method to generate a double unmarked *Mycobacterium tuberculosis tlyA plcABC* mutant by gene replacement. *Microbiology, 146*, 1969–1975.

Sassetti, C. M. (2008). Inducible expression systems for mycobacteria. In T. Parish, & A. C. Brown (Eds.), *Mycobacteria protocols* (pp. 255–264). Totowa, NJ: Humana Press.

Seeliger, J. C., Topp, S., Sogi, K. M., Previti, M. L., Gallivan, J. P., & Bertozzi, C. R. (2012). A riboswitch-based inducible gene expression system for mycobacteria. *PLoS One, 7*, e29266.

Singh, A. K., & Reyrat, J. M. (2009). Laboratory maintenance of *Mycobacterium smegmatis*. *Current Protocols in Microbiology, Chapter 10*, Unit 10C.1.

Stover, C. K., de la Cruz, V. F., Fuerst, T. R., Burlein, J. E., Benson, L. A., Bennett, L. T., et al. (1991). New use of BCG for recombinant vaccines. *Nature, 351*, 456–460.

Topp, S., Reynoso, C. M., Seeliger, J. C., Goldlust, I. S., Desai, S. K., Murat, D., et al. (2010). Synthetic riboswitches that induce gene expression in diverse bacterial species. *Applied and Environmental Microbiology, 76*, 7881–7884.

Van Kessel, J. C., & Hatfull, G. F. (2007). Recombineering in *Mycobacterium tuberculosis*. *Nature Methods, 4*, 147–152.

CHAPTER THIRTEEN

Controlling Expression of Genes in the Unicellular Alga *Chlamydomonas reinhardtii* with a Vitamin-Repressible Riboswitch

Silvia Ramundo[1], Jean-David Rochaix[2]
Departments of Molecular Biology and Plant Biology, University of Geneva, Geneva, Switzerland
[1]Current address: Department of Biochemistry & Biophysics, University of California, San Francisco, CA, USA.
[2]Corresponding author: e-mail address: jean-david.rochaix@unige.ch

Contents

Abstract

Chloroplast genomes of land plants and algae contain generally between 100 and 150 genes. These genes are involved in plastid gene expression and photosynthesis and in various other tasks. The function of some chloroplast genes is still unknown and some of them appear to be essential for growth and survival. Repressible and reversible expression systems are highly desirable for functional and biochemical characterization of these genes. We have developed a genetic tool that allows one to regulate the expression of any coding sequence in the chloroplast genome of the unicellular alga *Chlamydomonas reinhardtii*. Our system is based on vitamin-regulated expression of the nucleus-encoded chloroplast Nac2 protein, which is specifically required for the expression of any plastid gene fused to the *psbD* 5′UTR. With this approach, expression of the *Nac2* gene in the nucleus and, in turn, that of the chosen chloroplast gene artificially driven by the *psbD* 5′UTR, is controlled by the *MetE* promoter and *Thi4* riboswitch, which can be inactivated in a reversible way by supplying vitamin B_{12} and thiamine to

Methods in Enzymology, Volume 550
ISSN 0076-6879
http://dx.doi.org/10.1016/bs.mie.2014.10.035

the growth medium, respectively. This system opens interesting possibilities for studying the assembly and turnover of chloroplast multiprotein complexes such as the photosystems, the ribosome, and the RNA polymerase. It also provides a way to overcome the toxicity often associated with the expression of proteins of biotechnological interest in the chloroplast.

1. INTRODUCTION

A distinctive feature of land plants and algae is that their chloroplasts contain an autonomous genetic and protein-synthesizing system. Chloroplast genomes have generally a size comprised between 100 and 200 kb. They exist as circular and/or multimeric linear molecules and are present in multiple copies per cell. Chloroplast genomes from land plants and green algae contain between 100 and 150 genes of which the majority encodes components of the chloroplast gene expression system and of the photosynthetic apparatus. These include genes of the subunits of the chloroplast DNA-dependent RNA polymerase, ribosomal RNAs and proteins, tRNAs, and subunits of the photosynthetic complexes. These genomes also encode additional proteins involved in various functions such as proteolysis, heme attachment, and lipid metabolism. Finally, these genomes encode a set of genes of unknown function (for review, see Green, 2011).

Chloroplast transformation is feasible in the alga *Chlamydomonas reinhardtii* (Boynton & Gillham, 1993) and in some plants (Svab & Maliga, 1993). The *aadA* gene, which confers resistance to spectinomycin and streptomycin, has been used extensively as a selectable marker for chloroplast transformation (Goldschmidt-Clermont, 1991). Because homologous recombination occurs readily, it has been possible to perform specific chloroplast gene disruptions and site-directed mutagenesis for chloroplast genetic engineering (Boynton & Gillham, 1993; Rochaix, 1997). However, any known chloroplast genome exists in multiple copies (about 80 in the case of *Chlamydomonas*). Thus, disruption of a plastid locus usually requires several subcloning steps under selective conditions for obtaining homoplasmicity. Remarkably, photosynthetic function in *Chlamydomonas* is dispensable in the presence of acetate as reduced carbon source. Therefore, targeted deletion of many chloroplast genes involved in photosynthesis has been easily achieved in this model organism. In some cases, however, the heteroplasmicity persists even under prolonged selective pressure. This heteroplasmic state occurs when the disrupted gene is essential for cell growth. In this way, several chloroplast genes with essential function have been identified (Rochaix, 1995). They include all genes tested that encode components of the chloroplast gene expression system indicating that this process is essential for cell survival. However, one problem with this approach is that expression of the selectable marker itself depends on a functional chloroplast protein-synthesizing system. Hence, it is not clear whether the heteroplasmic state is due to the essential role of plastid protein synthesis or to the chloroplast protein synthesis-dependent expression of the selectable marker.

An alternative way to solve this problem was the development of a repressible chloroplast gene expression system in which any chloroplast gene of interest can be tightly downregulated in a temporal manner (Ramundo, Casero, et al., 2014; Ramundo, Rahire, Schaad, & Rochaix, 2014; Surzycki, Cournac, Peltier, & Rochaix, 2007). This method is described in this chapter and is based on the use of the vitamin-repressible *MetE* promoter and *Thi4* riboswitch from *Chlamydomonas* (Croft, Lawrence, Raux-Deery, Warren, & Smith, 2005; Croft, Moulin, Webb, & Smith, 2007).

Besides its use for elucidating the role of essential chloroplast genes, this method also provides new possibilities for studying the assembly of specific photosynthetic complexes in a preformed thylakoid membrane. It also allows one to examine the degradation of these complexes under controlled conditions (Dinc, Ramundo, Croce, & Rochaix, 2014). This system has already proven helpful to avoid the occurrence of a strong negative selective pressure against the constitutive expression of genes such as the chloroplast codon-optimized [FeFe] hydrogenase (cpHydA) (Reifschneider-Wegner, Kanygin, & Redding, 2014). It can also improve the production of proteins that are of biomedical interest but that are detrimental for *Chlamydomonas* viability. For example, expression of DILP-2, a growth promoter, could only be achieved using a repressible chloroplast gene expression system (Surzycki et al., 2009).

2. DESIGN OF THE REPRESSIBLE RIBOSWITCH SYSTEM ACTING ON CHLOROPLAST GENES

To overcome the lack of a tight repressible promoter in the chloroplast genome of *Chlamydomonas*, we took advantage of regulatory elements involved in nuclear gene expression. In particular, we tested the nuclear *Cyc6* promoter, which is tightly repressed in copper-replete medium and induced in the absence of copper (Merchant & Bogorad, 1987), the *MetE* promoter, whose activity is inhibited in the presence of vitamin B_{12} (Croft et al., 2005; Helliwell et al., 2014) and the thiamin pyrophosphate riboswitch in the 5′UTR of the nucleus-encoded *Thi4* mRNA, which drives a nonproductive splicing event in the presence of thiamine (Croft et al., 2007). The chloroplast repressible system we have developed is also based on the properties of some nucleus-encoded factors that are required for specific chloroplast posttranscriptional steps of gene expression such as RNA processing and translation. One of these factors is Nac2, which is specifically required for stabilization and translation of the chloroplast *psbD* mRNA encoding the D2 reaction center protein of PSII (Kuchka, Goldschmidt-Clermont, van Dillewijn, & Rochaix, 1989). Nac2 acts by interacting with the *psbD* 5′UTR, which is both necessary and sufficient to convey the *Nac2* dependence on any gene sequence to which this 5′UTR is fused (Nickelsen, van Dillewijn, Rahire, & Rochaix, 1994). By fusing the *MetE* promoter and *Thi4* riboswitch to the *Nac2* gene in the nuclear genome of a *nac2* mutant, it was possible to repress the expression of this gene by addition of vitamins B_{12} and thiamine to the growth medium (Ramundo, Rahire, et al., 2014). Binding of thiamine pyrophosphate to the *Thi4* riboswitch induces a conformational change which leads to alternative splicing and inclusion of a premature termination codon, thus preventing synthesis of the protein encoded by the downstream open reading frame (Fig. 1).

In turn, upon replacement of the endogenous 5′UTR of the plastid gene of interest with the *psbD* 5′UTR, it was possible to control expression of this target gene in a *Nac2*-dependent fashion. We have successfully applied this strategy to study several essential plastid genes present in *Chlamydomonas* and in many land plants such as *rpoA* encoding the α subunit of the RNA polymerase, *rps12*, encoding the ribosomal protein Rps12 (Ramundo, Rahire, et al., 2014) and *clpP1* encoding the catalytic subunit of the ATP-dependent ClpP protease (Ramundo, Casero, et al., 2014). With this system, the expression of any chloroplast gene of interest can be repressed by fusing its coding sequence to the *psbD* promoter and 5′UTR. For the necessary chloroplast DNA manipulations, it is most convenient to use the *aadA* gene as selectable marker that conveys spectinomycin/streptomycin resistance (Goldschmidt-Clermont, 1991). The construct containing *aadA* and the target gene fused to the *psbD* 5′UTR is then introduced into the chloroplast genome through biolistic chloroplast transformation using the A31 strain as host (Figs. 2 and 3). The A31 strain is derived from the *nac2* mutant transformed with the pRAM23.1 plasmid, which contains the chimeric gene

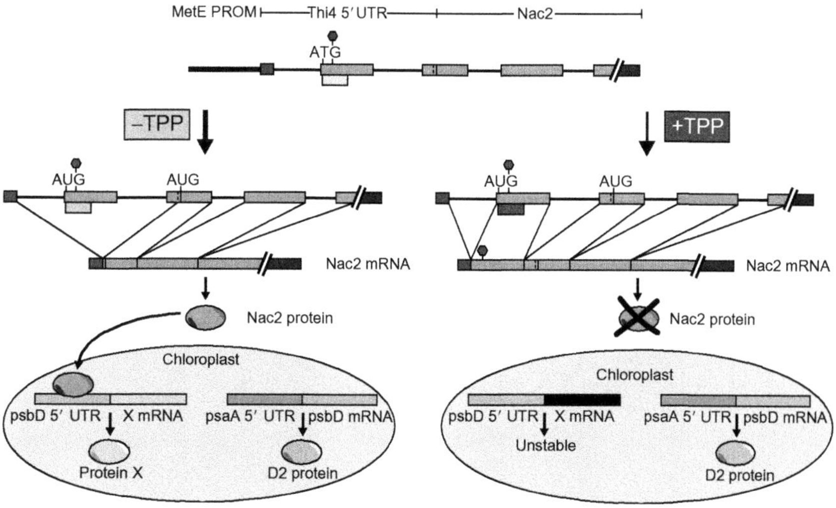

Figure 1 Reversible vitamin-mediated repression of chloroplast gene expression. Scheme of the vitamin-mediated repressible chloroplast gene expression system. The *Nac2* gene, which is specifically required for the accumulation of the chloroplast *psbD* mRNA, is fused to the *Thi4* 5′UTR containing the TPP-responsive riboswitch and the *MetE* upstream region is used as promoter. Addition of thiamine causes alternative splicing in the riboswitch region, which results in translation termination due to the inclusion of a stop codon (red flag) (Croft et al., 2007). The yellow box below the second exon indicates the genomic location of the TPP riboswitch. The change of the box color represents in a schematic way the conformational change of the riboswitch upon binding of TPP (green: no TPP binding; red: TPP binding). Because Nac2 acts specifically on the *psbD* 5′UTR, it is possible to render the expression of any chloroplast gene dependent on Nac2 by fusing its coding sequence to the *psbD* 5′UTR. To allow for phototrophic growth in the presence of the vitamins, the *psbD* 5′UTR of the *psbD* gene was replaced by the *psaA* 5′UTR, thus making *psbD* expression independent of Nac2. *Reproduced from Ramundo, Rahire, et al. (2014) with permission (Copyright American Society of Plant Biologists).* (See the color plate.)

consisting of the *MetE* promoter and *Thi4* riboswitch fused to the *Nac2* gene, so that expression of *Nac2* can be repressed with vitamins (Ramundo, Rahire, et al., 2014). Moreover, in the A31 strain, the promoter/5′UTR of the chloroplast *psbD* gene was replaced with that of *psaA*. Thus, expression of *psbD* is no longer dependent on *Nac2* and the strain grows photoautotrophically in the presence or absence of vitamins.

Besides conditional repression of chloroplast genes, this system can also be used for repressing nuclear genes. The only requirement is the availability of a mutant strain of *Chlamydomonas* that is deficient in the expression of the target gene. The wild-type sequence of the target gene can then be fused to

A

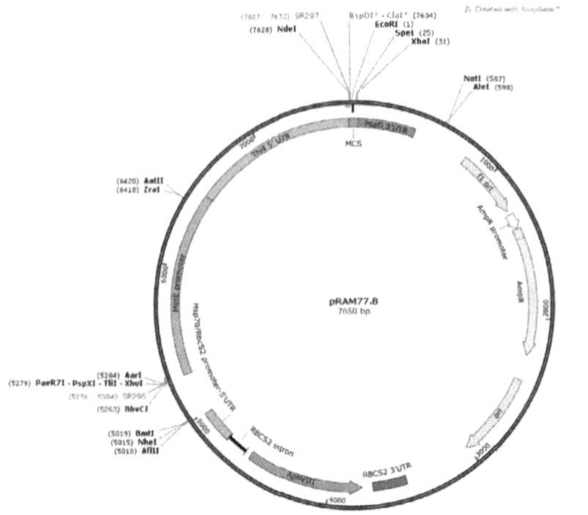

B

	DNA template	Primer Fw	Primer Rv
(psbD 5′) (amplicon 1)	Chloroplast genome	SR 186	SR185
(Rps12 - flanking region for HR) (amplicon 2)	Chloroplastgeonome	SR184	SR183
(psbD 5′-Rps12 – flanking region for HR) (OE-PCR amplicon)	Template 1 + Template 2	SR186	Sr183

C

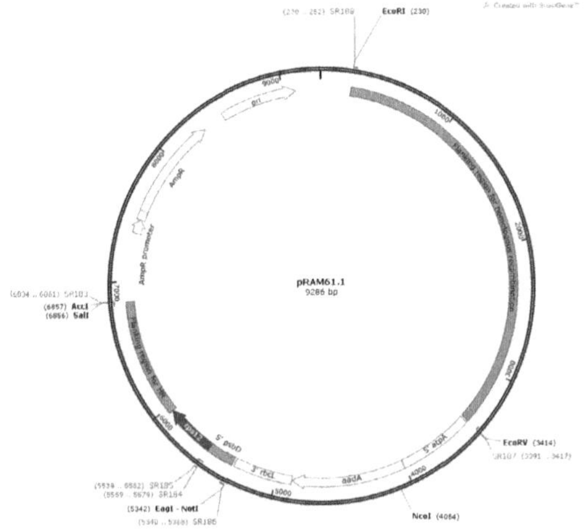

Figure 2 See legend on opposite page.

the *MetE* promoter and *Thi4* riboswitch by replacing the *Nac2* coding sequence with that of the target gene. To this end, we have generated a novel plasmid, named pRAM77.8 (Fig. 2A), in which the *PsaD* promoter/5′UTR was excised from pSL18 (S. Lemaire and J.D. Rochaix, unpublished) by digestion with *Xho*I and *Nde*I and replaced with the *MetE* promoter and *Thi4* 5′UTR by standard ligation. These plasmids allow one to express the nuclear gene of interest in a constitutive or repressible manner

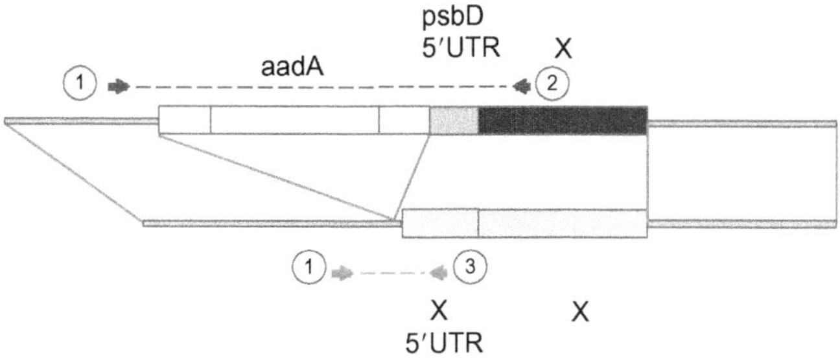

Figure 3 Test for homoplasmicity. The upper line shows the construct with the *aadA* cassette and the *psbD* 5′UTR-geneX inserted into the chloroplast genome through homologous recombination. The lower line represents the locus of gene X in the wild-type chloroplast genome. Primers 1 and 2 amplify a fragment specific for the inserted *psbD* 5′UTR-X construct (red-dashed) whereas primers 1 and 3 amplify a wild-type fragment (blue-dashed). A strain homoplasmic for the *psbD* 5′UTR construct will only amplify the red PCR product, whereas a heteroplasmic strain will amplify both the blue and the red PCR product. Since there are about 80 copies of the chloroplast genome per *Chlamydomonas* cell, to prove that the transformed strain is homoplasmic, it is highly recommended to perform a PCR with 80 times less wild-type DNA and ensure that a single copy of the wild-type gene is detectable with PCR under the same conditions. (See the color plate.)

Figure 2 (A) Map of pRAM77.8, the vector containing the *MetE* promoter and *Thi4* 5′UTR derived from pSL18; the illustrated primers in magenta SR296 (ccgctcgagTACTTCGTGCAGGTGTCTTA) and SR297 (ccgctcgagTACTTCGTGCAGGTGTCTTA) were used to amplify the *MetE* promoter and the *Thi4* 5′UTR from pRAM23.1; some relevant and unique restriction sites are indicated with a bold text string; MCS, Multicloning site. (B) Example of overlap extension polymerase chain reaction (OE-PCR) to fuse the *psbD* 5′UTR/promoter to the plastid gene *rps12*. (C) Map of pRAM61.1, the plasmid used to transform the A31 strain to generate a repressible expression system for the plastid gene *rps12*. In this case, the *aadA* gene was cloned in the same orientation as the *psbD5′-rps12* gene although no read-through transcription was observed. Some relevant and unique restriction sites are indicated with a bold text string. The sequence of the illustrated primers in magenta (Ramundo, Rahire, et al., 2014) and the map of the pUC-atpX-AAD plasmid are available (Goldschmidt-Clermont, 1991). (See the color plate.)

(pSL18 and pRAM77.8, respectively) and to use paromomycin resistance for selection of the transformants (Sizova et al., 1996). We and others have noticed that the *Thi4* 5′UTR is sufficient to achieve effective repression of the Nac2 protein, and it does not interfere with the specific regulation of the upstream promoter. Thus, when the *Thi4* 5′UTR is employed to regulate expression of a nuclear gene, it can be desirable to replace the *MetE* promoter with the endogenous promoter of the target gene. In this way, the gene will be expressed as in the wild type under nonrepressive conditions.

3. METHODS

3.1. Growth conditions

The *C. reinhardtii* wild-type, *nac2-26* and the other strains are usually maintained on Tris–acetate phosphate (TAP) or minimal (HSM) medium plates supplemented with 1.5% Bacto-agar (Gorman & Levine, 1966; Harris, 1989) at 25 °C, either under constant light (60–40 $\mu mol\ m^{-2}\ s^{-1}$)/dim light (10 $\mu mol\ m^{-2}\ s^{-1}$) or in the dark depending on their light sensitivity (Table 1).

Medium with vitamins is prepared in the following way: stock solutions 1000-fold concentrated of thiamine-HCl (20 m*M*) (Sigma Aldrich CAT. N. T4625) and vitamin B_{12} (20 mg mL^{-1}) (Sigma Aldrich CAT. N. V2006) are prepared in sterile, MilliQ water and stored at 4 °C in the dark. TAP medium is sterilized by autoclaving, then cooled before adding 1 mL of each vitamin stock solution per liter of medium to final concentrations of 20 μM thiamine-HCl, 20 μg mL^{-1} B_{12}. The TAP medium can be stored at room

Table 1 Properties of the vitamin-repressible strains of *Chlamydomonas reinhardtii*

Strain	Genotype	Nuclear transgene	Chloroplast*	TAP	TAP+Vit	HSM	HSM +Vit
WT	*Nac2*	–	*psbD-psbD*	+	+	+	+
Rep112	*nac2*	$VitB_{12}$/TRS-Nac2	*psbD-psbD*	+	+	+	–
A31	*nac2*	$VitB_{12}$/TRS-Nac2	*psaA-psbD*	+	+	+	+
nac2	*nac2*	–	*psbD-psbD*	+	+	–	–

Nac2 and *nac2* refer to the wild-type and mutant *Nac2* gene, respectively. $VitB_{12}$/TRS-Nac2 indicates the transgene Nac2 fused to the MetE promoter and Thi4 riboswitch. * Refers to the chloroplast *psbD* gene and its 5′UTR; *psaA-psbD* indicates that the *psaA* 5′UTR was fused to the *psbD* coding sequence. TAP, Tris–acetate medium; HSM, high salt minimal medium; Vit, vitamin; + and – indicate growth and loss of growth, respectively.

temperature until it is used. It is best not to store vitamin-containing TAP medium more than a few days and to add the vitamins to the medium shortly before it is used.

At the beginning of each experiment, cells are preinoculated from fresh plates into liquid TAP media with/without vitamins, and allowed to grow under continuous light at 25 °C on a rotary shaker at 150 rpm to a density of 2–4 $\times 10^6$ cells/mL (as determined with a hemocytometer). Subsequently, they are diluted to a concentration of 0.5–1 $\times 10^6$ cells/mL and allowed to grow to the desired cell concentration in medium with/without vitamins. To maintain cells in exponential growth during the time course, which can extend over several days, the cells are diluted to 0.5×10^6 cells/mL when they reach a concentration between 2 and 4×10^6 cells/mL and growth is continued.

For derepression experiments, i.e., when cells are transferred from vitamin-replete medium to medium lacking vitamins, cells are pelleted by centrifugation at 2500 $\times g$ for 5 min at room temperature and washed two to three times in medium lacking vitamins prior to inoculation. For the growth tests 7 μL containing 1×10^4, 5×10^4, 1×10^5, 5×10^5, 1×10^6 cells are spotted from serially diluted culture onto solid TAP or HSM medium with or without vitamins and subjected to continuous light during 5–6 days (60 μmol m^{-2} s^{-1}) or dark during 10–12 days.

3.2. Nuclear transformation

Nuclear transformation is usually performed using the method described by Shimogawara, Fujiwara, Grossman, and Usuda (1998). The nuclear gene of interest is cloned downstream of the *Thi4* 5′UTR that is present in the pRAM77.8 vector, between the *Nde*I restriction site and any of the downstream restriction sites available in the multicloning site (which is shared with the pSL18 vector) (Fig. 2A). This can be achieved using classical restriction enzyme-based subcloning or In-Fusion BioBrick assembly (Sleight, Bartley, Lieviant, & Sauro, 2010). Cells are grown in TAP medium under dim light, harvested in mid-log phase (2–4 $\times 10^6$ cells/mL), and treated with gamete autolysin for 1 h, then resuspended in TAP + 40 mM sucrose medium. Autolysin is prepared by mixing equal volumes of wild-type gametes of + and − mating type together and by gently shaking the culture under low light (40 μmol m^{-2} s^{-1}) for 1 h. During this period, the gametes secrete autolysin into the medium. The cells are then centrifuged and the supernatant is used as crude autolysin, which can be stored at −70 °C. To remove the cell wall of wild-type cells, they are resuspended in the autolysin medium and incubated for 1 h. Autolysin treatment is not necessary for cell wall-deficient strains. For each electroporation, 10^8 treated cells are incubated with 0.1–1 μg of linearized plasmid, then transformed by electroporation in a 0.2-cm electroporation cuvette (Biorad, USA) using the Biorad Gene Pulser II set to 0.75 kV, 25 μF, and no resistance. The transformants obtained are recovered in 1 mL fresh TAP + 40 m*M* sucrose + 0.4% PEG-8000 + 20% starch medium for 10 min, followed by plating on selective medium at 25 °C in constant light (40 μmol m^{-2} s^{-1}). This medium can be HSM if the transformation aims to complement a strain deficient in photosynthesis. If there is no significant growth phenotype, paromomycin resistance can be used as selectable marker (Sizova et al., 1996). The transformants are subsequently tested for their mutant phenotype when the medium is supplemented with vitamins (10 μ*M* thiamine-HCl/10 μg B_{12}/mL). On average, only 10% of the recovered transformants display a conditional vitamin-dependent phenotype.

3.3. Chloroplast transformation

Chloroplast transformation of *Chlamydomonas* cells can be easily achieved with a commercially available particle gun: 20×10^6 cells are spread evenly on agar plates containing 100 μg mL^{-1} spectinomycin (Sigma Aldrich S9007). As soon as the plates are dry, they are bombarded with 550 nm-diameter gold particles (Seashell Technology S550d) coated with 300 ng of plasmid DNA. The plasmid DNA is usually derived from the transformation vector pUC-atpX-AAD (Goldschmidt-Clermont, 1991). In this plasmid, the target chloroplast ORF should be fused to the *psbD* 5′UTR/promoter (210 bp) through overlap extension polymerase chain reaction (OE-PCR) (Fig. 2B) or In-Fusion BioBrick assembly (Higuchi, Krummel, & Saiki, 1988). Moreover, the transgene should be inserted as close as possible to the *aadA* selection cassette and preferably, but not necessarily, in the opposite direction to avoid futile recombination events and potential read-through transcripts, respectively (Figs. 2C and 3). The target gene can either be an authentic chloroplast gene or a foreign gene. To achieve an efficient integration at the chosen chloroplast target site, at least 2 kb of chloroplast DNA should be included in the transformation plasmid on both sides of the cassette and *psbD* 5′UTR. Obviously, to control the expression of a plastid endogenous gene, the 5′UTR/promoter region of this gene should be excluded. Given the potential formation of highly stable secondary structures, it may sometimes also be necessary to test different fusion constructs of the *psbD* 5′UTR with the target gene. These variants can be easily generated by introducing short linker sequences and/or a protein tag (i.e., FLAG, HA) after the start codon of the target gene. Colonies are usually detectable 7–8 days after transformation. Individual colonies (about 20–30) are picked up with toothpicks and restreaked on new plates containing a higher amount of spectinomycin (750 $\mu g\ mL^{-1}$) to enhance the selective pressure. At least three of these restreaking steps are required to achieve homoplasmicity, usually checked by PCR using primers that allow for specific amplification of the wild-type gene and of the chimeric *psbD* 5′UTR target gene (Fig. 3).

3.4. Screening for essential chloroplast genes

The transformants are first screened for their ability to grow on acetate agar plates with or without vitamins. Under these conditions, the parental A31 strain will not be affected by the presence of vitamins in the medium, whereas transformants in which the target gene is essential will gradually display a pale-green phenotype and die subsequently on medium supplemented with vitamins, as expected from the repression of a gene that is essential for cell survival. When the lethal phenotype is observed in the dark, it can be assumed that the target gene is essential for cell growth and survival. In particular, it can be excluded that cell death results from an indirect photooxidative effect. Before a given transformant is further characterized, at least two to three independently generated transformants should be examined by performing growth tests and their homoplasmicity should be assessed by semiquantitative PCR and/or Southern analysis of their genomic DNA. This is important as other genetic changes may occur during chloroplast transformation.

The level of expression of the target gene should be tested at the protein level under both nonrepressive and repressive conditions. We have had several cases where the mRNA level of a plastid gene does not match the level of the encoded protein (e.g., ClpP1, Rps12, RpoA) (Ramundo, Casero, et al., 2014; Ramundo, Rahire, et al., 2014). This phenomenon can be explained by assuming the existence of negative feedback loops regulating chloroplast gene expression. It also suggests that translational or posttranslational processes are probably the most important and limiting steps determining the steady-state levels of endogenous or heterologous proteins expressed from the chloroplast genome.

3.5. Effect of vitamins

Treatment of *Chlamydomonas* with vitamin B_{12} and thiamine-HCl at the indicated concentrations during the repression experiments does not have any apparent deleterious effect on the growth rate of this alga; but it does affect expression of a number of off-target genes. It is therefore necessary to compare changes in RNA expression between the strain in which a specific gene has been repressed and the A31 strain. As an example in the case of the repression of *clpP1*, comparison of changes in RNA expression between the two strains revealed that genes whose expression changes upon addition of vitamins in both strains comprise 26% and 23% of the up- and downregulated genes, respectively, in the *clpP1* repressible strain. This list is available in (Ramundo, Casero, et al., 2014). Further examination of the expression levels of these genes indicated that these values were similar in 25% of the cases, indicating that for this restricted set, the genes respond primarily to the addition of vitamins. Among the downregulated genes one finds as expected those involved in vitamin biosynthesis such as *MetE*, *Thi4a*, and *ThiC*.

4. CONCLUSIONS AND PERSPECTIVES

The vitamin-repressible system described is rather robust and can be used to repress any plastid gene of interest in a temporal and reversible manner. It is especially useful for the functional analysis of plastid genes that are essential (Ramundo, Casero, et al., 2014; Ramundo, Rahire, et al., 2014; Surzycki et al., 2007). Another potential use of this system is the expression of proteins that are toxic to the host cells and/or that are of biotechnological interest (Reifschneider-Wegner et al., 2014; Surzycki et al., 2009). Given its reversibility, this tool also opens interesting possibilities for investigating the biogenesis of chloroplast multiprotein complexes, such as the photosystems, the ribosome or the RNA polymerase. Finally, the availability of two repressible chloroplast gene expression systems in *Chlamydomonas* (one based on the copper-repressible *Cyc6* promoter and the other on the vitamin-repressible *MetE* promoter-*Thi4* riboswitch) offers the opportunity to repress genes together and independently from each other in time.

Although this system has been developed mainly for *Chlamydomonas*, a similar approach could be developed in land plants, provided that chloroplast transformation is feasible. This is the case for several dicot species (Maliga & Bock, 2011) including tobacco (Svab & Maliga, 1993), tomato (Ruf,

Hermann, Berger, Carrer, & Bock, 2001), petunia (Zubkot, Zubkot, van Zuilen, Meyer, & Day, 2004), potato (Valkov et al., 2011), soybean (Dufourmantel et al., 2007), lettuce (Kanamoto et al., 2006), and cabbage (Liu, Lin, Chen, & Tseng, 2007). The only other two important requirements for designing such a repressible expression system for a given plant species of interest are (1) the availability of highly inducible/repressible nuclear regulatory elements such as the dexamethasone-inducible (Aoyama & Chua, 1997) and tetracycline-repressible (Gatz, Frohberg, & Wendenburg, 1992) promoters in tobacco and (2) mutant lines for a nuclear gene functionally analogous to Nac2. Indeed, nuclear genes that act in a similar way as *Nac2* at a plastid posttranscriptional step have been identified in land plants (Jacobs & Kück, 2011). Thus, the characterized mutant lines lacking any of these genes could be used as a platform for establishing a similar repressible/inducible plastid gene expression system.

ACKNOWLEDGMENTS

We thank Nicolas Roggli for preparing the figures. The work in the authors' laboratory was supported by grant 31003A_133089/1 from the Swiss National Foundation.

REFERENCES

Aoyama, T., & Chua, N. H. (1997). A glucocorticoid-mediated transcriptional induction system in transgenic plants. *The Plant Journal, 11*, 605–612.

Boynton, J. E., & Gillham, N. W. (1993). Chloroplast transformation in *Chlamydomonas. Methods in Enzymology, 217*, 510–536.

Croft, M. T., Lawrence, A. D., Raux-Deery, E., Warren, M. J., & Smith, A. G. (2005). Algae acquire vitamin B12 through a symbiotic relationship with bacteria. *Nature, 438*, 90–93.

Croft, M., Moulin, M., Webb, M. E., & Smith, A. (2007). Thiamine biosynthesis in algae is regulated by riboswitches. *Proceedings of the National Academy of Sciences of the United States of America, 104*, 20770–20775.

Dinc, E., Ramundo, S., Croce, R., & Rochaix, J. D. (2014). Repressible chloroplast gene expression in *Chlamydomonas*: A new tool for the study of the photosynthetic apparatus. *Biochimica et Biophysica Acta, 1837*, 1548–1552.

Dufourmantel, N., Dubald, M., Matringe, M., Canard, H., Garcon, F., Job, C., et al. (2007). Generation and characterization of soybean and marker-free tobacco plastid transformants over-expressing a bacterial 4-hydroxyphenylpyruvate dioxygenase which provides strong herbicide tolerance. *Plant Biotechnology Journal, 5*, 118–133.

Gatz, C., Frohberg, C., & Wendenburg, R. (1992). Stringent repression and homogeneous de-repression by tetracycline of a modified CaMV 35S promoter in intact transgenic tobacco plants. *The Plant Journal, 2*, 397–404.

Goldschmidt-Clermont, M. (1991). Transgenic expression of aminoglycoside adenine transferase in the chloroplast: A selectable marker for site-directed transformation of *Chlamydomonas. Nucleic Acids Research, 19*, 4083–4089.

Gorman, D. S., & Levine, R. P. (1966). Cytochrome *f* and plastocyanin: Their sequence in the photoelectric transport chain. *Proceedings of the National Academy of Sciences, 54*, 1665–1669.

Green, B. R. (2011). Chloroplast genomes of photosynthetic eukaryotes. *The Plant Journal, 66*, 34–44.

Harris, E. H. (1989). *The Chlamydomonas source book: A comprehensive guide to biology and laboratory use.* San Diego, CA: Academic Press, Inc.

Helliwell, K. E., Scaife, M. A., Sasso, S., Araujo, A. P., Purton, S., & Smith, A. G. (2014). Unraveling vitamin B12-responsive gene regulation in algae. *Plant Physiology, 165*, 388–397.

Higuchi, R., Krummel, B., & Saiki, R. K. (1988). A general method of in vitro preparation and specific mutagenesis of DNA fragments: Study of protein and DNA interactions. *Nucleic Acids Research, 16*, 7351–7367.

Jacobs, J., & Kück, U. (2011). Function of chloroplast RNA-binding proteins. *Cellular and Molecular Life Sciences, 68*, 735–748.

Kanamoto, H., Yamashita, A., Asao, H., Okumura, S., Takase, H., Hattori, M., et al. (2006). Efficient and stable transformation of *Lactuca sativa L. cv. Cisco* (lettuce) plastids. *Transgenic Research, 15*, 205–217.

Kuchka, M. R., Goldschmidt-Clermont, M., van Dillewijn, J., & Rochaix, J. D. (1989). Mutation at the *Chlamydomonas* nuclear NAC2 locus specifically affects stability of the chloroplast psbD transcript encoding polypeptide D2 of PS II. *Cell, 58*, 869–876.

Liu, C. W., Lin, C. C., Chen, J. J., & Tseng, M. J. (2007). Stable chloroplast transformation in cabbage (*Brassica oleracea L. var. capitata L.*) by particle bombardment. *Plant Cell Reports, 26*, 1733–1744.

Maliga, P., & Bock, R. (2011). Plastid biotechnology: Food, fuel, and medicine for the 21st century. *Plant Physiology, 155*, 1501–1510.

Merchant, S., & Bogorad, L. (1987). Metal ion regulated gene expression: Use of a plastocyanin-less mutant of *Chlamydomonas reinhardtii* to study the Cu(II)-dependent expression of cytochrome c-552. *EMBO Journal, 6*, 2531–2535.

Nickelsen, J., van Dillewijn, J., Rahire, M., & Rochaix, J.-D. (1994). Determinants for stability of the chloroplast *psbD* RNA are located within its short leader region in *Chlamydomonas reinhardtii. EMBO Journal, 13*, 3182–3191.

Ramundo, S., Casero, D., Muhlhaus, T., Hemme, D., Sommer, F., Crevecoeur, M., et al. (2014). Conditional depletion of the *Chlamydomonas* chloroplast ClpP1 protease activates nuclear genes involved in autophagy and plastid protein quality control. *Plant Cell, 30*, 2201–2222.

Ramundo, S., Rahire, M., Schaad, O., & Rochaix, J. D. (2014). Repression of essential chloroplast genes reveals new signaling pathways and regulatory feedback loops in *Chlamydomonas. Plant Cell, 25*, 167–186.

Reifschneider-Wegner, K., Kanygin, A., & Redding, K. E. (2014). Expression of the [FeFe] hydrogenase in the chloroplast of *Chlamydomonas reinhardtii. Intlernational Journal of Hydrogen Energy, 30*, 1–9.

Rochaix, J. D. (1995). *Chlamydomonas reinhardtii* as the photosynthetic yeast. *Annual Review of Genetics, 29*, 209–230.

Rochaix, J.-D. (1997). Chloroplast reverse genetics: New insights into the function of plastid genes. *Trends in Plant Science, 2*, 419–425.

Ruf, S., Hermann, M., Berger, I. J., Carrer, H., & Bock, R. (2001). Stable genetic transformation of tomato plastids and expression of a foreign protein in fruit. *Nature Biotechnology, 19*, 870–875.

Shimogawara, K., Fujiwara, S., Grossman, A., & Usuda, H. (1998). High-efficiency transformation of *Chlamydomonas reinhardtii* by electroporation. *Genetics, 148*, 1821–1828.

Sizova, I. A., Lapina, T. V., Frolova, O. N., Alexandrova, N. N., Akopiants, K. E., & Danilenko, V. N. (1996). Stable nuclear transformation of *Chlamydomonas reinhardtii* with a *Streptomyces rimosus* gene as the selective marker. *Gene, 181*, 13–18.

Sleight, S. C., Bartley, B. A., Lieviant, J. A., & Sauro, H. M. (2010). In-Fusion BioBrick assembly and re-engineering. *Nucleic Acids Research, 38*, 2624–2636.

Surzycki, R., Cournac, L., Peltier, G., & Rochaix, J. D. (2007). Potential for hydrogen production with inducible chloroplast gene expression in *Chlamydomonas*. *Proceedings of National Academy of Science of the United States of America, 104*, 17548–17553.

Surzycki, R., Greenham, K., Kitayama, K., Dibal, F., Wagner, R., Rochaix, J. D., et al. (2009). Factors effecting expression of vaccines in microalgae. *Biologicals, 37*, 133–138.

Svab, Z., & Maliga, P. (1993). High-frequency plastid transformation in tobacco by selection for a chimeric *aadA* gene. *Proceedings of the National Academy of Sciences of the United States of America, 90*, 913–917.

Valkov, V. T., Gargano, D., Manna, C., Formisano, G., Dix, P. J., Gray, J. C., et al. (2011). High efficiency plastid transformation in potato and regulation of transgene expression in leaves and tubers by alternative 5' and 3' regulatory sequences. *Transgenic Research, 20*, 137–151.

Zubkot, M. K., Zubkot, E. I., van Zuilen, K., Meyer, P., & Day, A. (2004). Stable transformation of petunia plastids. *Transgenic Research, 13*, 523–530.

CHAPTER FOURTEEN

Conditional Control of Gene Expression by Synthetic Riboswitches in *Streptomyces coelicolor*

Martin M. Rudolph, Michael-Paul Vockenhuber, Beatrix Suess[1]
Fachbereich Biologie, Synthetische Biologie, Technische Universität Darmstadt, Darmstadt, Germany
[1]Corresponding author: e-mail address: bsuess@bio.tu-darmstadt.de

Contents

Abstract

Here we provide a step-by-step protocol for the application of synthetic theophylline-dependent riboswitches for conditional gene expression in *Streptomyces coelicolor*. Application of the method requires a sequence of only ~85 nt to be inserted between the transcriptional start site and the start codon of a gene of interest. No auxiliary factors are needed. All tested riboswitch variants worked well in concert with the promoters *galP2*, *ermEp1*, and SF14. Moreover, they allowed theophylline-dependent expression not only of the heterologous β-glucuronidase reporter gene but also of *dagA*, an endogenous agarase gene. The right combination of the tested promoters with the riboswitch variants allows for the adjustment of the desired dynamic range of regulation in a highly specific and

Methods in Enzymology, Volume 550
ISSN 0076-6879
http://dx.doi.org/10.1016/bs.mie.2014.10.036

dose-dependent manner and underlines the orthogonality of riboswitch regulation. We anticipate that any additional natural or synthetic promoter can be combined with the presented riboswitches. Moreover, this system should easily be transferable to other *Streptomyces* species, and most likely to any other genetically manipulable bacteria.

1. INTRODUCTION

In the last decade, it has become obvious that RNA plays an important role in the regulation of gene expression. This has led to a plethora of approaches aiming at the exploitation of the outstanding chemical properties of RNA to develop synthetic RNA-based regulators for conditional gene expression systems (Groher & Suess, 2014; Wittmann & Suess, 2012). Among them, synthetic riboswitches are very promising tools.

Their siblings, natural riboswitches—discovered in 2002—represent genetic regulatory elements that turned out to be widely distributed throughout the bacterial kingdom where they control numerous basic metabolic pathways (Breaker, 2012). Riboswitches are mostly located in the 5′ untranslated region of an mRNA and serve as molecular switches that modulate transcription, translation, or mRNA processing. They consist solely of RNA, sense their ligand in a highly specific binding pocket, and undergo a structural rearrangement upon metabolite binding, whereby one of the two mutual exclusive conformations efficiently interferes with gene expression. The specific characteristic of riboswitches is that RNA accomplishes both the sensor and the effector functions, demonstrating that a protein cofactor is not an obligate requirement for regulation.

One popular design principle when engineering synthetic riboswitches makes use of small molecule-binding RNA aptamers that can modulate the accessibility of the ribosomal binding site (SD, Shine–Dalgarno sequence) in a ligand-dependent manner (Fig. 1). The aptamer is placed close to the SD, and surrounding sequences are chosen in a way that a part of the aptamer sequence is involved in direct base pairing with the SD sequence. In absence of the ligand, this leads to the sequestration of both the ribosomal binding site and the start codon, thereby preventing the expression of the following gene. Ligand binding results in a structural rearrangement of the RNA, freeing the translation initiation region and allowing translation. By means of rational design combined with cell-based screenings, a set of six riboswitch constructs was developed, permitting inducible gene expression in a range of Gram-positive and -negative bacteria (Lynch, Desai, Sajja, & Gallivan, 2007; Topp & Gallivan, 2008; Topp et al., 2010).

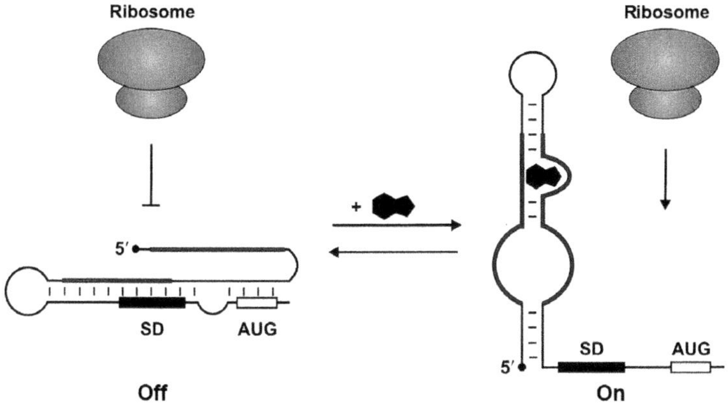

Figure 1 Regulation of translation initiation by engineered riboswitches in bacteria. In the absence of the ligand, parts of the aptamer domain pair with the Shine–Dalgarno (SD) sequence (black). The SD sequence and the translational start codon AUG (white) are not accessible for the binding of the 30S ribosomal subunit. As a consequence of ligand binding (black molecule), the restructuring of the aptamer domain (highlighted in gray) sets the SD sequence and AUG free. This event enables the ribosome to bind and renders translation possible.

Streptomyces coelicolor is a filamentous, GC-rich, soil-dwelling bacterium with a complex life cycle. It is a model representative of the Gram-positive actinomycetes that are known to exhibit an unparalleled metabolic and enzymatic diversity. They are the main producers of valuable compounds for pharmaceutical and industrial purposes, such as antibiotics, antiviral and anticancer drugs, immunosuppressive agents, herbicides, and diverse enzymes. What makes them even more attractive for biotechnological applications is their ability to express and secrete heterologous proteins in a properly folded and biologically active state (Anné, Maldonado, Van Impe, Van Mellaert, & Bernaerts, 2012).

At present, conditional gene expression in streptomycetes is driven by a few protein-based, ligand-inducible systems. These require the expression of accessory regulatory proteins whose genes have to be additionally provided on plasmids or on the chromosome. To circumvent the implementation and codon usage optimization of accessory protein factors and to create an easy and robust inducible system, we survey the possibility of RNA-based regulation. There is increasing evidence that RNA-based regulation plays an important role for regulation in streptomycetes, too, both by riboswitches (Borovok, Gorovitz, Schreiber, Aharonowitz, & Cohen, 2006; Pedrolli et al., 2012; Tezuka & Ohnishi, 2014) and noncoding RNAs (Heueis,

Vockenhuber, & Suess, 2014; Moody, Young, Jones, & Elliot, 2013; Vockenhuber et al., 2011; Vockenhuber & Suess, 2012). Therefore, we made use of the principal design of synthetic riboswitches developed by the Gallivan group and demonstrated their applicability in *S. coelicolor,* the model organism of the Gram-positive actinomycetes (Rudolph, Vockenhuber, & Suess, 2013).

We provide here a method to silence constitutive promoters in *S. coelicolor* and to turn on the riboswitch-controlled genes in a temporally defined and dose-dependent manner. Using synthetic theophylline-responsive riboswitches, the regulation of the expression of both a reporter and an endogenous gene was possible. The presented workflow should serve as a guideline to adapt the riboswitch-mediated control of expression for further heterologous or native genes, both in *S. coelicolor* and most likely in any other genetically manipulable bacteria.

2. CONSTRUCTION OF RIBOSWITCH-CONTROLLED EXPRESSION SYSTEMS

2.1. Vector

We used the integrative plasmid pGusT to assess the function of theophylline-dependent riboswitches in *S. coelicolor.* It is a pAR933a derivative (Rodriguez-Garcia, Combes, Perez-Redondo, & Smith, 2005) in which the *tetR* gene has been replaced with a β-glucuronidase gene (*gusA*). *gusA* was amplified from pSETGUS (Myronovskyi, Welle, Fedorenko, & Luzhetskyy, 2011) in an overlap PCR, whereby an *Age*I restriction site was introduced directly behind the start codon and an *Xba*I site 3′ of *gusA*. The vector carries the reporter gene *gusA* in the opposite direction relative to the remaining elements on the backbone to prevent read-through of the neighboring genes. Additionally, a synthetic *rho*-independent transcriptional terminator downstream of *gusA* (inserted at *Xba*I restriction site) ensures proper transcription termination.

The vector contains an integrase gene and *attP* site of the temperate *Streptomyces* phage phiC31. These elements enable a stable integration of one single copy per chromosome (Kuhstoss, Richardson, & Rao, 1991), avoiding possible problems associated with multicopy plasmids.

Taken together, pGusT carries the following genetic elements:

- combination of *gusA* reporter gene
- transcriptional terminator (*term*)
- origin of replication for maintenance in *Escherichia coli* (*oriV*)

- apramycin resistance gene for selection in *E. coli* and *S. coelicolor (acc(3)IV)*
- integrase gene and attachment site of phiC31 phage (*int* and *attP*)
- origin of transfer (*oriT*).

The vector map and a snapshot showing the riboswitch-containing region are shown in Fig. 2. The complete vector sequence can be provided upon request.

2.2. Riboswitch design

Six theophylline-responsive riboswitches (named A–E*, Topp et al., 2010) were cloned into pGusT via *Kpn*I and *Age*I (Fig. 2C). The expression of the riboswitch::*gusA* constructs was driven by the three different constitutive promoters *galP2*, *ermEp*1, or SF14, respectively, allowing to adjust different expression levels (Bibb, Janssen, & Ward, 1985; Fornwald, Schmidt, Adams, Rosenberg, & Brawner, 1987; Labes, Bibb, & Wohlleben, 1997). The promoters were inserted into the vector using the restriction sites *Bgl*II and *Acc*65I (Fig. 2B). Promoter and riboswitch were separated by a constant region (AATACGACTCACTATAGGTTCC), serving as a spacer between the two elements. For each promoter, a control plasmid was constructed lacking the riboswitch sequence but containing the core SD motif GGAGG located nine nucleotides upstream of the *gusA* start codon.

The vector pGusT can be used for modular cloning of different promoters and genes combined with one of the six riboswitches. The order of these elements and the exact nucleotide sequence of the whole 5′ UTR are both crucial for proper riboswitch function and should not be altered. When alternative promoters are used, it is advised to remove nonessential regulatory elements and to position only the first bases that are transcribed directly before the riboswitch sequence. As indicated in Fig. 2B, we cloned the transcription start sites of all three tested promoters directly upstream of the constant region in front of the riboswitch.

Given that AUG as translational start codon is retained, in principle every gene of interest can be regulated when cloned directly downstream of the riboswitch-containing 5′ UTR. It cannot be entirely excluded, however, that nucleotides of the coding region may base pair with the riboswitch. An *in silico* prediction of RNA secondary folding (e.g., with RNAfold WebServer) is therefore advisable before testing the system on new genes. Note that *in silico* RNA folding is only a prediction and several factors within the cell may influence local RNA folding. If the prediction hints an alternative folding, the sequence of the N-terminus should be revised (if possible) by means of silent mutations.

A

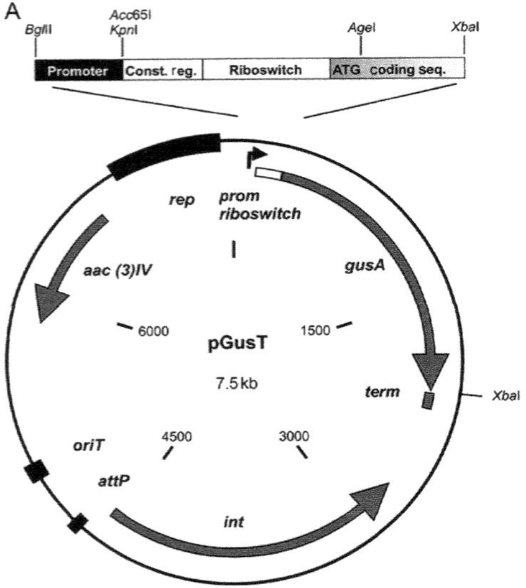

B

	Promoter sequence
galP2	*AGATCT*CTCCACCTGGAACTTTTCACTTCCGCCGTACGTCCGGCAAGCTGAAGTTCCTC**G***GGTACC*
ermEp1	*AGATCT*GTTGTGGGCTGGACAATCGTGCCGGTTGGTAGGATCCAGC**G***GGTACC*
SF14	*AGATCT*GACCTACGCCTTGACCTTGATGAGGCGGCGTGAGCTACAATCAATACT**C***GGTACC*

C

	Constant region
	AATACGACTCACTATAGGTTCC

D

	Riboswitch sequence
A	------------GGTGATACCAGCATCGTCTTGATGCCCTTGGCAGCACCC--TGAG**AAGGGG**----CAACAAGatg*ACCGGT*
B	------------GGTGATACCAGCATCGTCTTGATGCCCTTGGCAGCACCCGCTGCGC**AGGGGGT**ATCAACAAGatg*ACCGGT*
C	TGATAAGATAGGGGTGATACCAGCATCGTCTTGATGCCCTTGGCAGCACC-------**AAGGG**A----CAACAAGatg*ACCGGT*
D	------------GGTGATACCAGCATCGTCTTGATGCCCTTGGCAGCACCC--TGCT**AAGGT**AA---CAACAAGatg*ACCGGT*
E	------------GGTGATACCAGCATCGTCTTGATGCCCTTGGCAGCACCC--TGCT**AAGGAGGT**AACAACAAGatg*ACCGGT*
E*	------------GGTGATACCAGCATCGTCTTGATGCCCTTGGCAGCACCC--TGCT**AAGGAGG**---CAACAAGatg*ACCGGT*

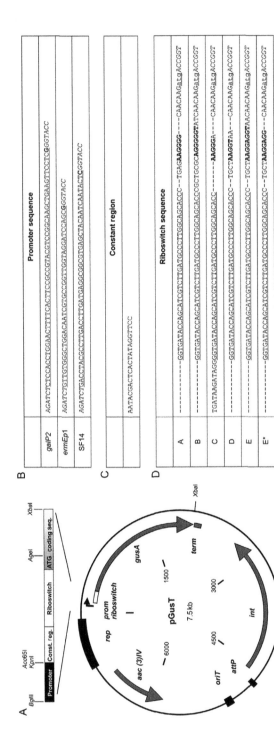

Figure 2 Vector pGusT and sequence of theophylline-dependent riboswitches. (A) pGusT, an integrative *gusA* reporter plasmid. *prom* = promoter; gusA = β-glucuronidase gene; *term* = synthetic terminator; *int* and *attP* = integrase gene and attachment site of phiC31 phage; *oriT* = origin of transfer; *acc (3)IV* = apramycin resistance; and rep = *E. coli* replication region. (B) Constitutive promoters used in this study. The weak (*galP2*), medium (*ermEp*1), and strong (SF14) promoters each were cloned via *Bgl*II (5′) and *Acc*65I (3′). The promoter sequence is underlined and the transcriptional start site is given in bold. (C) Constant region, separating promoter, and riboswitch sequences. (D) Riboswitch sequences. The sequence of the theophylline aptamer is underlined, the SD sequences are given in bold, and the translational start codon is in lower case. The start codon is followed by *Age*I restriction site, where the gene of interest is directly cloned.

2.3. Genetic manipulations in *S. coelicolor*

For all genetic manipulations, common laboratory *E. coli* DH5α strains can be used. All vectors are transferred to *S. coelicolor* by intergeneric conjugation. For our experiments, we used *S. coelicolor* M145, a prototrophic, plasmidless derivative of the model representative strain A3(2) (Kieser, Bibb, Buttner, Chater, & Hopwood, 2000). Since this strain possesses a methyl-specific restriction system, all vectors have to be passaged through the methylation-defective *E. coli* strain ET12567/puZ8002 (Flett, Mersinias, & Smith, 1997).

After verifying the correct insertion into the *S. coelicolor* genome by colony-PCR and sequence analysis, the riboswitch-regulated reporter gene expression can be visualized directly by chromogenic assays, as depicted hereafter.

3. MEASUREMENT OF RIBOSWITCH ACTIVITY

3.1. β-Glucuronidase measurement

The codon-optimized version of the β-glucuronidase gene allows the quantification of riboswitch activity (Myronovskyi et al., 2011). Its sensitivity and ease of detection make it a favorable reporter gene for monitoring the regulation of gene expression by the applied riboswitches. This reporter gene system also has been applied successfully in other *Streptomyces* species. β-Glucuronidase activity can be visualized in colonies grown on agar plates, but can also be measured accurately in cultures grown in liquid medium.

3.2. Detection on agar plates

- The *S. coelicolor* exconjugants are streaked out on agar plates (e.g., mannitol–soy flour (MS) agar, R2YE, or MM; Kieser et al., 2000) supplemented with or without 2 m*M* theophylline and 50 μg ml^{-1} apramycin.
- After 90 h of cultivation, the colonies are flooded with 2 ml of the chromogenic substrate X-Gluc (50 μg ml^{-1} in DMF); alternatively, the substrate can be added when the medium is poured, allowing for detection throughout the development.
- After incubation for approximately 2 h at 28 °C, the appearance of a blue precipitate (5,5′-dibromo-4,4′-dichloro-indigo) indicates GusA expression.

3.3. Measurement in liquid culture

- 10^8 Pregerminated spores of *S. coelicolor* strains are inoculated in 50 ml TSB medium supplemented with 50 μg ml^{-1} apramycin and varying concentrations of theophylline.

  Mycelium is harvested after up to 90 h of cultivation, washed with distilled water, and centrifuged again.
- For disruption, the mycelium is mixed with 200 μl glass beads (0.25–0.5 mm) in 50 m*M* $NaHPO_4$ pH 7.0, 0.1% Triton X-100, 5 m*M* DTT (Gus buffer).
- Disruption of cells is performed using a high-speed benchtop homogenizer FastPrep-24 instrument (MP Biomedicals, 6 × 30 s at 6.0 m s^{-1}).
- Mycelial debris is removed by centrifugation at 4 °C (17,000 × *g* for 30 min).
- Protein concentration of supernatant is determined by Bradford assay.
- 1–25 mg of total protein in a total volume of 750 μl Gus buffer is incubated at 28 °C for 15 min.
- Enzymatic reaction is started by addition of 80 μl of 0.2 *M* *p*-nitrophenyl-β-D-glucuronide (Glycosynth) and stopped with 300 μl of 1 *M* Na_2CO_3 upon appearance of a yellowish coloration.
- Absorption at 415 nm is measured and divided by the reaction time to calculate Gus activity [A_{415} min^{-1}], or further divided by the amount of protein used to calculate specific Gus activity [A_{415} mg^{-1} min^{-1}], which is used herein to represent Gus units.

4. CHARACTERIZATION OF RIBOSWITCH-CONTROLLED GENE EXPRESSION

4.1. Dynamic range of regulation can be adjusted by appropriate promoter–riboswitch pairing

The combination of the different riboswitches with promoters of different expression levels allows the adjustment of a desired regulatory window. Figure 3 summarizes the theophylline-dependent induction of β-glucuronidase activity for each combination of the riboswitches A–E* with a weak (*galP2*), medium (*ermEp*1), and strong (SF14) promoter monitored on agar plates (see Section 3.2).

Constitutive GusA expression can be observed for all controls (without the riboswitch) irrespective of theophylline supplementation. Thereby, the color depth correlates well with the promoter strength, indicating that

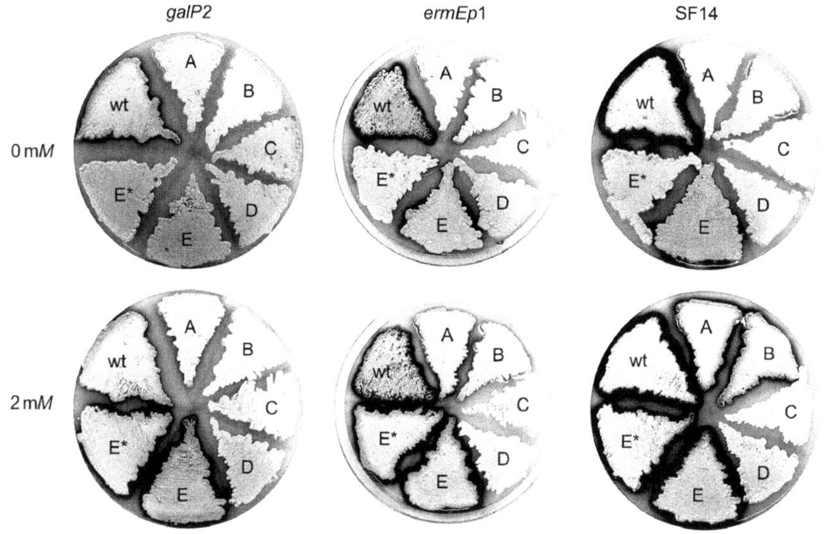

Figure 3 Analysis of different theophylline-dependent riboswitches in *S. coelicolor.* X-Gluc staining of *S. coelicolor* strains growing on MS agar expressing *galP2*-/*ermEp*1-/SF14-driven *gusA* under control of the theophylline riboswitches A, B, C, D, E, E*, or without riboswitch (wt). The medium was supplemented with 2 m*M* theophylline (theo) for induction. (See the color plate.)

increasing promoter strength from *galP2* over *ermEp*1 to SF14 directly correlates with an increased GusA production in the control strains.

The implementation of any riboswitch in the 5′ UTR leads to a strong reduction in promoter activity, only riboswitch E shows marginal GusA expression with the medium or strong promoter. Addition of the ligand theophylline then leads to a clear induction of GusA, albeit to a different extent. Strains with the riboswitches C and D show no induction by theophylline; hence, they silence every promoter regardless of inducer supplementation. Strains harboring riboswitches A, B, E, and E* showed an inducible GusA expression upon addition of theophylline with E and E* having the strongest effect.

To quantify the dynamic range of regulation, GusA expression of *ermEp*1-driven riboswitches A–E* was measured (see Section 3.3). The results given in Table 1 underscore the agar plate results. They confirm that the riboswitch C is most tightly regulated followed by A and D, whereas riboswitches B, E, and E* show leaky expression. The riboswitches A and E* have the highest activation ratios due to the lowest background activity.

Table 1 Measurement of β-glucuronidase (GusA) activity of different theophylline-dependent riboswitches

	GusA activity (GU) 0 m*M* theo	**GusA activity (GU) 4 m*M* theo**	**Activation ratio**
A	0.016 ± 0.002	0.500 ± 0.020	32
B	0.080 ± 0.001	0.260 ± 0.022	3
C	0.003 ± 0.003	0.007 ± 0.003	2
D	0.020 ± 0.004	0.040 ± 0.004	2
E	0.100 ± 0.003	1.260 ± 0.050	12
E*	0.040 ± 0.004	1.200 ± 0.140	30
wt	2.790 ± 0.180	2.890 ± 0.240	1

S. coelicolor strains of liquid cultures expressing *ermEp*1-driven gusA under the control of the synthetic riboswitches A–E* were grown for 90 h in the absence and presence of theophylline (theo). The cultures were supplemented with theophylline right from the beginning. Three independent cultures were measured in parallel. The measurements were repeated twice. Standard deviations are given ($n = 3$).

The variations in riboswitch activities may be rationalized by differing SD sequences and their respective spacing to the translational start codon. Translational efficiency strongly depends on complementarity between the SD sequence and the anti-SD sequence located on the 3′ end of the 16S rRNA. The strong complementarity (eight perfect matches) between the SD sequence of riboswitch E and the 3′ end of 16S rRNA may contribute to relatively high GusA expression observed in the absence of theophylline. In contrast, the SD sequence in riboswitch E* is located three nucleotides closer to the initiation codon and is predicted to form only seven base pair interactions with the anti-SD sequence. In riboswitch A, six to seven bases including one or two wobble base pairs are involved in the recognition by the *S. coelicolor* anti-SD sequence, leading to tight regulation but also to a lower overall expression level in the induced state.

In the presented experimental setup with a single copy per genome, riboswitch E* represents the best regulation device due to its high dynamic range concomitant with a very low basal expression. Figures 4 and 5 illustrate the proposed folding of riboswitch E* and its capability to repress and activate promoters with different strengths of transcription initiation. Whereas activation ratios of about 30-fold were achieved with *galP2* and *ermEp*1, a strikingly low basal expression in the ligand-free state leads to a ≥260-fold activation by the SF14-E* construct. Only the *galP2*-E* construct

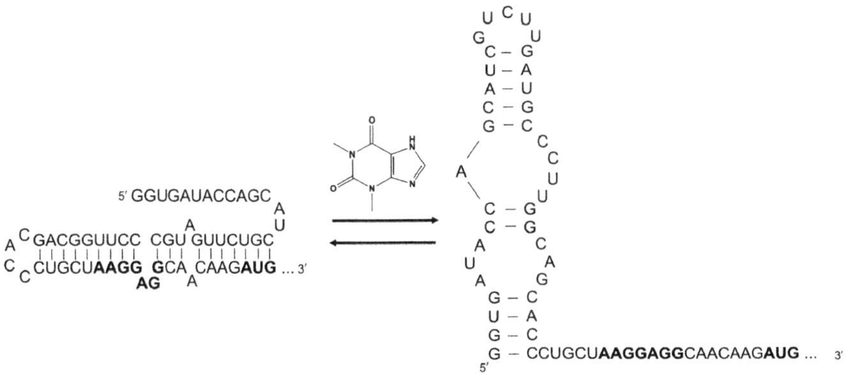

Figure 4 Proposed switching of the theophylline riboswitch E*. The SD sequence and the start codon are shown in black.

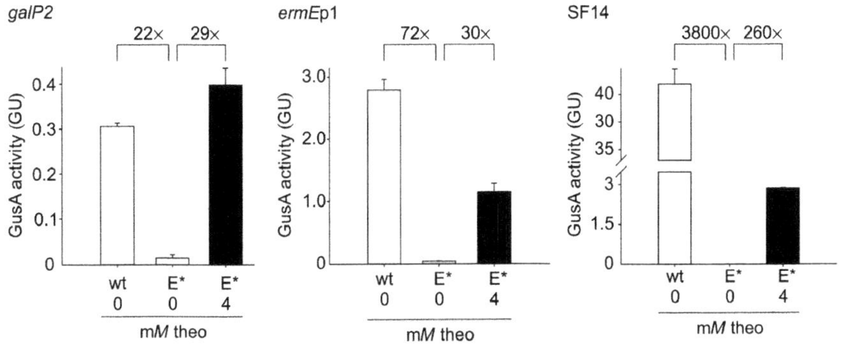

Figure 5 Repression of reporter gene expression driven by three different constitutive promoters mediated by riboswitch E*. Measurement of glucuronidase (GusA) activity from liquid cultures grown for 90 h in the absence (closed) and presence of 4 m*M* theophylline (open bars). Theophylline was present right from the initial inoculation. The maximal induction is given above the respective bars. *Reprinted from Rudolph et al. (2013), with permission from Society for General Microbiology.*

reached—upon induction—a similar expression level as the control plasmid without the riboswitch (Fig. 5).

In addition to the β-glucuronidase reporter gene, we also showed that the control of an endogenous gene is possible. Using the riboswitch E*, complete restoration of the activity of the gene for agarase *dagA* was possible after addition of theophylline (Rudolph et al., 2013). To that end, the *dagA* ORF was cloned into the plasmid pGusT with *ermE*p1 under the control of the E* riboswitch, thereby replacing the *gusA* gene. As a control, the same plasmid without E* was constructed. The two plasmids—named pTE*dagA

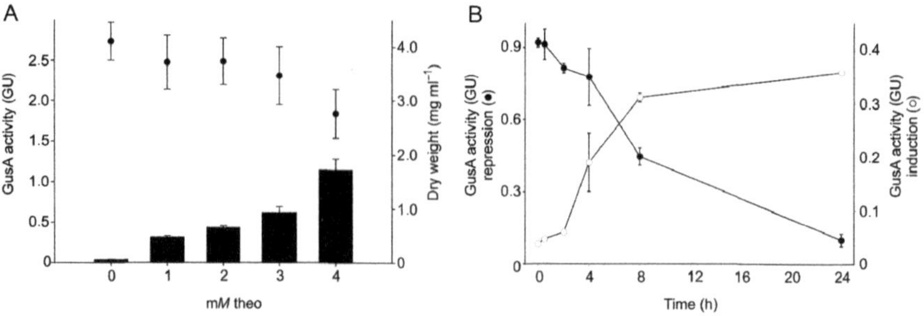

Figure 6 Dose dependence of the E* riboswitch. (A) Measurement of glucuronidase (GusA) activity (black bars) of liquid cultures grown for 90 h in the absence of theophylline or supplemented with increasing concentration of theophylline (theophylline was present right from the initial inoculation). The influence of theophylline on the growth of *S. coelicolor* was determined by measuring dry weight (black circles). (B) Cells were grown in liquid culture in the absence or presence of 4 m*M* of theophylline for 48 h. At time point 0, theophylline was added to the culture (white circles) or removed by harvesting and resuspending the cells in new medium without theophylline (black circles). Glucuronidase (GusA) activity was measured 0.5, 2, 4, 8, and 24 h after this point.

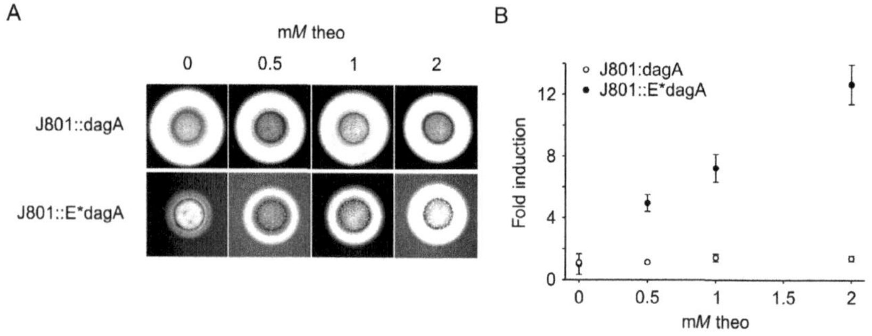

Figure 7 Regulation of *dagA* expression by the E* riboswitch. (A) Agarase activity assay (Lugol staining) of *S. coelicolor* strain J801 complemented with *ermE*p1-driven *dagA* growing on MBSM plates with agar and glycerol as only carbon sources supplemented with increasing amounts of theophylline. The area of clearing around a colony shows where agar has been utilized. Addition of theophylline has no effect on agarase expression if no riboswitch is present (J801::dagA). Addition of theophylline can almost completely restore agarase activity when *dagA* is under control of the riboswitch E* (J801::E*dagA). (B) Quantification of agarase induction. Given is the level of agarase expression at a given theophylline concentration divided by the expression level in the absence of theophylline (fold induction).

and pTdagA, respectively—were then integrated into the *attB* site of the *dag*$^{-}$ *S. coelicolor* strain J801 (Hodgson & Chater, 1981). Figure 7 shows an agarase assay of the constructed strains. All strains were exposed to a range of 0–2 m*M* theophylline. The control (J801::dagA) shows complementation of the *dag*$^-$ phenotype independent of the presence of theophylline. In contrast, no agarase activity is detectable in the construct carrying the riboswitch E* upstream of *dagA* in the absence of theophylline (J801::E*dagA). Upon induction with up to 2 m*M* theopylline, however, a dose-dependent increase in agarase activity is visible, reaching an expression level of ~76% when compared to the construct without riboswitch E*. This represents a ~12-fold, theophylline-dependent, induction of agarase activity.

Taken together, all tested promoters in our study could be regulated upon combination with the riboswitches. This led to the assumption that any natural or synthetic promoter can be regulated, confirming an apparently unrestricted applicability of the synthetic theophylline-dependent riboswitches. The smart combination of a certain promoter with one of the riboswitches will, therefore, allow for the adjustment of almost any regulatory window.

4.2. Dose dependence of riboswitch regulation

It is important to have an inducible system where the expression rate of proteins can be controlled within a large regulatory window; since some gene products may only be needed in minute amounts, while others must be highly expressed. To assess the dose dependence of riboswitch regulation, the *S. coelicolor* M145 strain with *ermEp*1-driven *gusA* controlled by the riboswitch E* was cultivated with increasing amounts of theophylline ranging from 1 to 4 m*M* for 90 h in liquid culture. GusA expression depicted in Fig. 6A provides evidence of the riboswitch acting in a dose-dependent manner. Utilization of increasing theophylline concentrations entailed higher levels of reporter gene expression. At the same time, however, a decrease in cell dry weight per milliliter was observed. Theophylline concentrations higher than 4 m*M* resulted in a reduction of cell growth that has already been reported for other Gram-positive and -negative bacteria (Desai & Gallivan, 2004).

Due to these limitations in the inducer concentration, a quantitative restoration of β-glucuronidase expression of the control strain lacking a riboswitch cannot be achieved using medium and strong promoters. Complete restoration was only possible for low expression levels (Fig. 5). To gain a high expression level, stronger promoters rather than increasing theophylline concentrations must be used.

Remarkably, the observed regulation is highly specific. In contrast to theophylline, the structurally very similar molecule caffeine was incapable of activating gene expression (Rudolph et al., 2013). Caffeine is methylated at position N-7, which impairs recognition by the aptamer. The binding affinity for theophylline is 10,000 times greater than for caffeine (Jenison, Gill, Pardi, & Polisky, 1994), which makes the utilized riboswitches highly specific and makes interference by other metabolites or media components extremely unlikely.

4.3. Assessing kinetics of induction and repression

In streptomycetes, a temporal coordination of gene expression is indispensable for the complex life cycle comprising morphological differentiation and physiological adaptation processes. To employ these bacteria as protein production systems or for studying gene function, it is important that individual genes can be turned on at a specific time point whereas other genes must be downregulated. For example, the production of some proteins that cause lethal phenotypes has to be triggered only after induction at a defined phase of growth. On the other hand, it could be necessary to switch off cellular functions. Here, conditional gene knockdowns are required. In both cases, a switching element with rapid and predictable kinetics of either activation or silencing is needed. The *ermEp*1-driven *gusA* controlled by the riboswitch E* was used to assess the theophylline-dependent induction/repression behavior of β-glucuronidase expression.

The induction profile is shown in Fig. 6B. An increase in GusA activity was detectable already 4 h after addition of theophylline to an exponentially growing culture. The maximum of reporter gene expression was reached after 8 h. The reversibility of theophylline induction, that is, a repression of reporter gene activity upon removal of the inducer, was also clearly demonstrated. Taken together, the induction/repression data indicate that the system can be rapidly induced and switched off in the logarithmic growth phase of *S. coelicolor*. It has to be noted, however, that the reversal of induction is severely dependent on the stability of the mRNA and protein of interest. The time needed to shut off expression again might vary dramatically for each tested gene.

5. CONCLUSION

The synthetic theophylline-responsive riboswitches described in this section considerably expand the genetic toolbox for controlling gene expression in *S. coelicolor*. The results indicate that synthetic riboswitches are an appropriate alternative to commonly used induction systems in these bacteria. The analyzed switches exhibit some outstanding properties:

- the only element necessary for regulation is a short RNA of ~85 nt inserted upstream of the SD of the target gene; therefore, no accessory protein factors are needed,
- the riboswitches may be combined with any promoter of interest, allowing the adjustment of the required regulatory window,
- the regulation is very specific for theophylline, which is nontoxic and nonmetabolizable, and the system can be switched on and off in a dose-dependent manner.

Among all constructs examined, riboswitches A and E* were the most appropriate to control gene expression tightly in *S. coelicolor.* Riboswitch E* leads to a comparatively high gene expression level upon addition of theophylline and can be used for applications where sheer protein yield is crucial. By contrast, construct A would be rather eligible when tight repression is required over high induction, for example, for the expression of toxic or otherwise detrimental proteins. For use with self-replicating (multicopy) plasmids, switch A might be the best choice in streptomycetes.

The successful combination of all tested promoters with the riboswitches underlines the orthogonality of riboswitch regulation. Therefore, we anticipate that any additional natural or synthetic promoters can be combined with the riboswitch not only in *S. coelicolor.* Moreover, recent publications report on the applicability of these switches not only for established model organisms like *E. coli* or *Bacillus subtilis* but also for further bacteria. Among those are species that serve as model for pathogenicity or diverse physiological or developmental processes, like *Francisella tularensis*, *Mycobacterium tuberculosis*, *Synechococcus elongatus*, *Agrobacterium tumefaciens*, *Acinetobacter baumannii*, and *Streptococcus pyogenes* (reviewed in Groher & Suess, 2014). This promising development suggests that these engineered riboswitches may be applied to a very broad spectrum of organisms. They even may—if adapted to the specific organism—prove themselves to work universally with any SD-mediated translation.

ACKNOWLEDGMENTS

The authors like to thank Julia E. Weigand for fruitful discussions and Justin Gallivan for providing us with plasmids. This work was funded by the DFG (SPP 1258 SU402/2-2 and SFB902, A2) and LOEWE CGT.

REFERENCES

Anné, J., Maldonado, B., Van Impe, J., Van Mellaert, L., & Bernaerts, K. (2012). Recombinant protein production and streptomycetes. *Journal of Biotechnology*, *158*(4), 159–167.

Bibb, M. J., Janssen, G. R., & Ward, J. M. (1985). Cloning and analysis of the promoter region of the erythromycin resistance gene (*ermE*) of *Streptomyces erythraeus*. *Gene*, *38*(1–3), 215–226.

Borovok, I., Gorovitz, B., Schreiber, R., Aharonowitz, Y., & Cohen, G. (2006). Coenzyme B12 controls transcription of the *Streptomyces* class Ia ribonucleotide reductase nrdABS operon via a riboswitch mechanism. *Journal of Bacteriology*, *188*(7), 2512–2520.

Breaker, R. R. (2012). Riboswitches and the RNA world. *Cold Spring Harbor Perspectives in Biology*, *4*(2), a003566.

Desai, S. K., & Gallivan, J. P. (2004). Genetic screens and selections for small molecules based on a synthetic riboswitch that activates protein translation. *Journal of the American Chemical Society*, *126*(41), 13247–13254.

Flett, F., Mersinias, V., & Smith, C. P. (1997). High efficiency intergeneric conjugal transfer of plasmid DNA from *Escherichia coli* to methyl DNA-restricting streptomycetes. *FEMS Microbiology Letters*, *155*(2), 223–229.

Fornwald, J. A., Schmidt, F. J., Adams, C. W., Rosenberg, M., & Brawner, M. E. (1987). Two promoters, one inducible and one constitutive, control transcription of the *Streptomyces lividans* galactose operon. *Proceedings of the National Academy of Sciences of the United States of America*, *84*(8), 2130–2134.

Groher, F., & Suess, B. (2014). Synthetic riboswitches—A tool comes of age. *Biochimica et Biophysica Acta*, *1839*(10), 964–973.

Heueis, N., Vockenhuber, M. P., & Suess, B. (2014). Small non-coding RNAs in streptomycetes. *RNA Biology*, *11*(5), 464–469.

Hodgson, D. A., & Chater, K. F. (1981). A chromosomal locus controlling extracellular agarase. *Journal of General Microbiology*, *124*(2), 339–348.

Jenison, R. D., Gill, S. C., Pardi, A., & Polisky, B. (1994). High-resolution molecular discrimination by RNA. *Science*, *263*(5152), 1425–1429.

Kieser, T., Bibb, M. J., Buttner, M. J., Chater, K. F., & Hopwood, D. A. (2000). *Practical* Streptomyces *genetics*. Norwich, UK: The John Innes Foundation.

Kuhstoss, S., Richardson, M. A., & Rao, R. N. (1991). Plasmid cloning vectors that integrate site-specifically in Streptomyces spp. *Gene*, *97*(1), 143–146.

Labes, G., Bibb, M., & Wohlleben, W. (1997). Isolation and characterization of a strong promoter element from the *Streptomyces ghanaensis* phage I19 using the gentamicin resistance gene (*aacC1*) of Tn*1696* as reporter. *Microbiology*, *143*(Pt. 5), 1503–1512.

Lynch, S. A., Desai, S. K., Sajja, H. K., & Gallivan, J. P. (2007). A high-throughput screen for synthetic riboswitches reveals mechanistic insights into their function. *Chemistry & Biology*, *14*(2), 173–184.

Moody, M. J., Young, R. A., Jones, S. E., & Elliot, M. A. (2013). Comparative analysis of non-coding RNAs in the antibiotic-producing Streptomyces bacteria. *BMC Genomics*, *14*, 558.

Myronovskyi, M., Welle, E., Fedorenko, V., & Luzhetskyy, A. (2011). Beta-glucuronidase as a sensitive and versatile reporter in actinomycetes. *Applied and Environmental Microbiology*, 77(15), 5370–5383.

Pedrolli, D. B., Matern, A., Wang, J., Ester, M., Siedler, K., Breaker, R., et al. (2012). A highly specialized flavin mononucleotide riboswitch responds differently to similar ligands and confers roseoflavin resistance to *Streptomyces davawensis*. *Nucleic Acids Research*, *40*(17), 8662–8673.

Rodriguez-Garcia, A., Combes, P., Perez-Redondo, R., & Smith, M. C. (2005). Natural and synthetic tetracycline-inducible promoters for use in the antibiotic-producing bacteria *Streptomyces*. *Nucleic Acids Research*, *33*(9), e87.

Rudolph, M. M., Vockenhuber, M. P., & Suess, B. (2013). Synthetic riboswitches for the conditional control of gene expression in Streptomyces coelicolor. *Microbiology*, *159*(Pt. 7), 1416–1422.

Tezuka, T., & Ohnishi, Y. (2014). Two glycine riboswitches activate the glycine cleavage system essential for glycine detoxification in *Streptomyces griseus*. *Journal of Bacteriology*, *196*(7), 1369–1376.

Topp, S., & Gallivan, J. P. (2008). Random walks to synthetic riboswitches—A high-throughput selection based on cell motility. *Chembiochem*, *9*(2), 210–213.

Topp, S., Reynoso, C. M., Seeliger, J. C., Goldlust, I. S., Desai, S. K., Murat, D., et al. (2010). Synthetic riboswitches that induce gene expression in diverse bacterial species. *Applied and Environmental Microbiology*, *76*(23), 7881–7884.

Vockenhuber, M. P., Sharma, C. M., Statt, M. G., Schmidt, D., Xu, Z., Dietrich, S., et al. (2011). Deep sequencing-based identification of small non-coding RNAs in Streptomyces coelicolor. *RNA Biology*, *8*(3), 468–477.

Vockenhuber, M. P., & Suess, B. (2012). Streptomyces coelicolor sRNA scr5239 inhibits agarase expression by direct base pairing to the dagA coding region. *Microbiology*, *158*(Pt. 2), 424–435.

Wittmann, A., & Suess, B. (2012). Engineered riboswitches: Expanding researchers' toolbox with synthetic RNA regulators. *FEBS Letters*, *586*(15), 2076–2083.

CHAPTER FIFTEEN

Engineering of Ribozyme-Based Aminoglycoside Switches of Gene Expression by *In Vivo* Genetic Selection in *Saccharomyces cerevisiae*

Benedikt Klauser, Charlotte Rehm, Daniel Summerer, Jörg S. Hartig[1]
Department of Chemistry, Konstanz Research School Chemical Biology, University of Konstanz, Konstanz, Germany
[1]Corresponding author: e-mail address: joerg.hartig@uni-konstanz.de

Contents

Abstract

Synthetic RNA-based switches are a growing class of genetic controllers applied in synthetic biology to engineer cellular functions. In this chapter, we detail a protocol for the selection of posttranscriptional controllers of gene expression in yeast using the *Schistosoma mansoni* hammerhead ribozyme as a central catalytic unit. Incorporation of a small molecule-sensing aptamer domain into the ribozyme renders its activity ligand-dependent. Aptazymes display numerous advantages over conventional protein-based transcriptional controllers, namely, the use of little genomic space for encryption, their modular architecture allowing for easy reprogramming to new inputs, the physical linkage to the message to be controlled, and the ability to function without protein cofactors. Herein, we describe the method to select ribozyme-based switches of gene expression in *Saccharomyces cerevisiae* that we successfully implemented to engineer neomycin- and theophylline-responsive switches. We also highlight how to adapt

Methods in Enzymology, Volume 550
ISSN 0076-6879
http://dx.doi.org/10.1016/bs.mie.2014.10.037

the protocol to screen for switches responsive to other ligands. Reprogramming of the sensor unit and incorporation into any RNA of interest enables the fulfillment of a variety of regulatory functions. However, proper functioning of the aptazyme is largely dependent on optimal connection between the aptamer and the catalytic core. We obtained functional switches from a pool of variants carrying randomized connection sequences by an *in vivo* selection in MaV203 yeast cells that allows screening of a large sequence space of up to 1×10^9 variants. The protocol given explains how to construct aptazyme libraries, carry out the *in vivo* selection and characterize novel ON- and OFF-switches.

1. THEORY

Naturally occurring riboswitches are modular RNA constructs consisting of a metabolite-binding aptamer domain connected to an adjoining expression platform. Changes in intracellular concentration of a respective metabolite are sensed via binding of the metabolite to the aptamer, which then induces structural reorganization of the expression platform thereby switching on or off expression of the downstream gene (Mandal, Boese, Barrick, Winkler, & Breaker, 2003; Winkler, Nahvi, & Breaker, 2002; Winkler, Nahvi, Sudarsan, Barrick, & Breaker, 2003). In the past decade, a growing number of synthetic RNA-based switches have been developed as signal transducers to trigger conditional gene expression. In our lab, we are engineering artificial riboswitches using the *Schistosoma mansoni* hammerhead ribozyme, a self-cleaving RNA motif, as expression platform (Ferbeyre, Smith, & Cedergren, 1998). Allosteric control of the cleavage reaction is achieved by coupling of a ligand-sensing aptamer motif to the HHR domain yielding a so-called hammerhead aptazyme (HHAz) (Tang & Breaker, 1997; Wieland & Hartig, 2008). Hammerhead ribozymes comprise three stems radiating from the catalytically active center as shown in Fig. 1. In principle, attachment of an aptamer domain is possible at any of the three stems (Ogawa & Maeda, 2008; Wieland, Benz, Klauser, & Hartig, 2009; Win & Smolke, 2007, 2008); however, preservation of stem I–II interactions is needed for enhanced cleavage rates at physiological Mg^{2+} levels (De la Pena, Gago, & Flores, 2003; Khvorova, Lescoute, Westhof, & Jayasena, 2003). In addition, connection of the HHAz to the mRNA of interest is possible in different fashions, as depicted by a type 1 HHR aptazyme (Fig. 1A) and a type 3 HHR aptazyme (Fig. 1B), the latter mimicking the architecture of naturally occurring riboswitches with the aptamer preceding the expression platform. Using these designs, we have successfully selected functional switches of gene expression responsive to

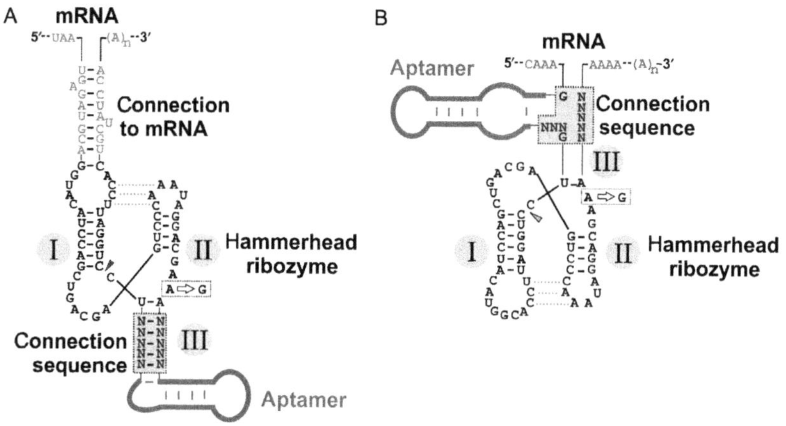

Figure 1 Insertion of the HHAz in the 3′-UTR of a yeast mRNA. (A) In the type 1 HHR, the HHAz is inserted into the 3′-UTR of an mRNA preceding the polyA tail via a connecting spacer (light gray) attached at stem I. The aptamer is attached at stem III via a randomized connection sequence (highlighted). (B) The HHAz is inserted into the mRNA via the randomized connection sequence (highlighted) at stem III in case of the type 3 HHR. The aptamer is attached within the connection sequence in stem III. In both designs, stem I–II interactions (dotted lines) are left intact. Inactivation point mutation A → G is shown (white box).

theophylline or neomycin, respectively (Klauser, Atanasov, Siewert, & Hartig, 2014).

Due to its modular architecture, the HHR has been used as a versatile platform to generate a variety of genetic switches. In bacteria, regulation of gene expression is achieved by sequestration of the ribosomal binding site (Wieland et al., 2009; Wieland & Hartig, 2008), while mRNA integrity is controlled in eukaryotic systems by detachment of either the 5′ cap structure (Auslander, Ketzer, & Hartig, 2010) or the 3′ polyA tail (Ketzer, Haas, Engelhardt, Hartig, & Nettelbeck, 2012; Ketzer et al., 2014; Klauser et al., 2014). In addition to regulation of gene expression via mRNAs, other RNA classes can be controlled, namely, transfer RNA to change amino acid identity (Berschneider, Wieland, Rubini, & Hartig, 2009; Saragliadis & Hartig, 2013) and ribosomal RNA (Wieland, Berschneider, Erlacher, & Hartig, 2010) in bacteria and microRNAs (Kumar, An, & Yokobayashi, 2009) in eukaryotes. The modular setup of HHAz also makes it possible to render the HHR responsive to different inputs such as the presence of *trans*-acting RNAs (Klauser & Hartig, 2013) or temperature sensing via an RNA thermometer (Saragliadis, Krajewski, Rehm, Narberhaus, & Hartig, 2013).

Conventional systems for posttranscriptional regulation of gene expression employing protein-based transcription factors (TFs) display several disadvantages with respect to artificial riboswitches: First, TFs are not readily programmable in terms of ligand input (Ketzer et al., 2012). Next, TFs require a large genomic space for encryption, and an additional TF-specific target site also needs to be incorporated upstream of the message to be controlled (Rohmer, Mainka, Knippertz, Hesse, & Nettelbeck, 2008). Finally, levels of the TFs may vary substantially in changing conditions or during the cell cycle (Fechner et al., 2003). In contrast, due to being part of the same RNA transcript, riboswitches are physically linked to the protein-coding sequence they control, ensuring at all times a 1:1 stoichiometry between the regulative element and the message to be controlled. In addition, these short motifs only need to be introduced at a single genomic site. Most importantly, the modular construction of HHAz allows the rearrangement or exchange of sensor domain, ribozyme, or output domain, thereby facilitating reprogramming of ligand dependency.

However, proper functionality of the connecting sequence between the catalytically active HHR domain and the input sensing aptamer domain is crucial for effective switching. This is achieved by randomization of the connecting sequence; screening for operable aptazymes is therefore indispensable. To test all possible sequence variants created by randomization of nucleotide positions, the sequence space needs to be oversampled various times (Reetz, Kahakeaw, & Lohmer, 2008). This limits the number of nucleotides that can be randomized when one wants to avoid prolonged screening periods for testing individual clones, extensive manual labor, or expensive technical equipment. In contrast, *in vivo* selection is a relatively time-saving and cheap method, as all clones can be processed in bulk. Coverage of large libraries is often only limited by transformation efficiency. Advanced yeast transformation methods have been reported (Benatuil, Perez, Belk, & Hsieh, 2010; Gietz & Schiestl, 2007) that theoretically make it possible to search yeast libraries with up to 1×10^9 variants.

In vivo selection in our lab is carried out with the yeast strain MaV203 (Invitrogen), often used in yeast two-hybrid systems, containing deletions of the endogenous *GAL4* and *GAL80* genes, which encode the Gal4 TF and its repressor. MaV203 cells carry three chromosomally encoded reporters: the auxotrophy markers *HIS3* and *URA3* used for positive and negative selection, respectively, and *lacZ* for quantification. All reporters are under control of Gal4-inducible promoters. As secondary structures in the 5′-UTR impair mRNA translation in yeast (Ringner & Krogh,

2005), the aptazyme is incorporated in the 3′-UTR of the Gal4 TF, which is expressed from a low-copy pBT3-based yeast plasmid under control of the constitutive CYC1 promoter. Activation of the HHAz leads to cleavage of the Gal4 mRNA, resulting in mRNA decay and therefore lower levels of the Gal4 TF. Contrarily, the mRNA remains intact upon inhibition of the intramolecular cleavage reaction leading to higher Gal4 levels (Fig. 2).

HIS3 encodes the imidazole glycerol phosphate dehydratase (IGPD), catalyzing a step in the histidine biosynthesis. Positive selection is carried out on SC medium lacking leucine, uracil, and histidine in the presence of 3-amino-1,2,4-triazol (3-AT), a competitive inhibitor of IGPD, thereby enriching cells that exceed a minimum threshold value of Gal4 expression (Vidal, Braun, Chen, Boeke, & Harlow, 1996). For negative selection, cells are grown on plates containing 5-fluoroorotic acid (5-FOA). *URA3* encodes the orotidine-5-phosphate decarboxylase (ODCase) that converts 5-FOA to the toxic product 5-fluorouracil (Vidal, Brachmann, Fattaey, Harlow, & Boeke, 1996). In this case, high levels of the Gal4 TF will lead to cell death. Finally, Gal4 expression is quantified by measuring Gal4-driven expression of LacZ in a β-galactosidase assay (Fig. 3).

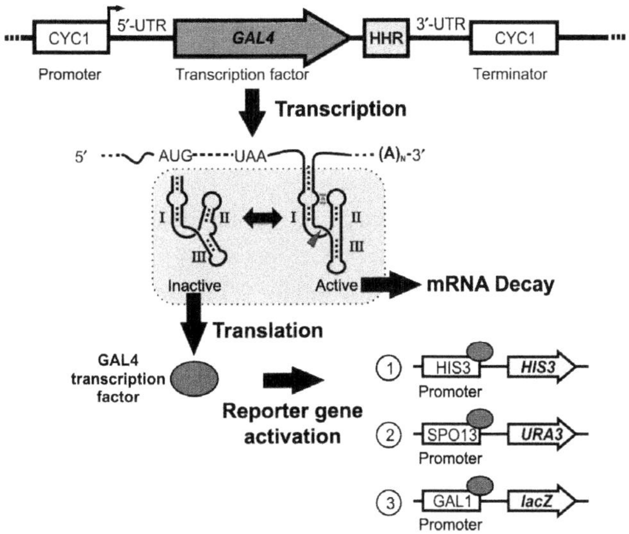

Figure 2 HHR-regulated transcription of Gal4 transcription factor. HHR is inserted in the 3′-UTR of the Gal4 mRNA. Transcription of the Gal4 mRNA results in autocatalytic cleavage of the ribozyme and subsequent mRNA decay. Inactivation of the HHR leads to translation of the Gal4 transcription factor, which in turn activates expression of the reporter genes *HIS3*, *URA3*, and *lacZ*.

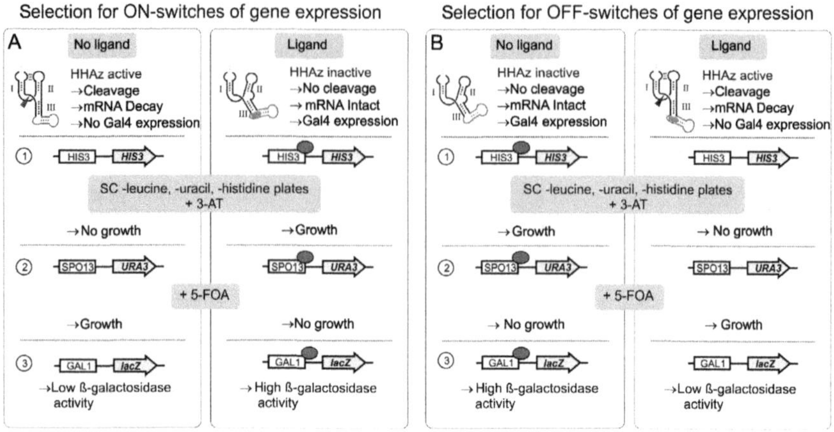

Figure 3 Selection strategies for ON- and OFF-switches of gene expression. (A) ON-switches of gene expression are achieved by inactivation of the HHAz self-cleavage by ligand binding. Cells are selected that in the presence of inducing ligand grow in the presence of 3-AT, do not grow in the presence 5-FOA, and exhibit a high β-galactosidase activity. (B) OFF-switches of gene expression are achieved by activation of the HHAz self-cleavage by ligand binding. Cells are selected that in the presence of inducing ligand do not grow in the presence of 3-AT, grow in the presence 5-FOA, and exhibit a low β-galactosidase activity.

Positive selection yields aptazymes with a low basal catalytic activity. When selecting for HHAz that switch on gene expression, the positive selection step is carried out in the presence of the aptazyme ligand under histidine-autotrophic conditions with 3-AT. In the negative selection step, aptazymes whose catalytic activity is not enhanced by the ligand are sorted out by growing cells in the absence of ligand on plates supplemented with 5-FOA (Fig. 3A). In contrast, selection for OFF-switches of gene expression is possible by growing cells first in the absence of ligand during positive selection. Afterward, a negative selection in the presence of ligand and 5-FOA is carried out (Fig. 3B).

2. EQUIPMENT AND MATERIAL

Use ultrapure water and analytical grade reagents for the preparation of solutions. All equipment that is required for bacterial and yeast cell culture needs to be sterilized by autoclaving or filtration. Reagents are stored at room temperature unless otherwise noted.

2.1. Molecular subcloning of the aptazyme library and propagation into *Escherichia coli*

1. Lab equipment
   - Tabletop centrifuge (e.g., Eppendorf Mini Spin®)
   - Centrifuge for 50 mL tubes (e.g., Eppendorf 5810 R)
   - Incubator
   - UV light table
   - Thermocycler for PCR amplification (e.g., Biometra TProfessional Thermocycler)
   - Electroporator (e.g., Eppendorf Electroporator 2510 and electroporation cuvettes: 1 mm for *E. coli* transformation and 2 mm for *Saccharomyces cerevisiae* transformation; e.g., Bio-Rad)
   - Plating beads
   - Microtiter plate incubation shaker (e.g., Heidolph Titramax T1000)
   - Luminescence plate reader (e.g., Tecan Infinite® M200)
   - Incubation shaker (e.g., Infors HT Ecotron)
2. Vectors
   - pBK164 and pBK129 (Hartig Lab, University of Konstanz) for cloning (Klauser et al., 2014)
   - pBT3-C (MoBiTec GmbH; negative control), pBK117 (positive control; encodes Gal4 protein), pBK147 (encodes active type 3 hammerhead ribozyme of *S. mansoni* in the 3′-UTR of Gal4 mRNA), pBK154 (encodes inactive type 3 hammerhead ribozyme of *S. mansoni* in the 3′-UTR of Gal4 mRNA) in the 3′-UTR of Gal4 mRNA) are required for the screening process as controls (Klauser et al., 2014). Transform plasmids into the *S. cerevisiae* MaV203 strain using standard lithium acetate procedure.
3. PCR
   - Phusion Hot Start II High-Fidelity DNA polymerase, 5 × HF buffer, and 100% DMSO (NEB). Store at −20 °C.
   - DNA oligonucleotides used for PCR amplification are ordered from your supplier of choice. Prepare stock solution of 100 μ*M*.
   - dNTP mix containing 2 m*M* of each dATP, dCTP, dGTP, and dTTP. Store at −20 °C.
4. Fragment preparation and purification
   - DNA Gel recovery kit (e.g., Gel DNA Recovery Kit; Zymo Research)
   - 5 × TBE buffer: Dissolve 54 g Trizma base, 27.5 g boric acid, 20 mL 0.5 *M* EDTA (pH 8.0) in 1 L of H_2O. Adjust to pH 8.3 with concentrated HCl. Dilute to 0.5 × in H_2O before use.

- Ethidium bromide solution (500 μg/L in H_2O). Solution can be reused when kept in the dark at room temperature.
- Ethidium bromide destaining solution: 400 mL H_2O. Can be kept at room temperature and reused.
- 6 × Agarose gel loading buffer: Dissolve 4 mL 50 m*M* Tris–HCl, pH 7.6, 12 mL 100% glycerol, 2.4 mL 0.5 *M* EDTA, 6 mg bromophenol blue, 6 mg xylene cyanol in 20 mL, prepare aliquots of 1 mL. Buffer can be stored at −20 °C for several months.
- Ready-to-use DNA size standard (e.g., GeneRuler 1 kb DNA Ladder and Ultra Low Range DNA Ladder; Thermo Scientific)
- DNA purification kit (e.g., DNA Clean and Concentrator™; Zymo Research)
- Agarose gels: dissolve agarose in 0.5 × TBE by boiling in a microwave oven.
- T4 DNA ligation kit (2,000,000 U/mL) and T4 DNA Ligase Reaction Buffer (NEB)
- Razor blade
- *Spe*1-HF (20,000 U/mL), *Not*1-HF (20,000 U/mL), and Cut smart buffer (NEB)
- Antarctic phosphatase (5000 U/mL; NEB)

5. Ligation and electroporation

- 1000 × Kanamycin stock solution (30 mg/mL): Prepare solution in H_2O and store at −20 °C.
- LB-agar plates supplemented with 30 μg/mL kanamycin (LB-Kan): Dissolve 10 g tryptone, 5 g yeast extract, 5 g NaCl, and 10 g agar agar in 1 L H_2O. Sterilize by autoclaving, cool down to 45 °C, and supplement solution with 30 mg/L kanamycin. Pour plates in 8.5-cm Petri dishes. Plates can be stored at 4 °C for up to 6 weeks.
- SOC medium: Dissolve 20 g tryptone, 5 g yeast extract, 0.6 g NaCl, 0.2 g KCl in 900 mL H_2O. Adjust pH to 6.8–7.0 using NaOH. Adjust volume to 960 mL and autoclave. Add 10 mL 1 *M* $MgCl_2$, 10 mL 1 *M* $MgSO_4$, and 20 mL 1 *M* glucose from sterile filtered stocks. Store 1 mL aliquots at −20 °C for several months.
- Electrocompetent *E. coli* XL10 gold (Stratagene) (genotype: *Tet^r; Δ(mcrA)183; Δ(mcrCB-hsdSMR-mrr)173; endA1; supE44; thi1; recA1; gyrA96; relA1; lac; Hte [F'; proAB; $lacI^q Z\Delta M15$; Tn10; (Tet^r); Amy; Cam^r]*. 80 μL aliquots frozen at −80 °C.
  - *Note*: A detailed protocol for the preparation of electrocompetent *E. coli* can be found at www.eppendorf.com. You can use another

E. coli strain, as long as the transformation efficiency is sufficiently high to cover the theoretical sequence space of the aptazyme library. A high transformation efficiency is supported by the Hte phenotype.
- Plasmid DNA recovery kit (e.g., Zyppy™ Plasmid DNA purification kit; Zymo research)

6. Transformation of the aptazyme library into yeast
   - *S. cerevisiae* MaV203 cells (Invitrogen™) (genotype: *MATα; leu2-3,112; trp1-901; his3Δ200; ade2-101; cyh2*^{R}*; can1*^{R}*; gal4Δ; gal80Δ; GAL1::lacZ; HIS3*$_{UASGAL1}$*::HIS3@LYS2; SPAL10::URA3.*)
   - YPDA liquid medium: Dissolve 50 g Yeast Extract-Peptone-Dextrose (YPD)-broth with 40 mg adenine hemisulfate in 1 L H_2O. Sterilize by autoclaving
   - Electroporation buffer: Aqueous solution containing 1 *M* sorbitol and 1 m*M* calcium chloride. Sterilize by autoclaving.
   - Aqueous solution of 0.1 *M* lithium acetate and 10 m*M* dithiothreitol. Sterilize by filtration.
   - 1:1 Mix of aqueous 1 *M* sorbitol and YPDA liquid medium

2.2. Selection and hit identification of the aptazyme library

1. Yeast cell culture
   - 0.9% Aqueous sodium chloride solution. Sterilize by autoclaving.
   - Aqueous 1 *M* 3-AT (84.1 g/L). Prepare freshly before use and sterilize by filtration.
   - 100 mg/mL 5-FOA in DMSO. Aliquots can be stored at −20 °C.
   - Aminoglycoside antibiotic stock solution, e.g., 50 mg/mL neomycin. Prepare sterile stock solution in H_2O and sterilize by filtration.
     - *Note*: 100 μg/mL Neomycin was used as final concentration for screening neomycin switches; however, screening concentrations for other ligands will have to be determined individually; more information is given in Section 3.4.
   - Synthetic complete (SC) medium without leucine (SC-Leu) liquid medium: Dissolve 6.7 g Yeast Nitrogen Base without amino acids (YNB; Sigma), 20 g glucose, 1.6 g Yeast synthetic drop-out supplements without leucine (Sigma), and 40 mg adenine hemisulfate in 1 L H_2O. Sterilize by autoclaving.
   - SC-Leu agar plates: These are prepared as SC-Leu liquid medium with additional 20 g/L agar agar. To increase solidity of agar, we recommend adding 0.3 g/L sodium hydroxide.

- Selection plates: Use 6.7 g/L Yeast Nitrogen Base without amino acids (YNB; Sigma), 2% (w/v) glucose, 2% (w/v) agar agar, 1.4 g/L Yeast synthetic drop-out supplements without histidine, leucine, tryptophane, and uracil (Sigma), and supplement as required with 2 g/L uracil, 7.6 g/L tryptophane, 7.6 g/L histidine, 2 g/L adenine hemisulfate, and aminoglycoside antibiotic.
  - *Note*: 3-AT, 5-FOA, and aminoglycosides are heat labile. Add after cooling down to 45 °C.

2. Screening and characterization of potential hits
   - Gal-Screen™ β-galactosidase reporter gene assay system for yeast cells (Life Technologies)
   - 24-well Microtiter plate
   - 96-well Microtiter plate
   - 96-well Half-area black microtiter plate

3. PROTOCOL

3.1. Aptazyme library construction

A pool of aptazyme mutants, as shown in Fig. 1B, is generated by PCR. Library diversity is obtained by using primer with unbiased random positions generated during solid-phase DNA synthesis using a 1:1:1:1 mixtures of nucleoside phosphoramidites. We recommend ordering primers HPLC purified or to purify primers by denaturing PAGE, for example, as described by Sambrook and Russell David (2001). Integration of the aptazyme library into the target vector pBK129 is performed using restriction endonucleases *Spe*I and *Not*I. The plasmid pool carrying the aptazyme library is propagated in the *E. coli* XL10 gold strain for amplification. Finally the recovered aptazyme library is transformed into the *S. cerevisiae* MaV203 strain.

1. Primers are designed according to following rules (see neomycin aptazyme example below):
   - Forward primer: Introduce *Spe*I endonuclease site and aptamer sequence (underlined)
   - Introduce randomized nucleotides at the bridge between the aptamer and the HHR catalytic core (highlighted gray)
   - Add A-rich linker sequence (10–15 nucleotides) between *Spe*I site and aptamer domain (bold)
   - Add additional nucleotides to the 5′ end of the restriction enzyme recognition site for efficient cleavage close to the end of DNA fragments (italic)

- Example: neomycin aptazyme
  5′-*ctcttc*ACTAGT**AAACAAACAAA**GCCGGCATAGCTTGTCCTTTAATGGTCCTATGTNNNGTCCTGGATTCCACGGTACATC-3′
- Reverse primer: Introduce *Not*I endonuclease site (underlined)
- Introduce randomized nucleotides (highlighted gray)
- Add T-rich linker sequence (10–15 nucleotides) between randomized nucleotides and *Not*I site (bold)
- Add additional nucleotides to the 5′ end for more complete restriction digest (italic)
- Example: neomycin aptazyme
  5′-*ctcttc*GCGGCCG**CTTTTTCTTTTT**NNNNNTTTCGTCCTATTTGGGACTC-3′

2. Prepare the reaction mixture and aliquot the mixture into PCR tubes (35 μL per tube)
   - *Note*: If you make use of another polymerase, the protocol needs to be changed according to the manufacturer's recommendations. We favor the application of a high-fidelity polymerase (Table 1).
3. Run PCR with Phusion Hot Start II High-Fidelity DNA polymerase according to following setup:
   - *Note*: We recommend to run a 2.5% (w/v) agarose gel to monitor successful PCR amplification (Table 2)

Table 1 PCR reaction mixture

Stock concentration	Material	Volume (μL)	Final concentration
5 ×	HF buffer	30	1 ×
2 m*M*	dNTP mix	15	200 μ*M*
100 μ*M*	Forward primer	0.9	600 n*M*
100 μ*M*	Reverse primer	0.9	600 n*M*
20 ng/μL	pBK164	1.5	30 ng
100% (v/v)	DMSO	4.5	3% (v/v)
2 U/μL	Phusion Hot Start II DNA polymerase	1.5	3 U
H_2O		95.7	
Total		150	

Table 2 PCR cycling conditions

Step	Stage	Temperature (°C)	Time	Go to step	Repeat
1	Initial denaturation	98	30 s		
2	Denaturation	98	10 s		
3	Annealing	60	30 s		
4	Extension	72	5 s	2	24
5	Final extension	72	7 min		

Table 3 Endonuclease DNA digestion mixture

		Insert (aptazyme library)	Vector (pBK129 200 ng/μL)
Stock concentration	Material	Volume (μL)	Volume (μL)
	PCR-Product/Vector	86	50
10 ×	CutSmart	10	25
20 U/μL	*Spe*1-HF	2	5
20 U/μL	*Not*1-HF	2	5
H_2O		66	165
Total volume		100	250

4. Purify PCR product using DNA Purification Kit. We recommend to increase the volume of DNA Binding Buffer to up to 7 Vol when purifying small fragments using DNA Clean and Concentrator kit; Zymo Research. Elute DNA with 86 μL H_2O.
5. Set up following reaction mixture for *Spe*1 and *Not*1 digestion. Incubate samples at 37 °C for 3 h (Table 3).
6. Purify digested pBK129 using DNA Purification Kit. Use multiple columns, if the capacity of a single column is not sufficient. Elute DNA in a total of 87 μL H_2O. Add 10 μL 10× antarctic phosphatase buffer and 3 μL antarctic phosphatase (5 U/μL) and incubate for 1 h at 37 °C.
7. Purify digested PCR-product (ca. 160 bp; depending on size of the aptamer sequence) and dephosphorylated pBK129 (9486 bp) by performing gel electrophoresis using a 2.5 % (w/v) and 0.8 % (w/v) agarose gel, respectively, in 0.5 × TBE. Prepare gels that provide a large well for

the PCR product and a small well for the size standard. Mix the sample with 6× agarose gel loading buffer before loading into the pocket. Use "ultra low range" size standard for the purification of the PCR product and 1 kb size standard for the purification of pBK129. Run the gel at 10 V cm^{-1} for 75–90 min.

8. Stain agarose gel in ethidium bromide staining solution for 10 min, then incubate for 10 min in H_2O. To avoid UV-light-induced mutations to the DNA, expose only part of the sample on a UV light table (excise the marker and a small portion of the size-fragmented PCR product/pBK129) to identify the correctly sized band.
9. Excise the respective bands and extract the DNA fragment using the DNA gel extraction kit according to the manufacturer's protocol. Elute with 20 μL H_2O.
   - *Note*: The band of the double-cut pBK129 vector is not distinguishable from the single cut band and is also excised. Therefore, it is important to perform the digestion to completion. A control ligation of dephosphorylated vector without insert, which is also transformed into *E. coli* XL10 gold, gives information about this issue.
10. Measure DNA concentrations and mix the necessary materials for ligation. Use a sevenfold molar excess of insert over digested pBK129. Incubate at 16 °C for 12 h (Table 4).
11. Purify the ligated DNA sample with a DNA purification kit to remove the ligation buffer for better transformation efficiency. Elute ligated DNA using 5 μL H_2O.

Table 4 Ligation mixture

Starting concentration	Material	Control ligation (μL)	Library ligation (μL)
50 ng/μL	*Spe*1/*Not*1-digested pBK129	1	4
10 ng/μL	*Spe*1/*Not*1-digested aptazyme library	–	2.4
10×	T4 ligase buffer	2	8
2000 U/μL	T4 ligase	0.2	0.8
H_2O		16.8	64.8
Total volume		20	80

3.2. Transformation of the aptazyme library into *E. coli* XL10 gold

1. For each transformation, preheat one aliquot of SOC medium at 37 °C, place four LB-Kan agar plates at 37 °C, and chill one 1-mm electroporation cuvette on ice. Thaw one aliquot of electrocompetent *E. coli* XL10 gold (80 μL) on ice.
2. Carefully mix 80 μL of the thawed *E. coli* cells with 2.5 μL ligated plasmid. Transfer the mixture into the chilled electroporation cuvette. Subsequently perform the transformation by applying a pulse (1800 V). Add 800 μL prewarmed SOC medium to the pulsed cells and transfer the mixture into a fresh 1.5-mL reaction tube. Incubate cells shaking at 37 °C for 45 min.
3. Centrifuge samples in a tabletop centrifuge at 13.4 k rpm for 15 s. Discard the supernatant and resuspend pellet in 150 μL SOC medium.
4. For the determination of the transformation efficiency, prepare serial dilutions of the transformed cells.
   - *Note*: We prepare serial dilutions by mixing 1.5 μL transformed cells with 148.5 μL 0.9% NaCl. From this initial dilution, we prepare two serial 1:10 dilutions.
5. Plate transformed cells on a prewarmed LB-Kan agar plate supplemented with kanamycin. Use plating beads to spread cells evenly. Incubate plates overnight at 37 °C.
6. Determine the transformation efficiency by counting the number of colonies on the plates with the serially diluted samples. The number of colonies of the sample with the control ligation should be reduced at least by a factor of 100 in comparison to the aptazyme library. The library size should exceed the theoretical sequence space at least threefold to obtain a pool coverage of greater 95% (Reetz et al., 2008).
7. Scrape the aptazyme library from the LB-Kan agar plates using a sterilized inoculating loop and transfer cells into 10 mL 0.9 % (w/v) aqueous NaCl solution. Centrifuge samples and isolate using Zymo Plasmid DNA Purification kit. Make sure to isolate greater than 10 μg of the aptazyme plasmid library for downstream applications.
   - *Note*: We recommend performing a sequence analysis of several single colonies to investigate the integrity of the plasmid library (sequencing primer: e.g., BAC2: 5′-CCCAGTCACGACGTTGTAAAAC-3′). In addition we recommend preparing a glycerol stock of the aptazyme library which can be stored at −80 °C for many years.

3.3. Generation of electrocompetent yeast cells

Herein, we describe a protocol for the electroporation of yeast cells. The method was previously described in literature and adapted for our purposes (Benatuil et al., 2010). There are other excellent protocols for the high-efficiency transformation of yeast cells described which are based on lithium acetate transformation (Gietz & Schiestl, 2007).

1. Use a sterile inoculating loop to scrape a 2-mm single colony of the MaV203 *S. cerevisiae* strain into 5 mL YPDA liquid medium. Resuspend well and transfer yeast cell suspension into 2-L Erlenmeyer flask with 1 L YPDA medium. Grow at 30 °C at 120 rpm until OD_{600} is in the range of 1.3–1.5 (takes about 16 h).
2. Use ten 50-mL centrifuge tubes to collect cells by centrifugation at 18,000 × *g* for 3 min. Discard supernatant and repeat once to collect remaining yeast cells.
3. Wash the pellet twice with 45 mL ice-cold H_2O and once with 45 mL ice-cold electroporation buffer (aqueous solution of 1 *M* sorbitol and 1 m*M* $CaCl_2$). Resuspend well, collect cells by centrifugation at 18,000 × *g* for 3 min, and discard supernatant after each washing step.
4. Condition yeast cell by resuspending each cell pellet in 25 mL 0.1 *M* LiAc/10 m*M* DTT. Incubate cell suspension vigorously shaking for 30 min at 30 °C.
5. Collect conditioned cells by centrifugation, discard supernatant, and resuspend each pellet in 17.5 mL ice-cold electroporation buffer. Reduce number of 50-mL centrifuge tubes to four by pooling cell suspensions.
6. Collect conditioned yeast cells by centrifugation, and resuspend each pellet in 45 mL of ice-cold electroporation buffer.
7. Collect conditioned cells by centrifugation and discard supernatant. Resuspend each cell pellet in 400 μL electroporation buffer. Keep cells on ice and move on to electroporation immediately.
8. Electroporation of yeast cells. Gently mix 400 μL electrocompetent MaV203 cells and 10 μg plasmid on ice. Transfer mixture immediately into prechilled 2-mm electroporation cuvette and pulse with 2500 V. Transfer electroporated cells into 5 mL of 1:1 mix of 1 *M* sorbitol and YPDA liquid medium prewarmed to 30 °C and incubate in an incubation shaker under vigorous shaking at 30 °C for 90 min. Prepare serial dilutions as described in Section 3.2 to determine transformation

efficiency. Plate yeast suspension onto SC-Leu plates. We usually obtain 200,000–800,000 cfu per transformation.

9. Collect the transformed yeast cells into 10 mL 0.9% (w/v) NaCl solution. We recommend preparing a glycerol stock of the yeast aptazyme library, which can be stored at −80 °C for many years.

3.4. Selection and screening of the yeast aptazyme library and "hit" identification

The concept of the yeast counterselection is described in Fig. 3. Depending on whether one selects for ON- or OFF-switches of gene expression, the aptazyme ligand is added to the agar plates supplemented with 3-AT (ON) or 5-FOA (OFF). We recommend investigating the toxicity of the aminoglycoside antibiotic to the yeast MaV203 cells prior to the selection. Toxicological profiling is based on the monitoring of the proliferation at increasing concentrations of the aminoglycoside antibiotic. In principle, an antibiotic concentration as high as possible should be applied in the selection setup (a minor impact on the proliferation pattern is usually well tolerated).

Successful selection is dependent on many factors. There are excellent reviews that give potential solutions to problems showing up in yeast-based selections (Stynen, Tournu, Tavernier, & Van Dijck, 2012). We obtained best results by applying 3-AT concentrations in the 30–40 m*M* range and incubating for less than 5 days. We recommend making use of the uracil auxotrophy of yeast MaV203 cells by omitting uracil on 3-AT agar plates to reduce the number of false-positive colonies growing up. We obtained best results when we performed the negative selection after the positive selection. Usually one does not know which applied 5-FOA concentration results in optimal screening results because growth on 5-FOA depends in part on the toxicity of the aptazyme ligand. Titration of the 5-FOA concentrations in the range of 0.03–0.075% (w/v) and conducting the negative selection with increasing stringency helps to overcome this issue.

1. For positive selection, plate 100 μL of yeast suspension ($OD_{600}=1$, corresponding to approximately 10^7 cfu/mL) on agar plates supplemented with 3-AT. We recommend exceeding the theoretical library size by a factor of at least 10. Do not increase the cell density to obtain full coverage of the aptazyme library, but instead use additional agar plates. Incubate at 37 °C for 3–5 days.
2. Collect colonies, wash with 0.9% (w/v) NaCl, and preincubate in 2 mL SC-Leu liquid medium supplemented with and without antibiotics

(final $OD_{600}=1$). Incubate under shaking in an incubation shaker at 30 °C overnight. Wash cells with aqueous 0.9% (w/v) NaCl and monitor OD_{600}.

- *Note*: Colonies harvested from plates containing 3-AT and antibiotic are preincubated in liquid medium without antibiotic. Accordingly, colonies collected from the agar plates without antibiotic are preincubated in medium containing antibiotic.

3. Negative selection is carried out on plates supplemented with 5-FOA according to the scheme shown in Fig. 3. For optimal selection results, prepare two types of selection plates supplemented with 5-FOA concentrations in the range of 0.03–0.075% (w/v). Prepare yeast suspensions with increasing cellular density (OD_{600} ranging from 0.01 to 1). Plate 100 μL yeast suspension on negative selection plates and spread evenly using glass beads. Incubate agar plates at 30 °C for 3–5 days.
4. Screening of single colonies of the negative selection is based on measuring the β-galactosidase activity of yeast cultures grown in the absence and presence of aptazyme ligand. Use 24-well microtiter plates to inoculate individual colonies in 1 mL SC-Leu liquid medium and grow yeast cultures shaking at 30 °C in an incubation shaker to stationary phase. Also include yeast cultures transformed with plasmids pBT3-C, pBK117, pBK147, and pBK154 as controls during the screening process. The next day 10 μL of yeast suspension are transferred into fresh SC-Leu liquid medium that is supplemented with the antibiotic ligand and without it. Grow yeast cultures by shaking at 30 °C to stationary phase and determine Gal4 expression levels by using the Gal-Screen™ β-galactosidase reporter gene assay system for yeast cells.
5. For measuring Gal4 expression levels, resuspend the yeast culture. Transfer 100 μL of resuspended culture into 96-well transparent microtiter plate for OD_{600} read-out and 25 μL into a 96-well half-area black microtiter plate for measuring β-galactosidase activity. Perform Gal-Screen™ β-galactosidase reporter gene assay according to the manufacturer's recommendations. Make sure to adjust volumes to the used volume of yeast culture and to vigorously shake the microtiterplate (1350 rpm for Heidolph Titramax 1000) after addition of the chemiluminescent substrate. Calculate Gal4 expression levels by performing an OD_{600} correction of the measures luminescence values and by subtracting Gal4 expression of the negative control. Identify potential aptazyme switches by calculating the ratio of Gal4 expression levels of cultures grown in the absence and presence of aptazyme ligand. Under optimal conditions, the

Gal4 expression levels of the control cultures is not influenced by the ligand.

6. Perform a sequence analysis of potential switches by isolating the plasmid from overnight culture and sending it for sequencing.
   - *Note*: We isolate yeast plasmids by performing lysis with 0.5-mm glass beads and then using a standard bacterial plasmid purification kit according to the manufacturer's recommendations. A fragment of the plasmid containing the aptazyme is amplified by standard Taq PCR using the primers pBK129-f 5′-ACGATGTGCAGCGTACC-3′ and pBK129-r 5′-GGCGATTAAGTTGGGTAACG-3′. The resulting ~2.1-kb fragment is purified by employing a DNA Purification Kit and is sent in for sequencing using the BAC2 primer.
7. Proof the aptazyme-dependent mechanism of the screened genetic switches by conducting additional experiments. For initial validation of potential hits, we suggest measuring the Gal4 expression levels of the most promising yeast culture as described above. Measure at least duplicates and use increasing concentrations of the aptazyme ligand. Furthermore, we recommend the generation of a catalytically inactive HHR variant which carries an A-to-G mutation within its catalytic core (see Fig. 1) by molecular subcloning to validate the ribozyme-dependent mechanism. Yeast cultures expressing the catalytically inactive aptazyme should display a comparable reporter gene expression level as the cultures transformed with pBK154 (encodes inactive type 3 hammerhead ribozyme of *S. mansoni* in the 3′-UTR of Gal4 mRNA) and should not respond to the aptazyme ligand anymore. In addition, we recommend the generation of an aptazyme variant that does not bind to its ligand anymore by inserting mutations which impair ligand binding.

REFERENCES

Auslander, S., Ketzer, P., & Hartig, J. S. (2010). A ligand-dependent hammerhead ribozyme switch for controlling mammalian gene expression. *Molecular BioSystems*, *6*(5), 807–814. http://dx.doi.org/10.1039/b923076a.

Benatuil, L., Perez, J. M., Belk, J., & Hsieh, C. M. (2010). An improved yeast transformation method for the generation of very large human antibody libraries. *Protein Engineering, Design & Selection*, *23*(4), 155–159. http://dx.doi.org/10.1093/protein/gzq002.

Berschneider, B., Wieland, M., Rubini, M., & Hartig, J. S. (2009). Small-molecule-dependent regulation of transfer RNA in bacteria. *Angewandte Chemie International Edition in English*, *48*(41), 7564–7567. http://dx.doi.org/10.1002/anie.200900851.

De la Pena, M., Gago, S., & Flores, R. (2003). Peripheral regions of natural hammerhead ribozymes greatly increase their self-cleavage activity. *The EMBO Journal*, *22*(20), 5561–5570. http://dx.doi.org/10.1093/emboj/cdg530.

Fechner, H., Wang, X., Srour, M., Siemetzki, U., Seltmann, H., Sutter, A. P., et al. (2003). A novel tetracycline-controlled transactivator-transrepressor system enables external control of oncolytic adenovirus replication. *Gene Therapy, 10*(19), 1680–1690. http://dx.doi.org/10.1038/sj.gt.3302051.

Ferbeyre, G., Smith, J. M., & Cedergren, R. (1998). Schistosome satellite DNA encodes active hammerhead ribozymes. *Molecular and Cellular Biology, 18*(7), 3880–3888.

Gietz, R. D., & Schiestl, R. H. (2007). High-efficiency yeast transformation using the LiAc/SS carrier DNA/PEG method. *Nature Protocols, 2*(1), 31–34. http://dx.doi.org/10.1038/nprot.2007.13.

Ketzer, P., Haas, S. F., Engelhardt, S., Hartig, J. S., & Nettelbeck, D. M. (2012). Synthetic riboswitches for external regulation of genes transferred by replication-deficient and oncolytic adenoviruses. *Nucleic Acids Research, 40*(21), e167. http://dx.doi.org/10.1093/nar/gks734.

Ketzer, P., Kaufmann, J. K., Engelhardt, S., Bossow, S., von Kalle, C., Hartig, J. S., et al. (2014). Artificial riboswitches for gene expression and replication control of DNA and RNA viruses. *Proceedings of the National Academy of Sciences of the United States of America, 111*(5), E554–E562. http://dx.doi.org/10.1073/pnas.1318563111.

Khvorova, A., Lescoute, A., Westhof, E., & Jayasena, S. D. (2003). Sequence elements outside the hammerhead ribozyme catalytic core enable intracellular activity. *Nature Structural Biology, 10*(9), 708–712. http://dx.doi.org/10.1021/sb500062p.

Klauser, B., Atanasov, J., Siewert, L. K., & Hartig, J. S. (2014). Ribozyme-based aminoglycoside switches of gene expression engineered by genetic selection in S. cerevisiae. *ACS Synthetic Biology*, http://dx.doi.org/10.1021/sb500062p.

Klauser, B., & Hartig, J. S. (2013). An engineered small RNA-mediated genetic switch based on a ribozyme expression platform. *Nucleic Acids Research, 41*(10), 5542–5552. http://dx.doi.org/10.1093/nar/gkt253.

Kumar, D., An, C. I., & Yokobayashi, Y. (2009). Conditional RNA interference mediated by allosteric ribozyme. *Journal of the American Chemical Society, 131*(39), 13906–13907. http://dx.doi.org/10.1021/ja905596t.

Mandal, M., Boese, B., Barrick, J. E., Winkler, W. C., & Breaker, R. R. (2003). Riboswitches control fundamental biochemical pathways in Bacillus subtilis and other bacteria. *Cell, 113*(5), 577–586.

Ogawa, A., & Maeda, M. (2008). An artificial aptazyme-based riboswitch and its cascading system in E. coli. *Chembiochem, 9*(2), 206–209. http://dx.doi.org/10.1002/cbic.200700478.

Reetz, M. T., Kahakeaw, D., & Lohmer, R. (2008). Addressing the numbers problem in directed evolution. *Chembiochem, 9*(11), 1797–1804. http://dx.doi.org/10.1002/cbic.200800298.

Ringner, M., & Krogh, M. (2005). Folding free energies of 5'-UTRs impact post-transcriptional regulation on a genomic scale in yeast. *PLoS Computational Biology, 1*(7), e72. http://dx.doi.org/10.1371/journal.pcbi.0010072.

Rohmer, S., Mainka, A., Knippertz, I., Hesse, A., & Nettelbeck, D. M. (2008). Insulated hsp70B' promoter: Stringent heat-inducible activity in replication-deficient, but not replication-competent adenoviruses. *The Journal of Gene Medicine, 10*(4), 340–354. http://dx.doi.org/10.1002/jgm.1157.

Sambrook, J., & Russell David, W. (2001). Purification of synthetic oligonucleotides by polyacrylamide gel electrophoresis. In J. Argentine (Ed.), *Molecular cloning: A laboratory manual: Vol. 2* (3rd ed., pp. 10.11–10.16). Cold Spring Harbor, NY: Cold Spring Harbor Laboratory Press, Chapter 10.

Saragliadis, A., & Hartig, J. S. (2013). Ribozyme-based transfer RNA switches for post-transcriptional control of amino acid identity in protein synthesis. *Journal of the American Chemical Society, 135*(22), 8222–8226. http://dx.doi.org/10.1021/ja311107p.

Saragliadis, A., Krajewski, S. S., Rehm, C., Narberhaus, F., & Hartig, J. S. (2013). Thermozymes: Synthetic RNA thermometers based on ribozyme activity. *RNA Biology*, *10*(6), 1010–1016. http://dx.doi.org/10.4161/rna.24482.

Stynen, B., Tournu, H., Tavernier, J., & Van Dijck, P. (2012). Diversity in genetic in vivo methods for protein-protein interaction studies: From the yeast two-hybrid system to the mammalian split-luciferase system. *Microbiology and Molecular Biology Reviews*, *76*(2), 331–382. http://dx.doi.org/10.1128/MMBR.05021-11.

Tang, J., & Breaker, R. R. (1997). Rational design of allosteric ribozymes. *Chemical Biology*, *4*(6), 453–459.

Vidal, M., Brachmann, R. K., Fattaey, A., Harlow, E., & Boeke, J. D. (1996). Reverse two-hybrid and one-hybrid systems to detect dissociation of protein-protein and DNA-protein interactions. *Proceedings of the National Academy of Sciences of the United States of America*, *93*(19), 10315–10320.

Vidal, M., Braun, P., Chen, E., Boeke, J. D., & Harlow, E. (1996). Genetic characterization of a mammalian protein-protein interaction domain by using a yeast reverse two-hybrid system. *Proceedings of the National Academy of Sciences of the United States of America*, *93*(19), 10321–10326.

Wieland, M., Benz, A., Klauser, B., & Hartig, J. S. (2009). Artificial ribozyme switches containing natural riboswitch aptamer domains. *Angewandte Chemie International Edition in English*, *48*(15), 2715–2718. http://dx.doi.org/10.1002/anie.200805311.

Wieland, M., Berschneider, B., Erlacher, M. D., & Hartig, J. S. (2010). Aptazyme-mediated regulation of 16S ribosomal RNA. *Chemical Biology*, *17*(3), 236–242. http://dx.doi.org/10.1016/j.chembiol.2010.02.012.

Wieland, M., & Hartig, J. S. (2008). Improved aptazyme design and in vivo screening enable riboswitching in bacteria. *Angewandte Chemie International Edition in English*, *47*(14), 2604–2607. http://dx.doi.org/10.1002/anie.200703700.

Win, M. N., & Smolke, C. D. (2007). A modular and extensible RNA-based gene-regulatory platform for engineering cellular function. *Proceedings of the National Academy of Sciences of the United States of America*, *104*(36), 14283–14288. http://dx.doi.org/10.1073/pnas.0703961104.

Win, M. N., & Smolke, C. D. (2008). Higher-order cellular information processing with synthetic RNA devices. *Science*, *322*(5900), 456–460. http://dx.doi.org/10.1126/science.1160311.

Winkler, W., Nahvi, A., & Breaker, R. R. (2002). Thiamine derivatives bind messenger RNAs directly to regulate bacterial gene expression. *Nature*, *419*(6910), 952–956. http://dx.doi.org/10.1038/nature01145.

Winkler, W. C., Nahvi, A., Sudarsan, N., Barrick, J. E., & Breaker, R. R. (2003). An mRNA structure that controls gene expression by binding S-adenosylmethionine. *Nature Structural Biology*, *10*(9), 701–707. http://dx.doi.org/10.1038/nsb967.

CHAPTER SIXTEEN

Kinetic Folding Design of Aptazyme-Regulated Expression Devices as Riboswitches for Metabolic Engineering

David Sparkman-Yager, Rodrigo A. Correa-Rojas, James M. Carothers[1]
Department of Chemical Engineering, Molecular Engineering and Sciences Institute, Center for Synthetic Biology, University of Washington, Seattle, WA, USA
[1]Corresponding author: e-mail address: jcaroth@uw.edu

Contents

Abstract

Recent developments in the fields of synthetic biology and metabolic engineering have opened the doors for the microbial production of biofuels and other valuable organic compounds. There remain, however, significant metabolic hurdles to the production of these compounds in cost-effective quantities. This is due, in part, to mismatches between the metabolic engineer's desire for high yields and the microbe's desire to survive. Many valuable compounds, or the intermediates necessary for their biosynthesis, prove deleterious at the desired production concentrations. One potential solution to these toxicity-related issues is the implementation of nonnative dynamic genetic control mechanisms that sense excessively high concentrations of metabolic intermediates and respond accordingly to alleviate their impact. One potential class of dynamic regulator is the riboswitch: *cis*-acting RNA elements that regulate the expression of

ISSN 0076-6879
http://dx.doi.org/10.1016/bs.mie.2014.10.038

downstream genes based on the presence of an effector molecule. Here, we present combined methods for constructing aptazyme-regulated expression devices (aREDs) through computational cotranscriptional kinetic folding design and experimental validation. These approaches can be used to engineer aREDs within novel genetic contexts for the predictable, dynamic regulation of gene expression *in vivo*.

1. INTRODUCTION

As the fields of synthetic biology and metabolic engineering have progressed, the ability to produce valuable chemical compounds through microbial biosynthesis has greatly expanded. It is becoming apparent, however, that the production of such compounds at commercially viable yields is not as simple as providing microbes with the DNA for enzymes to convert feedstock to product (Carothers, 2013; Carothers, Goler, & Keasling, 2009). Though the heterologous genes provide the essential catalytic functions that are required to construct a synthetic metabolic pathway, most of these components originate from disparate organisms and now lack their native control mechanisms. Natural biological systems have evolved layers of complex genetic control, dynamically regulated by physicochemical inputs from both their intra- and extracellular environments (Zhang, Carothers, & Keasling, 2012). This allows the microbes to prioritize survival over lesser cellular goals, redirecting metabolic flux down critical pathways when stress signals are recognized. At present, the process of engineering similar regulatory control circuits for engineered pathways continues to be challenging.

Novel metabolic pathways often contain key enzymes and intermediates with properties at odds with the primary goals of the host; metabolic pathway components can be cytotoxic, lost from the cell, deplete cellular nutrients, or interfere with essential cellular functions. In the future, the design of increasingly large metabolic pathways, able to produce sufficient product yields, will rely upon the implementation of sensor-regulator systems, capable of varying metabolic flux based on critical cellular concentrations. Here, we describe a methodology for engineering and implementing RNA-based devices for the dynamic control of gene expression.

Aptazymes are synthetic RNA structures that combine self-cleaving ribozyme sequences with *in vitro* selected, ligand-binding aptamer domains (Carothers, Goler, Kapoor, Lara, & Keasling, 2010). To date, aptazymes exhibiting ligand-dependent phosphodiester bond cleavage have been produced through combinations of *in vitro* selection, rational design, and computational and *in vivo* screening strategies (Goler, Carothers, & Keasling, 2014; Link et al., 2007; Penchovsky, 2013). Functional aptazymes have been successfully assembled into riboswitches that can dynamically control gene expression in bacteria, yeast, and mammalian cells (Ausländer, Ketzer, & Hartig, 2010; Wieland & Hartig, 2008; Win & Smolke, 2007). We have previously shown that aptazymes can be assembled into aptazyme-regulated expression devices (aREDs) with quantitatively predictable functions in *Escherichia coli* (Carothers, Goler, Juminaga, & Keasling, 2011). Programmable aREDs could have great utility as riboswitch-based biosensors to screen for production from engineered metabolic pathways, or as direct controllers of metabolic enzyme expression.

Here, we describe an approach to molecular engineering that combines cotranscriptional cleavage assays and kinetic RNA secondary structure folding simulations to construct novel riboswitches. The predictive power of RNA folding simulations allows the construction of genetic devices with increasing complexity from individual RNA components (parts) with known function. Cotranscriptional folding trajectories can have a large influence on RNA function within a cellular environment. While minimum free energy (MFE) simulations may be sufficient for predicting the folds of RNA molecules at equilibrium, or those of small sequences, they are not sufficient for the kinetic trajectories important for riboswitch, terminator, or aptazyme design (Cambray et al., 2013; Carothers et al., 2011; Wickiser, Winkler, Breaker, & Crothers, 2005).

Here, we focus on our own experience with the design and implementation of a *p*-aminophenylalanine (*p*-AF) aRED biosensor. *p*-AF is a metabolic intermediate in the production of *p*-AS, a vinyl aromatic monomer that can be formed into polymers with potential applications in biomedicine, photonics, and photolithography (Stevens & Carothers, 2014). The pathway for the production of *p*-AS provides an excellent test-bed in which to apply dynamic expression devices, as it contains several critical metabolic control problems (Fig. 1). Here, we provide an in-depth computational and experimental methodology (Fig. 2) for the implementation of aREDs that could be constructed to regulate the expression of metabolic enzymes in this pathway. In particular, we focus on the methods for the implementation of an existing aptazyme, *p*-AF-2 (Fig. 3) responsive to the pathway intermediate *p*-AF, within a cellular context.

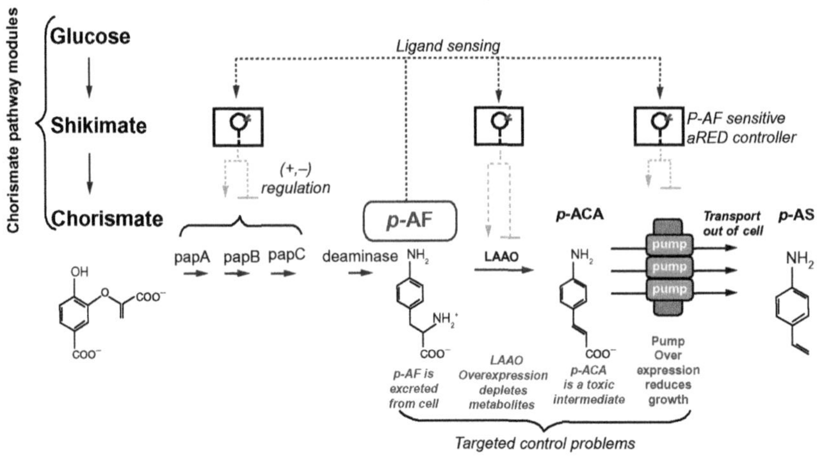

Figure 1 Schematic of the metabolic pathway for the production of *p*-aminostyrene, from glucose. Targeted control problems for the pathway are indicated. Dashed blue arrows represent the possible control points for a *p*-AF-sensing aptazyme. (See the color plate.)

2. *IN VITRO* CHARACTERIZATION

Notably, the kinetically determined RNA structures that result from cotranscriptional folding may differ from the global free energy minima that exist at thermodynamic equilibrium. Consequently, there may be significant differences in the structures and functions accessible to an RNA when folded cotranscriptionally, as compared to the more ubiquitous melt-and-anneal condition where aptazyme function has been commonly characterized (Carothers et al., 2011). In addition, aptazyme folding and catalysis are likely impacted by the presence (or absence) of divalent cations such as magnesium (Mg^{2+}). Therefore, we expect that assaying aptazyme function under physiologically relevant concentrations of free Mg^{2+} (<1 m*M*) will be valuable for predicting the function of engineered devices within a cellular environment.

This section is devoted to validating the activity of aptazymes within biologically relevant expression conditions. These procedures describe the steps needed to go from a synthesized aptazyme DNA oligo to a characterized device, ready to be employed within a cellular context (see below). Failure to display ligand-dependent differences in cleavage under the described assay conditions suggests that the biological implementation of the assayed aptazyme may prove unfruitful, regardless of other efforts to insulate it from the surrounding sequence context.

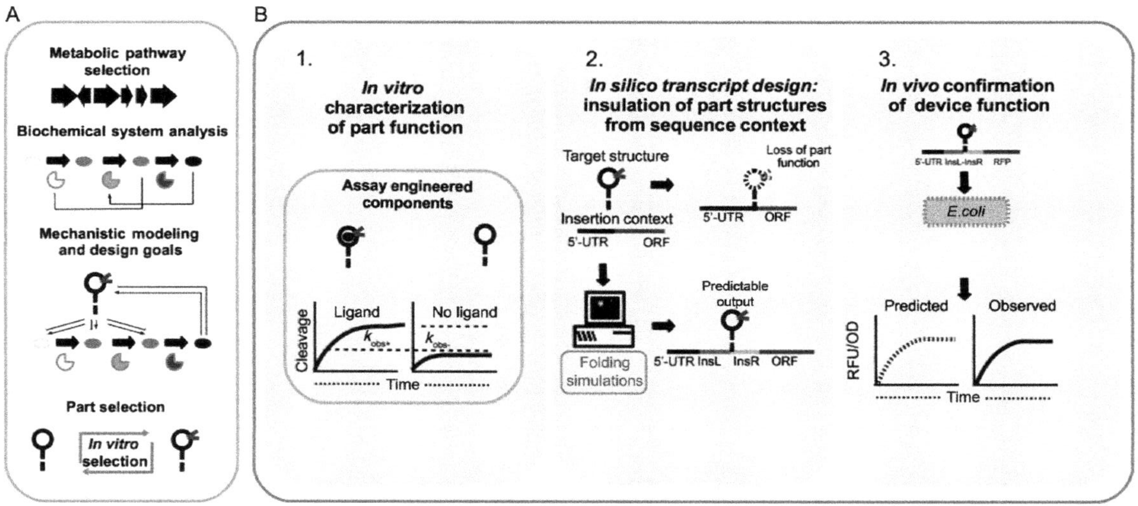

Figure 2 Overview of aRED biosensor design strategy. (A) Preliminary steps for the dynamic control of metabolic pathways. (B) Summary of the methods detailed in this publication. (1) Characterization of aptazyme rate constants, in presence and absence of effector molecule, through *in vitro* cotranscriptional cleavage assays. (2) Design of spacer sequences to maintain part function within novel contexts, via high-throughput cotranscriptional folding simulations. (3) *In vivo* analysis of genetic devices, using fluorescent readout.

A

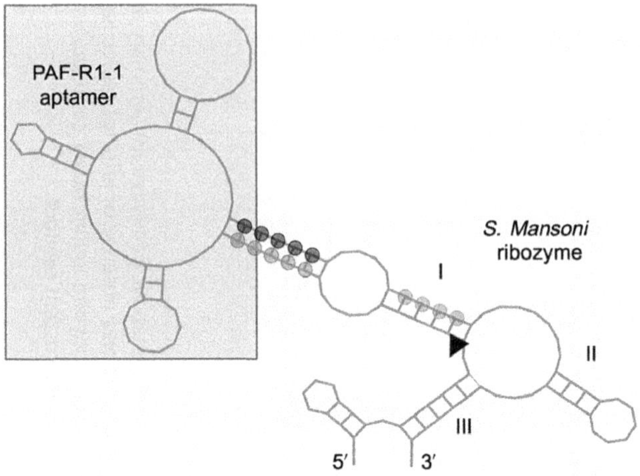

B

1

5′ GGGCGAAAGCCGGCGCGUCCUGGAUUCCAC
NNNNNAUGUCCCUACCAUACGGGAUUGCCCAGCUUC
GGCUGCCAUGCCGGCCAAACGGUAACCGGCCUACG
NNNNNGGUACANNNNGCUGAUGAGUCCCAAAUAGGA
CGAAACGCGCU 3′

2

5′ GGGCGAAAGCCGGCGCGUCCUGGAUUCCACC
AAGCAUGUCCCUACCAUACGGGAUUGCCCAGCU
UCGGCUGCCAUGCCGGCCAAACGGUAACCGGC
CUACGGGAGGGGUACACAACGCUGAUGAGUCC
CAAAUAGGACGAAACGCGCU 3′

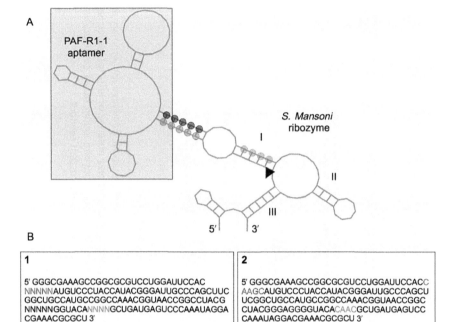

Figure 3 Structure and sequences of a *p*-AF-sensing aptazyme design pool. (A) Secondary structure diagram for the active state of the designed aptazymes. The artificial *p*-AF-R1-1 aptamer (gray box) is fused to the natural *S. mansoni* hammerhead self-cleaving ribozyme by a random linker (Carothers et al., 2010; Yen et al., 2004). (B) 1. Sequence of the aptazyme *in vitro* selection pool. Colored letters represent randomized nucleotides (N's) in selection pool. 2. Sequence of a functional aptazyme (*p*-AF-2) identified from the aforementioned *in vitro* selection pool. (See the color plate.)

2.1. Materials and equipment

Equipment

- 250 V constant power supply
- Vertical Slab Polyacrylamide gel System (C.B.S. Scientific)
- 1.5-mm gel spacers
- 20-well 1.5 mm Teflon combs
- Elutrap Electroelution System (Whatman)
- EluTrap BT1 and BT2 membranes for EluTrap system (Whatman)
- NanoDrop Spectrophotometer
- 400 nm UV transilluminator
- Handheld UV Lamp, 254 nm (UVP)
- Fluorescent PEI-Cellulose TLC Plate
- Orbital shaker

- Tabletop centrifuge
- Vortexer
- 25-ml pipette

Reagents

- UreaGel system (National Diagnostics), or 19:1 acrylamide:bisacrylamide mixture and urea
- Tris–borate–EDTA (TBE) buffer 5 ×
- Bromophenol blue 0.1% (w/v)
- Xylene cyanol 0.1% (w/v)
- KCl 3 *M*
- Ethanol, for molecular biology, ≥99.8%
- Sybr Gold 10,000 × Concentrate in DMSO (Life Technologies)
- Formamide, BioReagent, ≥99.5% (GC),
- Ethylenediaminetetraacetic acid (EDTA) 0.5 *M*
- T7 RNA polymerase 50,000 U/ml
- Thermostable inorganic pyrophosphatase (TIPP) 2000 U/ml
- RNase inhibitor 40,000 U/ml
- rNTP mixture (25 m*M* each of ATP, UTP, GTP, CTP)
- $MgCl_2$ 1 *M*
- Tris–HCl 1 *M*
- Dithiothreitol (DTT) 1 *M*
- Spermidine ≥99.0% (GC)
- Aptazyme Ligand 10 × (e.g., *p*-AF 50 m*M*)
- QIAprep PCR cleanup kit
- Phusion High-Fidelity DNA Polymerase 2000 U/ml
- DNA primers 100 μ*M*
- Sigmacote (Sigma–Aldrich)

2.2. Methods

- RNA is readily degraded by RNases, which are present on numerous lab surfaces. Proper precautions such as the use of gloves and physical separation of RNA from sources of RNAses, including bacteria, should be taken to maintain an RNase-free environment.

2.2.1 DNA oligo purification

1. Resuspend DNA oligos in RNase-free H_2O to a final concentration of 80 μ*M*. Usually, a 4 nmol synthesis is plenty for subsequent steps.
2. Clean the electrophoresis glass plates with ethanol and let them dry.

3. To aid in gel removal, treat gel plates with Sigmacote every 20–30 gels. Briefly cover or immerse the glass surface in the solution, and remove for later use. Allow the treated glass surface to air dry in a hood. Rinse the treated plates with water before use.
4. Set up gel apparatus following manufacturers recommendations. Cast an 8% PAGE gel containing a mixture of 19:1 acrylamide/bisacrylamide to a final concentration of 1 × TBE buffer, and 7.5 *M* of urea. To achieve sufficient resolution, cast gels of at least 24-cm long and 1.0-mm thick. Allow polymerization at room temperature for at least 2 h. While acrylamide stocks can be prepared from scratch, we recommend National Diagnostics' Ureagel system for consistency, and ease of use. If prepared properly, gels can be stored for up to 48 h: Keep comb in the gel, cover the top in a wet paper towel, wrap in plastic wrap, and store at 4 °C until use.
5. Mount gel on apparatus, fill both reservoirs using 0.5 × TBE buffer. Flush wells with buffer using an appropriate gauge syringe or pipette tip. Remove any bubbles caught under the bottom edge of the gel plates.
6. Pre-run the gel for 30 min at constant power (15 W).
7. Add 25 μl of 2 × loading buffer to 25 μl resuspended oligos, mix well. Freeze remaining oligo stock at −20 °C.
   - 2 × loading buffer (95% (v/v) formamide, 5% (v/v) 0.5 *M* EDTA)
8. Once again, flush wells, and immediately load samples (50 μl). Add 1 × loading buffer to empty wells to favor a uniform electric field and optimal resolution.
9. Load 10 μl of a 1:1:2 mixture of xylene cyanol, bromophenol blue, and 2 × loading buffer to one or more wells to track migration.
10. Run gel at constant power (15 W) until bromophenol blue dye has migrated to the bottom of the gel (~1–1.5 h). In an 8% polyacrylamide gel, xylene cyanol will migrate as though it were approximately 80 nucleotides, while bromophenol blue will migrate as though it were approximately 20 nucleotides.
11. Disassemble the apparatus. Unclip gel plates and place on a flat surface. Carefully work one of the spacers out from between the plates. Use the spacer to gently pry off the top plate from the gel.
12. Cover the exposed gel with a sheet of plastic wrap. Flip the gel so that the plate is on top and gently separate them. Cover exposed gel with another piece of plastic wrap.
13. Place the wrapped gel on a Fluorescent PEI-Cellulose TLC Plate (20 × 20 cm).

14. Wearing protective gear, place the gel under the handheld UV lamp, and turn it on. Quickly draw a box around the location of the desired bands using a permanent marker. Reduce UV exposure as much as possible. The desired bands should appear as a dark blob on a bright green background. Faint smears above and below correspond to incorrect synthesis products.
15. Return the sandwich to a glass plate. Using a clean razor blade, cut around marked box. Carefully extract the bands and place them in labeled microcentrifuge tubes.
16. Set up Elutrap elution chambers following manufacturers instructions.
17. Place each gel slice into its own elution chamber. Add enough 0.5 × TBE buffer to the box to reach halfway up the elution chamber entrance. Add just enough buffer into each elution chamber to cover the gel slice.
18. Electroelute at 250 V for 1.5 h.
19. Carefully withdraw solution from elution chambers and transfer to 2-ml microcentrifuge tubes.

2.2.2 Ethanol precipitation

1. Bring total volume of the eluted samples up to 400 μl with RNase-free water.
2. Add 1 ml of ethanol to the samples.
3. Add 150 μl of 3 M KCl to samples.
4. Place samples in a −80 °C freezer for at least 10 min. If possible use a refrigerated centrifuge. If none is available, keep samples in freezer for at least 1 h before spinning.
5. Centrifuge samples at 14,000 × *g* for 10 min.
6. Decant supernatant, making sure not to lose the white pellet.
7. Respin the pellets for a few seconds.
8. Use a micropipette to remove any remaining liquid.
9. Open tubes and allow pellets to air-dry for 10 min. A lyophilizer or gentle heating can be used to accelerate the drying process.
10. When dry, resuspend pellets in 50 μl of RNase free water.
11. Quantify DNA concentration using Nanodrop spectrophotometer.

2.2.3 Cotranscriptional cleavage analysis

1. Prepare *in vitro* transcription template by PCR amplifying previously purified DNA oligo with Phusion High-Fidelity DNA Polymerase. Design an appropriate 3′ reverse primer and a 5′ forward primer containing a standard T7 promoter sequence. T7 transcription efficiency is

greatly improved for transcripts containing two 5′ G's. These G's should be added to the primer if not already present in the template oligo.

2. Due to the large amount of secondary structure within aptazyme sequences, PCR amplification can prove challenging. Additives such as DMSO and betaine can help alleviate these issues. Additionally, PCR amplification using >20 cycles can result in significant off-target amplification products. Calculate the approximate number of cycles to completion based on initial moles of template and moles of primer. Good rule of thumb: 10-fold amplification occurs every four cycles. If PCR amplification yields only the desired product, purify DNA using a PCR cleanup kit, otherwise Gel purify.
3. Perform *in vitro* transcription reactions in PCR tubes. The final reaction volume per sample is 50 μl. Prepare two reactions per construct: one in the presence of saturating ligand concentrations and one without ligand. Include a no-polymerase control.
4. 50 μl *in vitro* transcription reaction conditions:
   - 5 pmol DNA
   - 3 m*M* each NTP
   - 12 m*M* $MgCl_2$
   - 40 m*M* Tris–HCl
   - 10 m*M* DTT
   - 2 m*M* spermidine
   - 0.1 U TIPP
   - 4 U RNase inhibitor (Ribolock)
   - 0–5 m*M* *p*-AF
   - 500 U T7 RNA polymerase

Notes: (1) While T7 RNA polymerase is commercially available through multiple vendors, it is quite expensive. When possible, in-lab production of the polymerase from a plasmid is preferable. Once purified, the produced polymerase can be diluted to match the activity of the commercial enzyme. (2) Based on the literature value for the dissociation of Mg-ATP (Gupta, Gupta, Yushok, & Rose, 1983), the described mixture should contain ~0.75 m*M* free Mg^{2+}.

5. Place capped reaction tubes into a prewarmed thermocycler. Incubate for 5 min without polymerase at 37 °C.
6. Add polymerase to the reaction tubes without removing them from the thermocycler. Quickly mix by pipetting up and down, recap and begin timing.

7. At intervals ranging from 5 min to 1 h carefully withdraw 10 μl of the reaction. Stop the reaction by adding it to a PCR tube containing 10 μl of ice-cold 2× loading buffer, mix well and place them on ice. Quenched aliquots can be stored at −20 °C or used immediately for PAGE.
8. Prior to loading, heat samples at 70 °C for 5 min and then chill them on ice immediately.
9. Cast, load, and run 8% Urea PAGE gel, as in Oligo purification. When loading, include your no-polymerase control sample as reference. This will help to differentiate template DNA bands from the RNA bands. A single-stranded RNA ladder (such as NEB's Low Range ssRNA Ladder) is highly recommended, to allow identification of cleavage products.
10. Disassemble apparatus, remove glass plate and carefully transfer gel into Sybr Gold staining solution (10× Sybr Gold in 0.5× TBE buffer).
11. Place the container on an orbital shaker and gently shake (100 rpm), protected from light, for 15 min.
12. Photograph gel on a 400 nm UV transilluminator. Ensure exposure time is sufficient to yield a nonsaturated image.
13. Quantify band intensity using ImageJ or other gel analysis software. Normalize band intensities to the length of RNA molecule.
14. Determine the first order cleavage rate constant k_{obs} for reactions with, and without, ligand by fitting fraction of transcript uncleaved over time to the following equation (Long & Uhlenbeck, 1994).

$$f_{uncleaved} = \frac{1 - e^{-k_{obs}t}}{k_{obs}t}$$

3. *IN SILICO* TRANSCRIPT DESIGN

To maintain the function of an aptazyme within a novel biological context, it is essential that the structures responsible for desired RNA function are conserved. For the described system, in which an aptazyme is inserted into the 5′-UTR of a bacterial messenger RNA, the critical components to preserve are the aptazyme and the ribosome binding site (RBS). The RBS is the sequence of RNA that mediates interaction between the ribosome and the transcript, allowing proper translation initiation. It has been shown that the secondary structure of the RBS is a significant factor in the determination of protein expression levels (de Smit & van Duin,

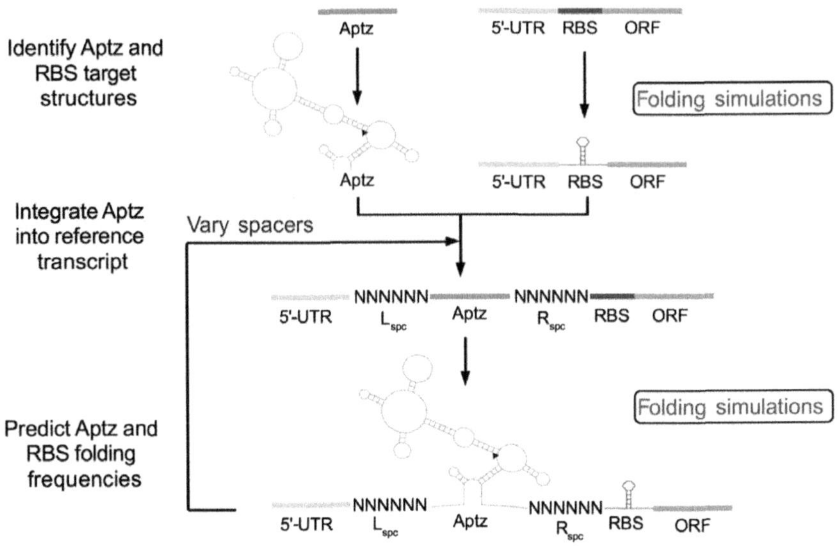

Figure 4 *In silico* workflow for the design of genetic devices with predictable outputs. First, the native secondary structures for the aptazyme and ribosome binding site are determined through stochastic folding simulations. Second, the aptazyme sequence is inserted upstream of the RBS, flanked by random spacer hexamers. Next, the generated sequences are folded and evaluated on the structural agreement of their component parts with their target structures. Spacer sequences are varied until the desired folding frequencies are achieved. (See the color plate.)

1990). Thus, to maintain similar basal levels of protein expression between a control device (without aptazyme) and the aptazyme-regulated device, the RBS structure should be the same in both cases.

The goal of this *in silico* transcript design method is to identify a pair of randomly generated spacer sequences that will allow both the aptazyme and RBS to fold properly within a novel context, resulting in predictable protein expression levels. Here, we explain a series of kinetic, cotranscriptional folding simulations designed to achieve this goal. First, the target part structures are identified within their native sequence contexts. Second, a library of random insulating sequences of specified length are produced *in silico*. Third, the aptazyme and spacer sequences are inserted into the reference device, immediately upstream of the RBS. Finally, each assembled transcript is then folded *in silico* and screened for the presence of the native component structures (Fig. 4). For a step-by-step procedure for establishing a framework for the submission of cotranscriptional folding simulations, see (Thimmaiah, Voje & Carothers, n.d.):

3.1. System and submission guidelines

System

- Multinode scalable computer cluster
- kinefold_long_static binary (Xayaphoummine, Bucher, & Isambert, 2005)
- Job submission framework
- Kinefold output processing scripts

Submission guidelines

1. We suggest running all folding simulations without considering pseudoknots (or entanglements). Pseudoknots are higher-order nucleic-acid structures, and folding simulations that consider them take dramatically longer to complete. As their consideration has not been shown to be critical (except in cases, where the target structure has been experimentally shown to include pseudoknots), they can generally be ignored.
2. We perform folding simulations with a helix MFE value of 6.346 kcal/mol.
3. Kinefold can perform two kinds of folding simulations: renaturation and cotranscriptional folds. Renaturation folds assume that the entire RNA sequence has been transcribed prior to folding. While this type of simulation is likely sufficient for small sequences, such as an aptazyme, it is not sufficient to predict the structures of full transcripts accurately.
4. Due to variation of polymerase elongation rates (k_{pol}), it is suggested to perform all cotranscriptional folding simulations at both a low rate (97 nt/s for T7, 25 nt/s for *E. coli*), and a fast rate (200 nt/s for T7, 55 nt/s for *E. coli*).
5. As Kinefold yields stochastic folding data, it is essential to fold each sequence many times to predict the distribution of structures possible *in vivo* accurately. We recommend performing at least 100 folding simulations at both high and low k_{pol} values, using random seeds, before choosing a given set of spacer sequences to test *in vivo*.
6. When determining whether a part has folded correctly, it is considered to agree with the reference construct if the part structure matches any of the ensemble of consensus structures. By default, each reference part has a folding frequency of 1.00.
7. To determine the overall success of a pair of spacer sequences, we evaluate it based on its distance score:

$$d = 1 - \sqrt{\frac{1}{2}\left((1 - f_{aptz})^2 + (1 - f_{rbs})^2\right)}$$

T7 promoter 5′UTR Lspc
pAF-2 Aptazyme Rspc RBS
5′ end of DsRed (RFP)

5′ GTCTAATACGACTCACTATAGGGACGACGACAGGCACCCGAACTCNNNNNNGGGCGA
AAGCCGGCGCGTCCTGGATTCCACCAAGCATGTCCCTACCATACGGGATTGCCCAGCTT
CGGCTGCCATGCCGGCCAAACGGTAACCGGCCTACGGGAGGGGTACACAACGCTGATG
AGTCCCAAATAGGACGAAACGCGCTNNNNNNGAAAGAGGAGAAATACTAGATGGCTTCC
TCCGAAGACGTTATCAAAGAGTTCATGCGTTTCAAAGTTCGTATGGAAGGTTCCGTTAACG
GTCACGAGTTCGAAATCGAAGGTGAAGGTGAAGGTCGTCCGTACGAAGGTACCCAGACC
GCTAAACTGAAAGTTACCAAAGGTGGTCCGCTGCCGTTCGCTTGGGACATCCTGTCCCC
GCAGTTCCAGTACGGTTCCAAAGCTTACGTTAAACACCCGGCTGACATCCCGGACTACCT
GAAACTGTCCTTCCCGGAAGGTTTCAAATGGGAACGTGTTATGAAC...3′

Figure 5 Sequence of a *p*-AF biosensor used for transcript design example. The *p*-AF-2 aptazyme is inserted downstream of the T7 promoter, and upstream of the RBS and coding sequence for RFP. The RBS target structure is evaluated within the given context, without the presence of the aptazyme or spacer sequences. (See the color plate.)

f_{aptz} and f_{rbs} are the observed frequencies of correct folding for the given part, defined by the number of correct folds divided by the number of folding simulations for the sequence (in this case 100).

8. When performing the folding simulations outlined below, the number of downstream nucleotides (from the center of the RBS) considered should be equal to the polymerase rate multiplied by 1 s, as 1 s approximates the minimum amount of time required for productive RBS–ribosome interaction (Fig. 5).

3.2. Computational methods

1. Fold the desired riboswitch part 100 × to determine the consensus target structure(s) for subsequent screening. We recommend performing a renaturation fold with a total simulation time of 60 s.
2. Fold the sequence of a previously characterized reference construct, into which the aptazyme will be inserted. Perform 100 simulations at both the high and low polymerase rates for the polymerase of interest. Determine the consensus RBS target structure(s) for subsequent screening.
3. Generate a library of randomly generated left (Lspc) and library right (Rspc) spacer hexamers. For our example, we generated 50 hexamers for each.
4. Create a sequence library, wherein the aptazyme sequence is placed immediately upstream of the RBS, flanked by every possible combination of spacer sequences.

L_{spc}	R_{spc}	d (k_{pol} low)	d (k_{pol} high)	Mean d
CUUUGA	CACGUU	0.311	0.286	0.298
UUGUUA	CACGUU	0.298	0.293	0.296
UAUGCG	GUCGUA	0.302	0.286	0.294
GAUCCC	GGCGUC	0.286	0.298	0.292
CCUAUG	AAAAGA	0.295	0.287	0.291
GACACU	CACGUU	0.291	0.279	0.285
UUGUUA	ACGUAU	0.291	0.275	0.283
CGGUCA	GUCGUA	0.277	0.273	0.275
UGAGGA	CGUUUA	0.250	0.284	0.267
GAGACA	CGUUUA	0.259	0.232	0.246
CUGUAU	ACCGUC	0.246	0.243	0.245
GAUCCC	AGCAGA	0.251	0.236	0.244
GAUCCC	ACGUAU	0.263	0.200	0.232
GAGACA	GGGCAA	0.228	0.207	0.217

Figure 6 Performance of selected insulating sequences. Distance score, a measure of the correct folding of the parts, is defined above. Fourteen of the best-performing insulating sequences are shown.

5. Fold every member of the library 10 ×. This allows the identification of spacer sequences that perform well, without the exhaustive use of computational resources. To accelerate the process, a single k_{pol} can be used for initial screening.
6. Select the top performing spacer pairs, based on the distance score for the initial folding, and resubmit each for 100 new folding simulations.
7. Calculate the distance scores for the new folding simulations, averaging the d values for the simulations performed at high and low k_{pol} values.
8. If the active secondary structure of the aptazyme is known, a scan for the unwanted presence of the active state (i.e., without ligand) along the folding path can be included as a negative selection criteria for the insulating sequences.
9. Select the best performing spacer sequences to build for *in vivo* validation (Fig. 6).

4. *IN VIVO* VALIDATION

Thus far, we have described the steps to characterize aptazyme function *in vitro* and a computational method to place this sequence in a context of cellular relevance without compromising individual part functions. Here, we show how using a *p*-AF-dependent self-cleaving ribozyme (aptazyme) in the 5′-UTR of a bacterial mRNA can control gene expression of a red fluorescent protein *in vivo*. Our regulatory mechanism of choice was inspired by the idea that the chemical composition of 5′ termini of a transcript can, either positively or negatively affect its half life (Mackie, 1998). One example is the mechanism of action of the *glmS* riboswitch of *Bacillus subtilis*. This natural riboswitch selectively increases the rate of degradation of the *glmS* transcript by cleaving upon binding of glucosamine 6-phosphate. Cleavage by the *glmS* riboswitch produces bacterial mRNAs terminated by a 2′,3′-cyclic phosphate and a 5′-hydroxyl group. RNA transcripts containing a 5′-hydroxyl group are rapidly degraded in *B. subtilis* by RNase J1. As a consequence, transcript expression is downregulated (Collins, Irnov, Baker, & Winkler, 2007). Placing an aptazyme into the 5′-UTR region of bacterial mRNA also produces transcripts with 5′-hydroxyl groups. In *E. coli*, however, these processed transcripts are degraded in an RppH-independent manner, which is significantly slower (Celesnik, Deana, & Belasco, 2007; Deana, Celesnik, & Belasco, 2008) than conventional RppH degradation. As a consequence, in our system, aptazyme cleavage increases mRNA abundance, and therefore protein expression of the downstream gene. It is important to note that the mechanism of choice depends on both the design goals of the engineered metabolic pathway and the nature of the host cell.

In the following method, we provide an approach to assess the effectiveness of the *in silico* transcript design *in vivo*, by producing a metabolite biosensor with a readable output. The biosensor (designed in Section 3) consists of an aptazyme cloned upstream of the gene for a fluorescent protein. Thus, the success of device implementation can be assessed by observing increases in fluorescence when cells are grown in media containing the target metabolite.

Device assembly was carried out using standard Gibson Assembly followed by transformation of *E. coli* DH10B, and the pertinent sequences confirmed. Analysis of gene expression was carried out using an *E. coli* strain (BL21(DE3)) that has been optimized for protein expression. This method could be extended to test the function of other devices; however, optimization of the conditions will depend on the activator ligand and the underlying mechanism of genetic regulation.

4.1. Equipment and reagents

Equipment

- Infinite M1000 Microplate reader (Tecan)
- Incubator with shaker

Reagents

- *E. coli* BL21(DE3) cells (Novagen, Inc.)
- *E. coli* DH10B (Life Technologies, Inc.)
- Luria-Bertani broth (LB)
- Kanamycin
- Isopropyl β-D-1-thiogalactopyranoside (IPTG)
- *p*-AF
- Black polystyrene 96-well plates with lid, flat bottom (Corning)
- Plasmid constructs

4.2. Methods

Note: For plasmid assembly, we recommend scarless methods, such as Gibson or CPEC cloning, due to the potential impact of restriction sites on part folding. For DNA manipulations and plasmid amplification, we recommend using *E. coli* DH10B cells.

Assembly plasmids (prRED0-T7-ref aRED1, prRED0-T7-ref aRED9) for aREDs are available from Addgene.

1. Transform plasmids into competent *E. coli* BL21(DE3) cells. Follow transformation method recommended by the supplier.
2. Plate recombinant cells on LB plates containing a final concentration of 50 μg/ml of kanamycin. Incubate, cover-side down, in a 37 °C incubator for 15–18 h.
3. Pick transformed colonies and inoculate 5-ml starter cultures of LB with a final concentration of 50 μg/ml of kanamycin.
4. Incubate liquid cultures shaking at 200 rpm, at 37 °C for 15–18 h.
5. Dilute starter cultures 1:100 into 5 ml of fresh LB. Incubate, with 200 rpm shaking, at 30 °C.
6. Induce protein expression with 1 m*M* IPTG. Add saturating amounts of ligand (5 m*M*) to media alongside transcript induction.
7. After 18–24 h of incubation, measure RFP fluorescence (excitation = 557 nm, emission = 579 nm) as a function of optical density at 380 nm.

5. FUTURE DIRECTIONS

We have laid out an experimental and computational workflow to develop aptazyme-regulated expression devices for the dynamic control of gene expression. Though the described procedure utilizes an aptazyme selected to respond to the metabolite *p*-AF, the method can, in theory, be extended to any system for which an aptazyme can be designed. The functional aRED designed using the above method has three main immediate applications: as a biosensor for metabolite production, as part of a positive feedback circuit to decrease rise-time to steady state, or as part of a negative feedback circuit to minimize the buildup of toxic intermediates (Fig. 7; Carothers et al., 2009). By replacing the fluorescent gene with that of a metabolic enzyme (e.g., the gene for the enzyme whose substrate is the aptazyme's inducer), the same process of part insulation (Section 3) can be performed to ensure predictable expression of the new target gene. As the library of publicly available riboswitches expands, the ability to design and assemble increasingly complex,

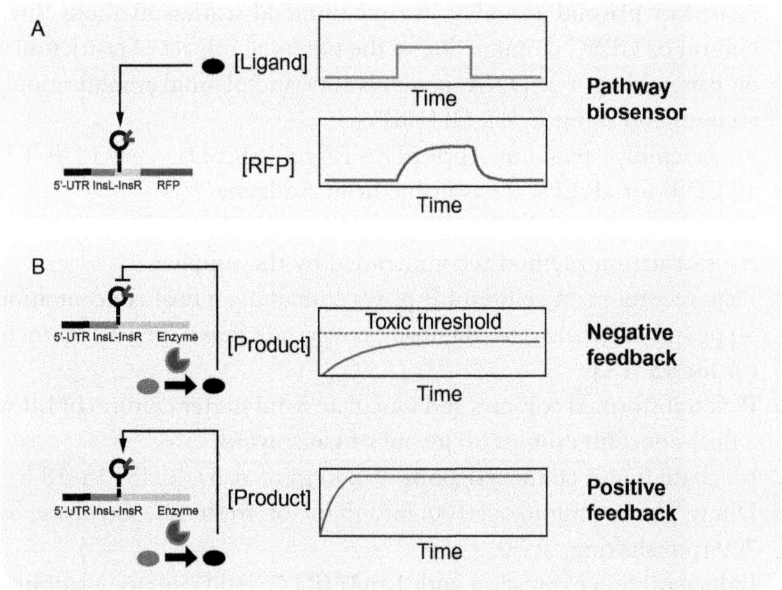

Figure 7 Possible applications of aptazyme-regulated expression devices. (A) Metabolite biosensors with fluorescent output. (B) Use of aREDs to achieve design goals in metabolic engineering (e.g., as part of a negative feedback motif to minimize the buildup of toxic intermediates or as part of a positive feedback motif to decrease rise-time to steady state).

self-regulated metabolic pathways should allow the economically viable microbial production of ever-increasing number chemicals (Fig. 7).

ACKNOWLEDGMENT

We thank W. M. Voje, Jr. and C. R. Burke for helpful discussions and comments. Work in the authors' laboratory was supported by the Molecular Engineering & Sciences Institute and the University of Washington. J. M. C. is a fellow of the Alfred P. Sloan Foundation.

REFERENCES

Ausländer, S., Ketzer, P., & Hartig, J. S. (2010). A ligand-dependent hammerhead ribozyme switch for controlling mammalian gene expression. *Molecular BioSystems*, *6*(5), 807–814. http://dx.doi.org/10.1039/b923076a.

Cambray, G., Guimaraes, J. C., Mutalik, V. K., Lam, C., Mai, Q. A., Thimmaiah, T., et al. (2013). Measurement and modeling of intrinsic transcription terminators. *Nucleic Acids Research*, *41*(9), 5139–5148. http://dx.doi.org/10.1093/nar/gkt163.

Carothers, J. M. (2013). Design-driven, multi-use research agendas to enable applied synthetic biology for global health. *Systems and Synthetic Biology*, 7(3), 79–86. http://dx.doi.org/10.1007/s11693-013-9118-2.

Carothers, J. M., Goler, J. A., Juminaga, D., & Keasling, J. D. (2011). Model-driven engineering of RNA devices to quantitatively program gene expression. *Science (New York, N.Y.)*, *334*(6063), 1716–1719. http://dx.doi.org/10.1126/science.1212209.

Carothers, J. M., Goler, J. A., Kapoor, Y., Lara, L., & Keasling, J. D. (2010). Selecting RNA aptamers for synthetic biology: Investigating magnesium dependence and predicting binding affinity. *Nucleic Acids Research*, *38*(8), 2736–2747. http://dx.doi.org/10.1093/nar/gkq082.

Carothers, J. M., Goler, J. A., & Keasling, J. D. (2009). Chemical synthesis using synthetic biology. *Current Opinion in Biotechnology*, *20*(4), 498–503. http://dx.doi.org/10.1016/j.copbio.2009.08.001.

Celesnik, H., Deana, A., & Belasco, J. G. (2007). Initiation of RNA Decay in *Escherichia coli* by 5′ Pyrophosphate Removal. *Molecular Cell*, *27*(1), 79–90. http://dx.doi.org/10.1016/j.molcel.2007.05.038.

Collins, J. A., Irnov, I., Baker, S., & Winkler, W. C. (2007). Mechanism of mRNA destabilization by the glmS ribozyme. *Genes & Development*, *21*(24), 3356–3368. http://dx.doi.org/10.1101/gad.1605307.

Deana, A., Celesnik, H., & Belasco, J. G. (2008). The bacterial enzyme RppH triggers messenger RNA degradation by 5′ pyrophosphate removal. *Nature*, *451*(7176), 355–358. http://dx.doi.org/10.1038/nature06475.

de Smit, M. H., & van Duin, J. (1990). Secondary structure of the ribosome binding site determines translational efficiency: A quantitative analysis. *Proceedings of the National Academy of Sciences of the United States of America*, *87*(19), 7668–7672.

Goler, J. A., Carothers, J. M., & Keasling, J. D. (2014). Dual-selection for evolution of in vivo functional aptazymes as riboswitch parts. *Methods in Molecular Biology (Clifton, N.J.)*, *1111*, 221–235. http://dx.doi.org/10.1007/978-1-62703-755-6_16.

Gupta, R. K., Gupta, P., Yushok, W. D., & Rose, Z. B. (1983). Measurement of the dissociation constant of MgATP at physiological nucleotide levels by a combination of 31P NMR and optical absorbance spectroscopy. *Biochemical and Biophysical Research Communications*, *117*(1), 210–216.

Link, K. H., Guo, L., Ames, T. D., Yen, L., Mulligan, R. C., & Breaker, R. R. (2007). Engineering high-speed allosteric hammerhead ribozymes. *Biological Chemistry, 388*(8), 779–786. http://dx.doi.org/10.1515/BC.2007.105.

Long, D. M., & Uhlenbeck, O. C. (1994). Kinetic characterization of intramolecular and intermolecular hammerhead RNAs with stem II deletions. *Proceedings of the National Academy of Sciences of the United States of America, 91*(15), 6977–6981.

Mackie, G. A. (1998). Ribonuclease E is a 5′-end-dependent endonuclease. *Nature, 395*(6703), 720–723. http://dx.doi.org/10.1038/27246.

Penchovsky, R. (2013). Computational design and biosensor applications of small molecule-sensing allosteric ribozymes. *Biomacromolecules, 14*(4), 1240–1249. http://dx.doi.org/10.1021/bm400299a.

Stevens, J. T., & Carothers, J. M. (2014). Designing RNA-based genetic control systems for efficient production from engineered metabolic pathways. *ACS Synthetic Biology*. http://dx.doi.org/10.1021/sb400201u.

Thimmaiah, T., Voje, W. E., Jr, & Carothers, J. M (n.d.). Computational design of RNA parts, devices, and transcripts with kinetic folding algorithms implemented on multiprocessor clusters. Methods in Molecular Biology (Clifton, N.J.) Accepted.

Wickiser, J. K., Winkler, W. C., Breaker, R. R., & Crothers, D. M. (2005). The speed of RNA transcription and metabolite binding kinetics operate an FMN riboswitch. *Molecular Cell, 18*(1), 49–60. http://dx.doi.org/10.1016/j.molcel.2005.02.032.

Wieland, M., & Hartig, J. S. (2008). Improved aptazyme design and in vivo screening enable riboswitching in bacteria. *Angewandte Chemie (International Ed. in English), 47*(14), 2604–2607. http://dx.doi.org/10.1002/anie.200703700.

Win, M. N., & Smolke, C. D. (2007). A modular and extensible RNA-based gene-regulatory platform for engineering cellular function. *Proceedings of the National Academy of Sciences of the United States of America, 104*(36), 14283–14288. http://dx.doi.org/10.1073/pnas.0703961104.

Xayaphoummine, A., Bucher, T., & Isambert, H. (2005). Kinefold web server for RNA/DNA folding path and structure prediction including pseudoknots and knots. *Nucleic Acids Research, 33*(Web Server issue), W605–W610. http://dx.doi.org/10.1093/nar/gki447.

Yen, L., Svendsen, J., Lee, J.-S., Gray, J. T., Magnier, M., Baba, T., et al. (2004). Exogenous control of mammalian gene expression through modulation of RNA self-cleavage. *Nature, 431*(7007), 471–476. http://dx.doi.org/10.1038/nature02844.

Zhang, F., Carothers, J. M., & Keasling, J. D. (2012). Design of a dynamic sensor-regulator system for production of chemicals and fuels derived from fatty acids. *Nature Biotechnology, 30*(4), 354–359. http://dx.doi.org/10.1038/nbt.2149.

CHAPTER SEVENTEEN

Riboselector: Riboswitch-Based Synthetic Selection Device to Expedite Evolution of Metabolite-Producing Microorganisms

Sungho Jang*, Jina Yang*, Sang Woo Seo*, Gyoo Yeol Jung*,†,1
*Department of Chemical Engineering, Pohang University of Science and Technology, Pohang, Gyeongbuk, Republic of Korea
†School of Interdisciplinary Bioscience and Bioengineering, Pohang University of Science and Technology, Pohang, Gyeongbuk, Republic of Korea
[1]Corresponding author: e-mail address: gyjung@postech.ac.kr

Contents

Abstract

Many successful metabolic engineering projects have utilized evolutionary approaches, which consist of generating phenotypic diversity and screening for desired phenotype. Since conventional screening methods suffer from low throughput and limited target metabolites, a universal high-throughput screening platform for selection of improved strains should be developed to facilitate evolution of metabolite high producer. Recently, riboswitches have received attention as attractive sensor–actuator hybrids that can control gene expression in response to intracellular metabolite concentration.

Methods in Enzymology, Volume 550
ISSN 0076-6879
http://dx.doi.org/10.1016/bs.mie.2014.10.039

Our group developed a riboswitch-based selection device called "Riboselector" which can give a growth advantage to metabolite-overproducing strains by modulating expression of a selectable marker gene. We applied the device to expedite evolution of lysine producing *Escherichia coli*, and the selected strain showed a dramatic improvement of lysine production compared to its parental strain. Moreover, a tryptophan-responsive Riboselector was also developed using synthetic tryptophan aptamer. In this chapter, we provide a step-by-step overview of developing synthetic RNA devices comprising a riboswitch and a selection module that specifically sense inconspicuous metabolites and enrich high producer strains out of library.

1. INTRODUCTION

Metabolic engineering of microbes for production of valuable chemicals mostly relies on rational engineering that attempts redirection and amplification of metabolic flux toward target metabolite (Keasling, 2008; Nakagawa et al., 2011; Peralta-Yahya et al., 2011). However, it is often required to optimize metabolic pathways to achieve highly efficient production strains, which is difficult for rational engineering (Pfleger, Pitera, Smolke, & Keasling, 2006). On the other hand, combinatorial or evolutionary engineering can circumvent hurdles in metabolic pathway optimization by screening of improved variants from large and diversified library (Alper, Moxley, Nevoigt, Fink, & Stephanopoulos, 2006; Isaacs et al., 2011; Santos & Stephanopoulos, 2008; Wang et al., 2009). Considering that the vast majority of metabolites are not optically detectable, an alternative screening or selection technology for enhanced metabolite production is needed (Dietrich, McKee, & Keasling, 2010).

Riboswitches have been employed in many applications which demand detection of specific molecules and regulation of gene expression in response to the molecular signal. Recently, our group has shown that a riboswitch-based selection device called a Riboselector can efficiently optimize metabolic pathways (Fig. 1) by linking the intracellular concentration of a particular metabolite to survival of the cell under selection pressure (Yang et al., 2013). Detection of the target chemical does not depend on visual signal in this system. Moreover, throughput is extremely high because all unimproved variants perish and only superior ones can survive. It is now possible to sense a wide array of ligands by the riboswitch since various kinds of natural riboswitches are known (Breaker, 2011; Henkin, 2008), and synthetic riboswitches for novel ligands can easily be developed by numerous

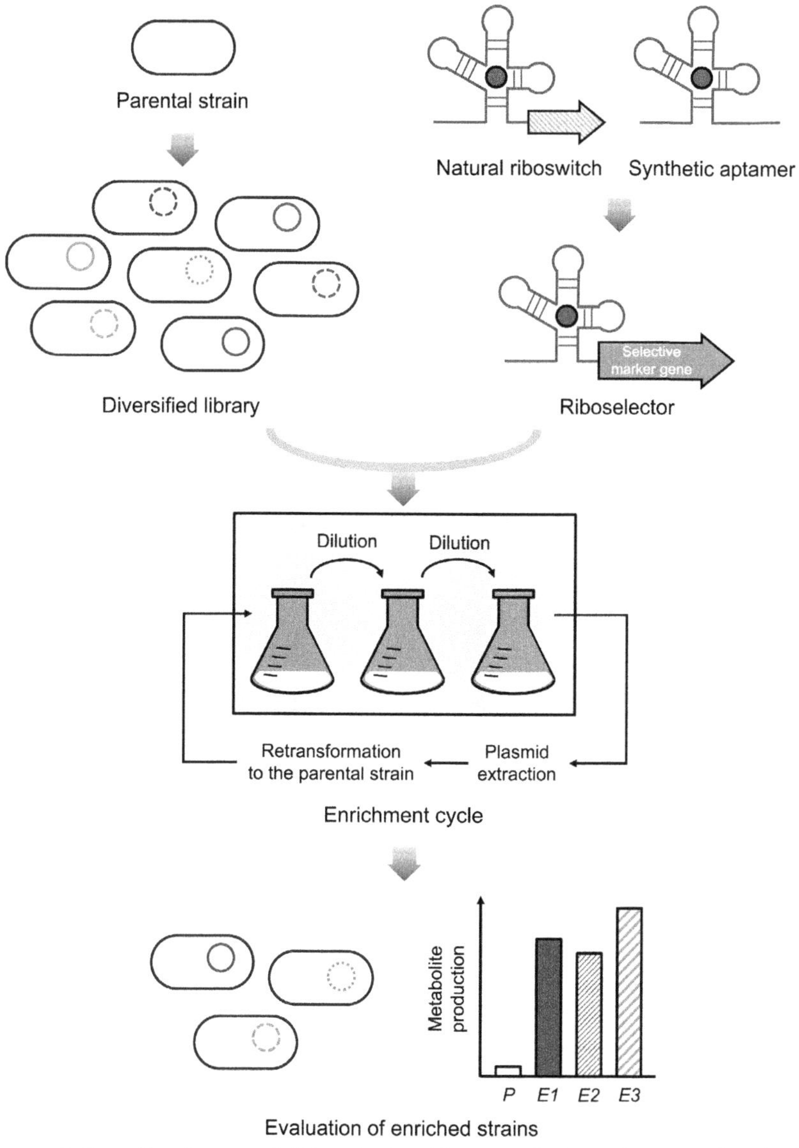

Figure 1 Overall scheme of Riboselector-driven evolution of metabolite high-producing microbe is depicted here. Library of phenotypically diversified strains is generated. Then a Riboselector which responds specifically to target molecule is constructed from either natural riboswitch or synthetic aptamer. The Riboselector is introduced to the strain library, and serial culture dilution is conducted in the presence of appropriate selection pressure. After several enrichment cycles, individual strains are isolated on agar plate and metabolite production is evaluated.

methods (Lynch & Gallivan, 2009; Muranaka, Sharma, Nomura, & Yokobayashi, 2009; Sinha, Reyes, & Gallivan, 2010). Therefore, we expect that Riboselectors for the ligands of interest could easily be created for optimization of any metabolic pathway.

In this chapter, we describe an experimental protocol for metabolic pathway optimization utilizing Riboselector. We introduce detailed procedures for fabrication of Riboselector, creation of microbial variant library, and pathway optimization through serial culture in the presence of the Riboselector and selection pressure.

2. MATERIALS

2.1. Equipment

Micropipettes, micropipette tips, microtubes, PCR tubes, PCR machine, electroporator, electroporation cuvettes, incubator (37 °C), shaking incubator (37 °C, 200 rpm), flasks, test tubes, Petri dishes, UV spectrophotometer, and fluorescence detector.

2.2. Materials

PCR master mix, restriction enzymes, T4 DNA ligase, cloning vectors (pACYCDeut (Novagen), pMD20 (Takara)), pKD46 (Datsenko & Wanner, 2000), pCP20 (Cherepanov & Wackernagel, 1995), LB medium, M9 minimal medium, agar plate, tetracycline hydrochloride, nickel chloride, antibiotics for plasmid maintenance, electrocompetent cell (*Escherichia coli* W3110, DH5α, MachT1^{R}), and target ligand.

2.3. Oligonucleotides

See Table 1

3. CONSTRUCTION AND VALIDATION OF RIBOSELECTOR: RIBOSWITCH-BASED SYNTHETIC SELECTION DEVICES

3.1. Riboselector based on a natural riboswitch

The Riboselector can be readily designed from natural riboswitches. There are many riboswitches discovered over decades (Baker et al., 2012; Weinberg et al., 2010). In this section, the experimental procedures for the development of a selection device based on a natural riboswitch will be described (Fig. 2A). The natural riboswitch is linked to a testing or

Table 1 Primers used in this study

Name	Sequence (5′–3′)[a–c]
Pro-LUTR-F	attgacggctagctcagtcctaggtacagtgctagcgtactacctgcgctagcgca
*Kpn*I-Pro-F	aGGTACCttgacggctagctcagtcctaggtacagtg
LUTR-R	aactacctcgtgtcaggggatccat
L-sgfp-F	ctcttcccttgtgccaaggctgaaaatggatcccctgacacgaggtagttatggctagcaagggcgaggagctgt
sgfp-R	acaaaaaacccctcaagacccgtttagaggc
T-tetA-F	aGGTACCaaggagcatctatgaaatctaacaatgcgctcatcgtca
T-tetA-R	gctcccaccgccactcccaccgccggacccaccgcccgacccaccgccggtcgaggtggcccggct
T-sgfp-F	ggcggtgggtcgggcggtgggtccggcggtgggagtggcggtgggagcgctagcaagggcgaggagct
T-sgfp-R	aGAGCTCtcacttgtacagctcgtccatgcc
ph-T-Library-F	atatttgacggctagctcagtcctaggtacagtgctagcctggacgacggggacgccactggactaggtaagccagga ccgtacgtcgggagccgtcagaataNNNNNNNNNNaaggagcatctatgaaatctaacaatgcgct
ph-T-Library-R	ttaaggtaccgcgcaacgcaattaatgtaagtta
D-lysC-F	gactttggaagattgtagcgccagtcacagaaaaatgtgatggttttagtgcgatgcctcatccgcttctc
D-lysC-R	gacaagaaaatcaatacggcccgaaatatagcttccaggccatacagtatgcaacgcagtagctggagtc
P-lysC-F	gactttggaagattgtagcgccagtcacagaaaaatgtgatggttttagtgcgttgctcctgacatggctc
P-lysC-R	gacaagaaaatcaatacggcccgaaatatagcttccaggccatacagtatgcatccaacgcgttgggagctcc

Continued

Table 1 Primers used in this study—cont'd

Name	Sequence (5′–3′)
P-dapA-F	ggaaagcataaaaaaaacatgcatacaacaatcagaacggttctgtctgcatgggaattagccatggtccatatg
P-dapA-R	atcgcgacaatacttcccgtgaacatgggccatcctctgtgcaaacaagtgctagcactgtacctaggactgagc
P-dapB-F	gtaacctgtcacatgttattggcatgcagtcattcatcgactcatgccatgggaattagccatggtcc
P-dapB-R	ggatgtttgcatcatgcatagctattctcttttgttaatttgcatagaccgctagcactgtacctaggactgagc
P-lysA-F	gccattagcgctctctcgcaatccggtaatccatatcatttttgcatagagaatatcctccttagttcctattccgaag
P-lysA-R	agattttcggcggtgagatcggtatcggtgctgaacagtgaatgtggcatatgtatatctccttcttaaagttaaacaa
S-ddh-F	aCATATGaccaacatccgcgtagctatcgtgg
S-ddh-R	aCTCGAGctaaattagacgtcgcgtg
P-ddh-F	ttaaactgacgattcaactttataatctttgaaataatagtgcttatcccgtctatttcgttcatccgaatatcctcc
P-ddh-R	gcggtatggcatgatagcgcccggaagagagtcaattcagggtggtgaatgagctccaaaaaacccctcaagac
D-met-F	atgagtgtgattgcgcaggcaggggcgaaaggtcgtcagctgcataaattctagtgctggagcgaactgc
D-met-R	aaaccataaacccgaaaacatgagtaccgggcattattaaatttctgaaaggagtactcgcggttgactg
D-thr-F	gcgtacaggaaacacagaaaaaagcccgcacctgacagtgcgggcttttttgttagcccgtctgtcccaa
D-thr-R	tgattcatcatcaatttacgcaacgcagcaaaatcggcgggcagattatgcatactcgctcttgggtcgg
D-iclR-F	ctgtggtaaaagcgaccaccacgcaacatgagatttgttcaacattaactcgtagcaccgagtcgtaccag
D-iclR-R	gcattccaccgtacgccagcgtcacttccttcgccgctttaatcaccatccggtactggcctaacgcact

D-ppc-F	ccagtgccgcaataatgtcggatgcgatacttgcgcatcttatccgaccgttagcccgtctgtcccaa
D-ppc-R	gcagacagaaatatattgaaaacgagggtgttagaacagaagtaTcatactcgctcttgggtcgg
S-ppc-F	aGGTACCacaggttcagagttctacagtccgacatgagcaaaggtttcagtaggaggaaagaacaatgaacgaa caatattccgcattgcg
S-ppc-R	aGAGCTCgagggtgttagaacagaagtatttc
ph-UTR-ppc-F	gaattcgtatgccgtcttctgcttgtcgaggaggatcccaatgaacgaacaatattccgcattgcg
ph-UTR-ppc-R	tcggactgtagaactctgaacctgtgg
ph-J23N-F	gctagcNNNNNNcctaggactgagctagcNNNNNNcatgtcggactgtagaactctgaacc
ph-J23N-R	ttcgtatgccgtcttctgcttgtcg

[a]Capital letters indicate restriction sites.
[b]Underlined letters indicate homologous sequences for recombination and overlap PCR.
[c]Underlined and capital letters indicate randomized sequences for library construction.

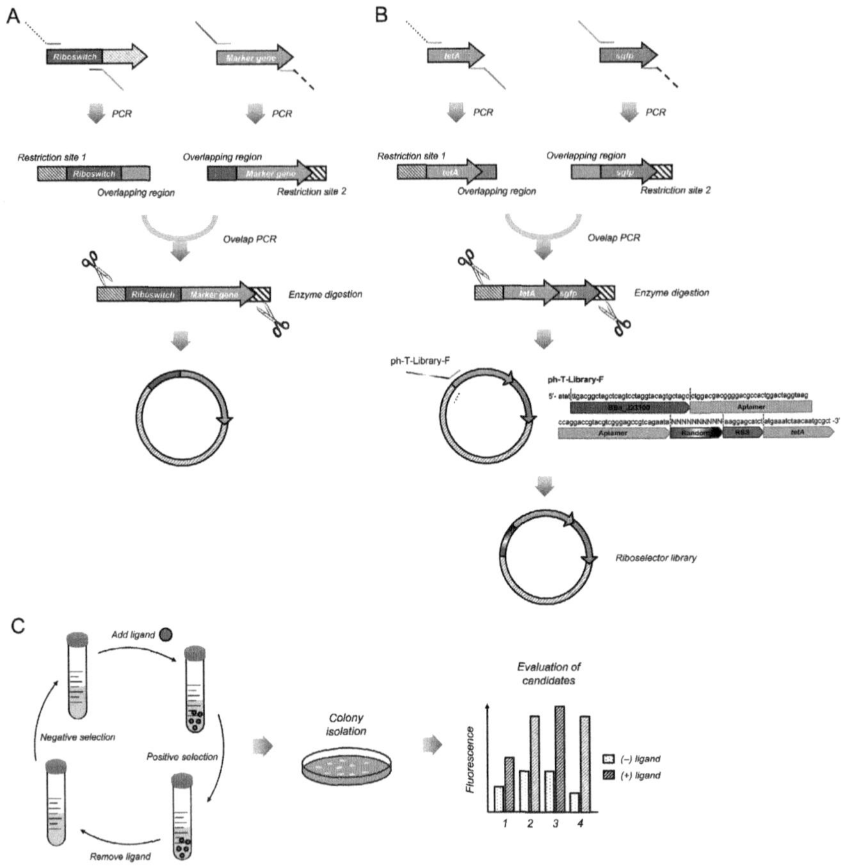

Figure 2 Development procedure of Riboselector. (A) Cloning strategy for fabrication of Riboselector from natural riboswitches. Riboswitch part and marker gene are PCR amplified, and the products are joined by overlap PCR. The whole construct is subcloned into vector to produce Riboselector plasmid. (B) Scheme of Riboselector library construction from artificially developed aptamers. Selective marker gene (e.g., *tetA*) and reporter gene (e.g., *sgfp*) are PCR amplified and they joined together. The PCR product is subcloned into vector. The whole plasmid is PCR amplified by 5′-phosphorylated primer set. The forward primer contains promoter, aptamer, UTR, and randomized sequence. (C) Iterative positive and negative selection enriches functioning Riboselectors from the plasmid library. Single colonies are isolated and tested for their performance. *Panel (C): Adapted from Muranaka et al. (2009) with permission.*

selection module composed of a fluorescent protein (e.g., *sgfp*) or dual selective marker gene (e.g., *tetA*) so that we can evaluate selection devices. Before starting the development of a selection device, it is desirable to check two requirements of the target system. First, the sequence of a riboswitch that binds specifically to the target molecule should be prepared. You can consult review paper (Serganov & Nudler, 2013) or well-organized databases (Burge et al., 2013) and (RibEx: Riboswitch Explorer, http://132.248.32.45/cgi-bin/ribex.cgi) that provide lists of published naturally occurring riboswitches. However, they can also be generated by combination of routine SELEX and selection procedures, which will be covered by the following sections. Second, cloning vector for selection device library should not contain the same antibiotic resistance gene as the marker for plasmid maintenance. Note that every growth medium throughout following protocols should always contain appropriate antibiotics to maintain plasmids.

3.1.1 Construction of a device for reporting its operation

The ligand-specific reporting device can be constructed by assembling the riboswitch and a reporter gene. The example in Fig. 2A illustrates assembly of the leader region of *lysC* for lysine with the gene for a selective marker or fluorescent protein *sgfp*. Fluorescent protein can report the performance of the riboswitch, while selective marker gene can be utilized in actual pathway optimization that is described in Section 4.2. If you are interested in construction of Riboselector based on synthetic aptamer, see Section 3.2.

i. Amplify riboswitch (e.g., leader region of *lysC*) with PCR. Forward primer (Pro-LUTR-F and *Kpn*I-Pro-F) contains restriction enzyme site and promoter sequence at its 5′-end, while reverse primer (LUTR-R) contains overlapping sequence at its 5′-end which will connect riboswitch with reporter gene.

ii. Amplify fluorescent protein (e.g., *sgfp*) with PCR. Forward primer (L-sgfp-F) contains the overlapping sequence with the cognate riboswitch at its 5′-end, and reverse primer (sgfp-R) contains restriction enzyme site at its 5′-end.

iii. Gel purify the PCR products from steps i and ii and mix them together to conduct overlap PCR. The overlapping sequence in the PCR products will hybridize to each other, and the reporting device sequence will be generated during overlap PCR.

iv. Gel purify the PCR product from step iii and digest it by restriction enzymes.

v. Digest target vector plasmid by same restriction enzymes.

vi. Ligate DNA fragments from steps iv and v together using T4 DNA ligase and transform it into competent cell.

vii. Check several colonies to confirm the sequence of the designed reporting system in the ligated plasmid.

3.1.2 Characterization of the constructed reporting device

The characterization of the reporting device can be conducted by two different ways depending on the properties of the ligands. If the ligand can easily penetrate the cell membrane, the fluorescence of the cells harboring the device in the presence or absence of the ligand in culture medium can be measured. If not, various strains with different ligand productivity should be constructed and intracellular concentration of them should also be measured with a proper method (see Note [1]).

i. Inoculate cell to 3 ml of M9 medium at final OD_{600} of 0.05 and culture for 24 h at 37 °C with vigorous shaking.

ii. Dilute culture broth to 3 ml of fresh M9 medium at final OD_{600} of 0.05 and culture for 8 h at 37 °C with vigorous shaking until reaching exponential growth phase.

iii. In case of ligand with cell membrane permeability, dilute culture broth at final OD_{600} of 0.05 to two vials of 3 ml of M9 medium with or without the ligand and culture for 12 h at 37 °C with vigorous shaking. If not, dilute each culture (numbers depending on the variants with different abilities of producing target ligand) broth at final OD_{600} of 0.05–3 ml of M9 medium without ligand and culture for 12 h at 37 °C with vigorous shaking.

iv. Measure cell density of the culture at wavelength of 600 nm using spectrophotometer.

v. Measure fluorescence intensity of the culture using fluorescence detector equipped with appropriate filters (e.g., 486 nm excitation filter and 535 nm emission filter for *sgfp*).

vi. Activation ratio of device in each colony could be calculated using the following equation:

[(fluorescence in the presence of the ligand)/(OD_{600} in the presence of the ligand)]/

[(fluorescence in the absence of the ligand)/(OD_{600} in the absence of the ligand)].

The confirmed device can be used for constructing a Riboselector to expedite the evolution of corresponding metabolite-producing microorganism by exchanging the fluorescent protein to a selective marker gene (e.g., *tetA*).

[1] Some ligands are not readily transported into the cell when transporter for the ligands does not exist. Then you need to construct several bacterial variants that contain the ligand of different intracellular concentrations. For example, an important gene of a biosynthetic pathway (e.g., *lysC* for lysine) can be deleted or overexpressed to make strains with different ligand productivity. In case of using diverse strains with different productivity for the target ligand, it is advisable to measure the intracellular concentration of the target molecule with proper sampling and detection methods to confirm the effect of the genetic changes that you have made.

3.2. Riboselector based on an artificial riboswitch from synthetic aptamer

The Riboselector can also be designed from an artificial riboswitch. There are many synthetic aptamers for sensing of specific molecules which are generated by SELEX (Stoltenburg, Reinemann, & Strehlitz, 2007; Szeto et al., 2013). In this section, the experimental procedures for the development of a selection device based on an artificial riboswitch from synthetic aptamer will be described (Fig. 2B and C). In brief, the synthetic RNA aptamer sequence is linked to the selection/screening module via randomized sequences which constitute selection device library. The selection/screening module, which is a fusion protein composed of a dual selective marker gene (e.g., *tetA*) and a fluorescent protein (e.g., *sgfp*), enriches and evaluates functioning selection devices from the library. Before starting the development of a selection device, it is desirable to check two requirements of target system. First, the sequence of an RNA aptamer that binds specifically to target molecule should be prepared. You can consult review papers (Hermann & Patel, 2000; Patel & Suri, 2000) or well-organized database (Aptamer Base, http://aptamerbase.semanticscience.org) that provide lists of published synthetic aptamers. In addition, aptamers can also be generated by routine SELEX procedure. Second, as above, the cloning vector for selection device library should not contain the same antibiotic resistance gene as marker for plasmid maintenance. Note that every growth medium throughout following protocols should always contain appropriate antibiotics to maintain plasmids.

3.2.1 Library generation for selection of synthetic device

Library of selection device candidates is composed of aptamer sequence and selection/screening module which are linked by randomized sequence. We utilized PCR and blunt-end ligation to construct the library. Synthetic aptamer and randomized sequence are incorporated at the overhang of the primer (ph-T-Library-F) for amplification of the plasmid that harbors the selection/screening module (Fig. 2B). You can also adopt other popular DNA assembly methods such as Gibson assembly and In-Fusion cloning for library generation. Note that this experimental procedure is used also for construction of *ppc* promoter library plasmid in Section 4.1.2.

i. Amplify dual selectable marker gene (e.g., *tetA*) with PCR. Forward primer (T-tetA-F) contains restriction enzyme site and SD sequence at its 5′-end, while reverse primer (T-tetA-R) contains four consecutive repeats of Gly-Gly-Gly-Ser linker sequence at its 5′-end which will connect the marker gene with reporter gene. The linker sequence facilitates independent folding of linked proteins.

ii. Amplify fluorescent protein (e.g., *sgfp*) with PCR. Forward primer (T-sgfp-F) contains the (Gly-Gly-Gly-Ser) × 4 linker sequence at its 5′-end, and reverse primer (T-sgfp-R) contains restriction enzyme site at its 5′-end.

iii. Gel purify the PCR products from steps i and ii and mix them together to conduct overlap PCR. The linker sequence in the PCR products will hybridize to each other, and the fusion protein sequence will be generated during overlap PCR.

iv. Gel purify the PCR product from step iii and digest it by restriction enzymes.

v. Digest target vector plasmid by the same restriction enzymes.

vi. Ligate DNA fragments from steps iv and v together using T4 DNA ligase and transform it into competent cell.

vii. Check several colonies to confirm the sequence of the fusion protein in the ligated plasmid.

viii. Amplify the whole plasmid with PCR (~5 kb). High-fidelity DNA polymerase such as Phusion or Q5 from NEB should be used to avoid any unwanted mutations to occur. Forward primer (Ph-T-Library-F) binds to SD sequence of the fusion protein and contains promoter, aptamer, and random sequences (about 10 nt) in its overhang. And reverse primer (Ph-T-Library-R) binds directly upstream of SD sequence in the plasmid. Both primers should be phosphorylated at their 5′-end.

ix. Ligate the PCR product from step viii using T4 DNA ligase and purify it.

x. Transform the ligated product from step ix into electrocompetent cell to maintain sequence diversity in the randomized region by highly efficient electroporation.

xi. The selection device library can be recovered from plated cells by various plasmid preparation protocols.

xii. Transform the selection device library plasmid into the target cell in which the selection device will function.

3.2.2 Iterative positive and negative selection of synthetic device

The randomized sequence that connects aptamer and RBS of marker gene in the selection device library is designed to modulate expression of marker gene by posttranscriptional mechanism in response to binding of ligand to the aptamer. Since we have used a dual selectable marker gene, ligand-responsive variants can be screened from the library by iterative positive and negative selections (Fig. 2C). Only activated variants survive during positive selection in the presence of the ligand and selection pressure (e.g., tetracycline). In contrast, addition of negative selection pressure (e.g., nickel chloride) in the absence of the ligand can select inactivated variants, since expression of *tetA* causes increased permeability to nickel salts, which is toxic to the cell. Concentrations of the ligand and of the components used to apply positive or negative selection pressure should be determined carefully, since they directly affect the success of whole selection procedure. As mentioned earlier, expression of *tetA* makes the cell resistant to tetracycline while sensitive to nickel ion because the protein encoded by *tetA* increases permeability of nickel ion to the cell. In this protocol, the concentration of $NiCl_2$ for negative selection was decided as 0.2 m*M* at which W3110 wild type could survive, while pBR322-harboring strain (expressing *tetA* marker gene) vanished. Concentration of tetracycline for positive selection depends on basal expression level of the selection device library, which is determined by promoter strength and copy number of the plasmid. For a plasmid with a p15A origin (~10–12 copies per cell) and strong constitutive promoter, 40 μg/ml of tetracycline was enough to discriminate activated state from inactivated state. Finally, ligand concentration in medium for positive selection should be far above usual intracellular concentration.

i. Scrape library colonies from agar plate and wash them three times using M9 medium.

ii. Inoculate the library to 3 ml of M9 medium at final OD_{600} of 0.05 and culture for 12 h at 37 °C with vigorous shaking (200–250 rpm).

iii. Inoculate culture broth to 3 ml of M9 medium containing 0.2 m*M* $NiCl_2$ at final OD_{600} of 0.005 and culture for 48 h at 37 °C with vigorous shaking (negative selection).

iv. Harvest cells by centrifugation at 13,000 rpm for 1 min and wash them three times using M9 medium.

v. Inoculate the library to 3 ml of M9 medium containing proper concentration of ligand and culture for 12 h at 37 °C with vigorous shaking.

vi. Inoculate culture broth to 3 ml of M9 medium containing ligand of proper concentration and 40 μg/ml of tetracycline at final OD_{600} of 0.005 and culture for 24 h at 37 °C with vigorous shaking (positive selection).

vii. Repeat steps from ii to vi two or three times more.

viii. Streak the culture broth onto LB agar plate and incubate at 37 °C to obtain isolated, individual colonies.

3.2.3 Characterization of potential synthetic device

Potential synthetic devices harbored by isolated single colonies should be tested whether they can activate expression of the *tetA–sgfp* fusion protein in response to the target ligand. Characterization of the potential devices follows the same procedure that was introduced in Section 3.1.2. The resulting synthetic devices can be used directly as a Riboselector to expedite the evolution of corresponding metabolite-producing microorganism.

3.3. Characterization and validation of the constructed Riboselector

To utilize Riboselector for enriching bacterial strains that produce high levels of the triggering metabolite, the intracellular concentration of the target ligand should affect the growth rate of cells in the presence of the Riboselector. The variants having different capacity to produce target metabolite should be constructed and tested to confirm that the constructed Riboselector specifically responds to the intracellular concentration of target metabolite by converting the nongrowth-associated phenotype into the selective phenotype.

3.3.1 Growth rate characterization with Riboselector under selective pressure

First, growth rates of variants should be measured under the selection pressure in the presence or absence of the Riboselector to establish whether the Riboselector can transform the metabolite concentration signal into the selective phenotype.

i. Inoculate cells (variants with Riboselector or control plasmid without Riboselector) to 3 ml of M9 medium at final OD_{600} of 0.05 and culture for 24 h at 37 °C with vigorous shaking.
ii. Dilute culture broth to 3 ml of fresh M9 medium at final OD_{600} of 0.05 and culture for 8 h at 37 °C with vigorous shaking until reaching exponential growth phase.
iii. Dilute culture broth to a final OD_{600} of 0.05–3 ml of fresh M9 medium with selection pressure and culture at 37 °C with vigorous shaking.
iv. Measure cell density of the culture at wavelength of 600 nm using spectrophotometer during the course of the culture.

3.3.2 *Growth competition with Riboselector under selective pressure*

Since the growth rates of variants are different depending on the metabolite concentration due to the Riboselector, the higher metabolite producers should outcompete the lower metabolite producers.

- **i.** Inoculate cells (variants with Riboselector or Control system) to 3 ml of M9 medium at final OD_{600} of 0.05 and culture for 24 h at 37 °C with vigorous shaking.
- **ii.** Dilute culture broth to 3 ml of fresh M9 medium to a final OD_{600} of 0.05 and culture for 8 h at 37 °C with vigorous shaking until reaching exponential growth phase.
- **iii.** Dilute culture broth again to a final OD_{600} of 0.05–3 ml of fresh M9 medium with selection pressure and culture at 37 °C with vigorous shaking until reaching exponential growth phase.
- **iv.** Mix the culture broth (variants with Riboselector or variants with Control system) at final OD_{600} of 0.05 for each variant to 3 ml of fresh M9 medium with selection pressure and culture at 37 °C with vigorous shaking until reaching exponential growth phase.
- **v.** Dilute mixed culture broth at final OD_{600} of 0.05–3 ml of fresh M9 medium with selection pressure and culture at 37 °C with vigorous shaking until reaching exponential growth phase. Repeat this step for two or three times more.
- **vi.** Amplify the genetic marker for each variant from the enriched culture broth by PCR to confirm that the higher producer is enriched.

Once the ability of Riboselector to translate the metabolite concentration into the selective phenotype has been confirmed, the device is now ready to be used for optimizing metabolic pathway of the target metabolite production.

4. APPLICATION OF RIBOSELECTOR FOR PATHWAY ENGINEERING

4.1. Metabolic pathway engineering for phenotypic diversification

The application of a synthetic selection device is demonstrated for the optimization of lysine synthesis pathway from physiologically diversified microbes (Fig. 3). However, the device can be applied to many other metabolites. In the case of lysine, we engineered lysine synthesis pathway and further optimized the flux of anaplerotic pathway to find an optimal point that balances the distribution of carbon flux between glycolytic and anaplerotic pathways around phosphoenolpyruvate node.

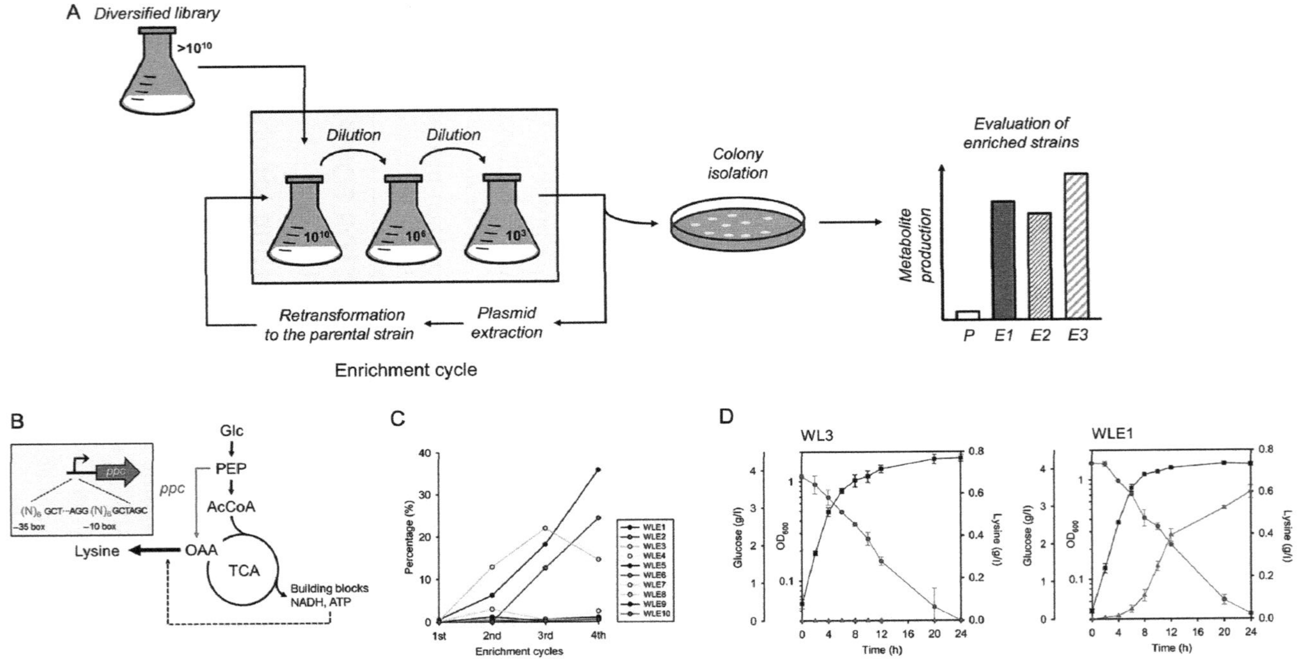

Figure 3 See figure legend on opposite page.

4.1.1 Pathway engineering to enhance lysine production

In the case of the lysine synthesis pathway, we have engineered eight genomic locations to exchange native promoters for synthetic constitutive promoter [BBa_J23100; *lysC*^{fbr} (D-lysC-F, D-lysC-R, P-lysC-F, P-lysC-R), *dapA* (P-dapA-F, P-dapA-R), *dapB* (P-dapB-F, P-dapB-R), and *lysA* (P-lysA-F, P-lysA-R)]; heterologous gene expression [*ddh* from *Corynebacterium glutamicum* (S-ddh-F, S-ddh-R, P-ddh-F, P-ddh-R)]; and knockout [*metL* (D-metL-F, D-metL-R), *thrA* (D-thrA-F, D-thrA-R), and *iclR* (D-iclR-F, D-iclR-R)] using lambda-Red recombination system encoded in pKD46 plasmid.

i. Prepare each PCR product for recombination using suitable oligonucleotides.
   a. Oligonucleotides name beginning with a "P" indicate primers used to replace promoter sequence and with a "D" indicate primers used to delete target genes.
   b. Oligonucleotides name beginning with a "S" indicate primers used to amplify gene.

ii. Transform pKD46 to electrocompetent cell to be engineered.

iii. Dilute the overnight culture at final OD_{600} of 0.05–5 ml of LB medium with the appropriate antibiotics at 30 °C with vigorous shaking.

Figure 3 Application of Riboselector for pathway engineering. (A) Program for optimization of metabolic pathway by means of Riboselector is illustrated here. Phenotypically diversified library is cultured in the presence of the Riboselector and corresponding selection pressure. Repetitive dilution should shift the microbial population to metabolite high-producing strains. Single colonies are isolated after enrichment cycles, and each variant is tested for metabolite production by flask culture and proper quantification method. (B) Schematic metabolic pathway toward lysine in *E. coli*. Expression level of *ppc* was diversified to construct the library of lysine producing *E. coli* variants with different productivity by randomizing sequences of −35 and −10 boxes of promoter (BBa_J23100) that drives transcription of *ppc* gene encoded in plasmid. Glc, glucose; PEP, phosphoenolpyruvate; AcCoA, acetyl-CoA; OAA, oxaloacetate. (C) Population of the library was analyzed using next-generation sequencing of promoter region of *ppc*. The *y*-axis represents the percentage of each variant in the total population. The *x*-axis represents the enrichment cycle depicted in (A), and each enrichment cycle corresponds to three serial cultures and one plasmid preparation. (D) Comparison of physiology between the parental strain (WL3) and one of the enriched strains (WLRE1) that occupied the population up to 40% after four enrichment cycles. The left *y* offset and right *y*-axis represent concentration (g/l) of glucose (green circles) and lysine (red triangles), respectively. The left *y*-axis represents optical density (black rectangles) at 600 nm in log scale. The *x*-axis represents the culture time (h). The experiments were replicated twice. *Panels (B–D): Adapted from Yang et al. (2013) with permission.* (See the color plate.)

iv. Grow cells at 30 °C until reaching OD_{600} of 0.3 and transfer the culture to 37 °C with vigorous shaking after adding arabinose at final concentration of 10 m*M* (induction of Red recombinase) for 1 h.
v. Rapidly chill the culture in ice for 5–10 min.
vi. Transfer cultures to the chilled microtubes and centrifuge 1 min at 8000 rpm at 4 °C.
vii. Pour off the supernatant and add 1 ml ice-cold sterile distilled H_2O to the cell pellet and gently suspend cells. Centrifuge 1 min at 8000 rpm at 4 °C. Repeat this step two more times.
viii. Pour off the supernatant and add 1 ml ice-cold sterile 10% (w/v) glycerol to the cell pellet and gently suspend cells. Centrifuge 1 min at 8000 rpm at 4 °C.
ix. Carefully remove the supernatant and suspend cells in 50 μl sterile ice-cold 10% (w/v) glycerol and add the PCR fragment for recombination (~1 μg).
x. Pipette 50 μl of the electrocompetent cells into a chilled cuvette.
xi. According to the manufacturer's instruction, perform the electroporation.
xii. Quickly add 1 ml of LB for recovery and transfer the entire volume to a new microtube.
xiii. Grow the cell for 3 h at 37 °C and plate them on the agar plate containing appropriate antibiotics. Confirm the recombination by colony PCR after the colonies are appeared.

We have repeated these genomic engineering steps to change all eight locations for amplifying lysine synthesis pathway.

4.1.2 Library generation for optimization of ppc *expression level*

To avoid unwanted influence of the innate *ppc* gene, the *ppc* gene of parental strain was knocked out using the same method described above using primers (D-ppc-F, D-ppc-R). Then we constructed a plasmid library by randomizing the −35 and −10 boxes of promoter of plasmid-encoded *ppc* gene to generate various expression levels (Fig. 3B). Construction of promoter library for *ppc* gene follows the steps that were introduced in Section 3.2.1 using different primer sets [amplification of *ppc* (S-ppc-F, S-ppc-R), addition of synthetic UTR for *ppc* (ph-UTR-ppc-F, ph-UTR-ppc-R), and addition of promoter library (ph-J23N-F, ph-J23N-R)]. Prepared promoter library plasmid is transformed into the target cell in which lysine production pathway will be optimized.

4.2. Pathway optimization using synthetic selection device

To screen the highest producing strains from the library using Riboselector, the appropriate concentration of components used to apply positive or negative selection pressure should be decided by investigating growth rates of parental strain in various concentrations of selection pressure. We should set a selection pressure at which growth rate of parental strain is severely impeded when the selection device is implemented to eliminate parental strain in enrichment steps.

4.2.1 Setting an appropriate selection pressure by measuring growth rates

i. Inoculate cells (parental strain with Riboselector or Control system) to 3 ml of M9 medium at final OD_{600} of 0.05 and culture for 24 h at 37 °C with vigorous shaking.
ii. Dilute culture broth to 3 ml of fresh M9 medium at final OD_{600} of 0.05 and culture for 8 h at 37 °C with vigorous shaking until reaching exponential growth phase.
iii. Dilute culture broth at final OD_{600} of 0.05 to several 3 ml of fresh M9 medium with different concentrations of selection pressure and culture at 37 °C with vigorous shaking.
iv. Measure cell density of the culture at wavelength of 600 nm using spectrophotometer during the course of the culture.
v. Select the concentration where growth rate of parental strain is severely hampered when Riboselector is introduced.

4.2.2 Enrichment experiment and analysis

The higher lysine producer will be enriched from the library when cells are serially diluted and cultured in the presence of selection pressure since only the higher lysine producer can repress the expression of the negative selection marker (*tetA*) of the Riboselector and survive in the presence of nickel salt (Fig. 3A). The cells should be diluted around mid-exponential phase to efficiently enrich superior strains. Although the long-term incubation under selection pressure allows cells to acquire resistance to selection pressure by unexpected mutations, we can prevent this situation by introducing harvested plasmids into freshly prepared parental strain.

i. Inoculate library cells carrying Riboselector to 3 ml of M9 medium at final OD_{600} of 0.05 and culture for 24 h at 37 °C with vigorous shaking.

ii. Dilute culture broth to 3 ml of fresh M9 medium at final OD_{600} of 0.05 and culture for 8 h at 37 °C with vigorous shaking until reaching exponential growth phase.

iii. Dilute culture broth to a final OD_{600} of 0.05–3 ml of fresh M9 medium containing appropriate selection pressure and culture at 37 °C with vigorous shaking.

iv. Dilute the culture broth again to a final OD_{600} of 0.05–3 ml of fresh M9 medium with appropriate selection pressure when it reached mid-exponential phase and culture at 37 °C with vigorous shaking. Repeat this step two times more.

v. Harvest plasmids from enriched library and retransform them to the fresh parental strain.

vi. Repeat steps from i to v three times or more to enrich the higher producer.

vii. Plate the enriched cell on the agar plate and analyze several colonies by sequencing and measuring concentration of target metabolite (Fig. 3C and D).

5. CONCLUDING REMARKS

An expanding number of riboswitch classes have been discovered from natural sources, and synthetic riboswitches can be easily constructed from *in vitro*-selected aptamers. Moreover, by subjecting the synthetic Riboselector to a series of selection processes, kinetic components of the device, such as binding affinity and dynamic range, can be changed to obtain an RNA device that is functional under physiological conditions. Thus, our synthetic Riboselector should be used as a universal high-throughput screening platform for inconspicuous-metabolite overproduction through accelerated evolution. Once the Riboselector is constructed, repetitive strain improvement can be accomplished by simply modulating the selection cutoff levels during enrichment process and using recently developed techniques for genome-wide alterations of cellular phenotypes.

ACKNOWLEDGMENTS

This work was supported by the Basic Science Research Program (2012R1A2A2A01009868) and the Advanced Biomass R&D Center (ABC-2010-0029800) through the National Research Foundation of Korea funded by the Ministry of Education, Science and Technology, Korea and the Marine Biomaterials Research Center of the Marine Biotechnology Program funded by the Ministry of Land, Transport and Maritime Affairs, Korea.

REFERENCES

Alper, H., Moxley, J., Nevoigt, E., Fink, G. R., & Stephanopoulos, G. (2006). Engineering yeast transcription machinery for improved ethanol tolerance and production. *Science, 314*, 1565–1568.

Baker, J. L., Sudarsan, N., Weinberg, Z., Roth, A., Stockbridge, R. B., & Breaker, R. R. (2012). Widespread genetic switches and toxicity resistance proteins for fluoride. *Science, 335*, 233–235.

Breaker, R. R. (2011). Prospects for riboswitch discovery and analysis. *Molecular Cell, 43*, 867–879.

Burge, S. W., Daub, J., Eberhardt, R., Tate, J., Barquist, L., Nawrocki, E. P., et al. (2013). Rfam 11.0: 10 years of RNA families. *Nucleic Acids Research, 41*, D226–D232.

Cherepanov, P. P., & Wackernagel, W. (1995). Gene disruption in Escherichia coli: TcR and KmR cassettes with the option of Flp-catalyzed excision of the antibiotic-resistance determinant. *Gene, 158*, 9–14.

Datsenko, K. A., & Wanner, B. L. (2000). One-step inactivation of chromosomal genes in Escherichia coli K-12 using PCR products. *Proceedings of the National academy of Sciences of the United States of America, 97*, 6640–6645.

Dietrich, J. A., McKee, A. E., & Keasling, J. D. (2010). High-throughput metabolic engineering: Advances in small-molecule screening and selection. *Annual Review of Biochemistry, 79*, 563–590.

Henkin, T. M. (2008). Riboswitch RNAs: Using RNA to sense cellular metabolism. *Genes & Development, 22*, 3383–3390.

Hermann, T., & Patel, D. J. (2000). Adaptive recognition by nucleic acid aptamers. *Science, 287*, 820–825.

Isaacs, F. J., Carr, P. A., Wang, H. H., Lajoie, M. J., Sterling, B., Kraal, L., et al. (2011). Precise manipulation of chromosomes in vivo enables genome-wide codon replacement. *Science, 333*, 348–353.

Keasling, J. D. (2008). Synthetic biology for synthetic chemistry. *ACS Chemical Biology, 3*, 64–76.

Lynch, S. A., & Gallivan, J. P. (2009). A flow cytometry-based screen for synthetic riboswitches. *Nucleic Acids Research, 37*, 184–192.

Muranaka, N., Sharma, V., Nomura, Y., & Yokobayashi, Y. (2009). An efficient platform for genetic selection and screening of gene switches in Escherichia coli. *Nucleic Acids Research, 37*, e39.

Nakagawa, A., Minami, H., Kim, J. S., Koyanagi, T., Katayama, T., Sato, F., et al. (2011). A bacterial platform for fermentative production of plant alkaloids. *Nature Communications, 2*, 326.

Patel, D. J., & Suri, A. K. (2000). Structure, recognition and discrimination in RNA aptamer complexes with cofactors, amino acids, drugs and aminoglycoside antibiotics. *Journal of Biotechnology, 74*, 39–60.

Peralta-Yahya, P. P., Ouellet, M., Chan, R., Mukhopadhyay, A., Keasling, J. D., & Lee, T. S. (2011). Identification and microbial production of a terpene-based advanced biofuel. *Nature Communications, 2*, 483.

Pfleger, B. F., Pitera, D. J., Smolke, C. D., & Keasling, J. D. (2006). Combinatorial engineering of intergenic regions in operons tunes expression of multiple genes. *Nature Biotechnology, 24*, 1027–1032.

Santos, C. N., & Stephanopoulos, G. (2008). Combinatorial engineering of microbes for optimizing cellular phenotype. *Current Opinion in Chemical Biology, 12*, 168–176.

Serganov, A., & Nudler, E. (2013). A decade of riboswitches. *Cell, 152*, 17–24.

Sinha, J., Reyes, S. J., & Gallivan, J. P. (2010). Reprogramming bacteria to seek and destroy an herbicide. *Nature Chemical Biology, 6*, 464–470.

Stoltenburg, R., Reinemann, C., & Strehlitz, B. (2007). SELEX—A (r)evolutionary method to generate high-affinity nucleic acid ligands. *Biomolecular Engineering*, *24*, 381–403.

Szeto, K., Latulippe, D. R., Ozer, A., Pagano, J. M., White, B. S., Shalloway, D., et al. (2013). RAPID-SELEX for RNA aptamers. *PLoS One*, *8*, e82667.

Wang, H. H., Isaacs, F. J., Carr, P. A., Sun, Z. Z., Xu, G., Forest, C. R., et al. (2009). Programming cells by multiplex genome engineering and accelerated evolution. *Nature*, *460*, 894–898.

Weinberg, Z., Wang, J. X., Bogue, J., Yang, J., Corbino, K., Moy, R. H., et al. (2010). Comparative genomics reveals 104 candidate structured RNAs from bacteria, archaea, and their metagenomes. *Genome Biology*, *11*, R31.

Yang, J., Seo, S. W., Jang, S., Shin, S. I., Lim, C. H., Roh, T. Y., et al. (2013). Synthetic RNA devices to expedite the evolution of metabolite-producing microbes. *Nature Communications*, *4*, 1413.

CHAPTER EIGHTEEN

Fluorescence Assays for Monitoring RNA–Ligand Interactions and Riboswitch-Targeted Drug Discovery Screening

J. Liu, C. Zeng, S. Zhou, J.A. Means, J.V. Hines[1]
Department of Chemistry & Biochemistry, Ohio University, Athens, Ohio, USA
[1]Corresponding author: e-mail address: hinesj@ohio.edu

Contents

Abstract

Riboswitches and other noncoding regulatory RNA are intriguing targets for the development of therapeutic agents. A significant challenge in the drug discovery process, however, is the identification of potent compounds that bind the target RNA specifically and disrupt its function. Essential to this process is an effectively designed cascade of screening assays. A screening cascade for identifying compounds that target the T box riboswitch antiterminator element is described. In the primary assays, moderate to higher throughput screening of compound libraries is achieved by combining the sensitivity of fluorescence techniques with functionally relevant assays. Active compounds are then validated and the binding to target RNA further characterized in secondary assays. The cascade of assays monitor ligand-induced changes in the steady-state

Methods in Enzymology, Volume 550
ISSN 0076-6879
http://dx.doi.org/10.1016/bs.mie.2014.10.040

fluorescence of an attached dye or internally incorporated 2-aminopurine; the fluorescence anisotropy of an RNA complex; and, the thermal denaturation fluorescence profile of a fluorophore-quencher labeled RNA. While the assays described have been developed for T box riboswitch-targeted drug discovery, the fluorescence methods and screening cascade design principles can be applied to drug discovery efforts targeted toward other medicinally relevant noncoding RNA.

1. INTRODUCTION

1.1. Noncoding RNAs and drug discovery

Riboswitches and other noncoding RNAs serve essential roles in many biological processes and consequently are important targets for drug discovery. Therapeutic agents that specifically target these RNAs have historically been limited to antibacterial agents which target ribosomal RNA, but recent efforts have focused on expanding the scope to targeting other noncoding RNA elements.(Guan & Disney, 2012) Riboswitches are intriguing potential antibacterial targets for which there are no known therapeutic agents. Antibacterial drug resistance is a significant global health concern (Boucher et al., 2009; CDC, 2013; WHO, 2014) resulting in an urgent need to identify new antibacterial agents. Riboswitch regulatory elements structurally respond to metabolite levels to turn on or off transcription, or translation of essential genes (Smith, Fuchs, Grundy, & Henkin, 2010). The T box riboswitch, for example, is found primarily in Gram positive bacteria (Gutierrez-Preciado, Henkin, Grundy, Yanofsky, & Merino, 2009; Vitreschak, Mironov, Lyubetsky, & Gelfand, 2008) and regulates the production of many aminoacyl tRNA synthetase, amino acid biosynthesis, and amino acid transport genes by responding to the amount of nonaminoacylated tRNA (Green, Grundy, & Henkin, 2010). The tRNA anticodon base pairs with a specifier sequence in the 5′-untranslated region, while the tRNA elbow (D/T region) forms tertiary contacts (Grigg & Ke, 2013) and the nonaminoacylated tRNA acceptor end base pairs with nucleotides in the bulge of an antiterminator element (Green et al., 2010) (Fig. 1A). This latter base pairing stabilizes the antiterminator element and enables complete transcription of the gene (Grundy, Moir, Haldeman, & Henkin, 2002). In the absence of uncharged tRNA (e.g., when there is a competing excess of aminoacylated tRNA), a more stable terminator element forms (Fig. 1B) and transcription termination occurs upstream of the translation start site (Green et al., 2010). The T box antiterminator RNA element is a high impact target for antibacterial drug

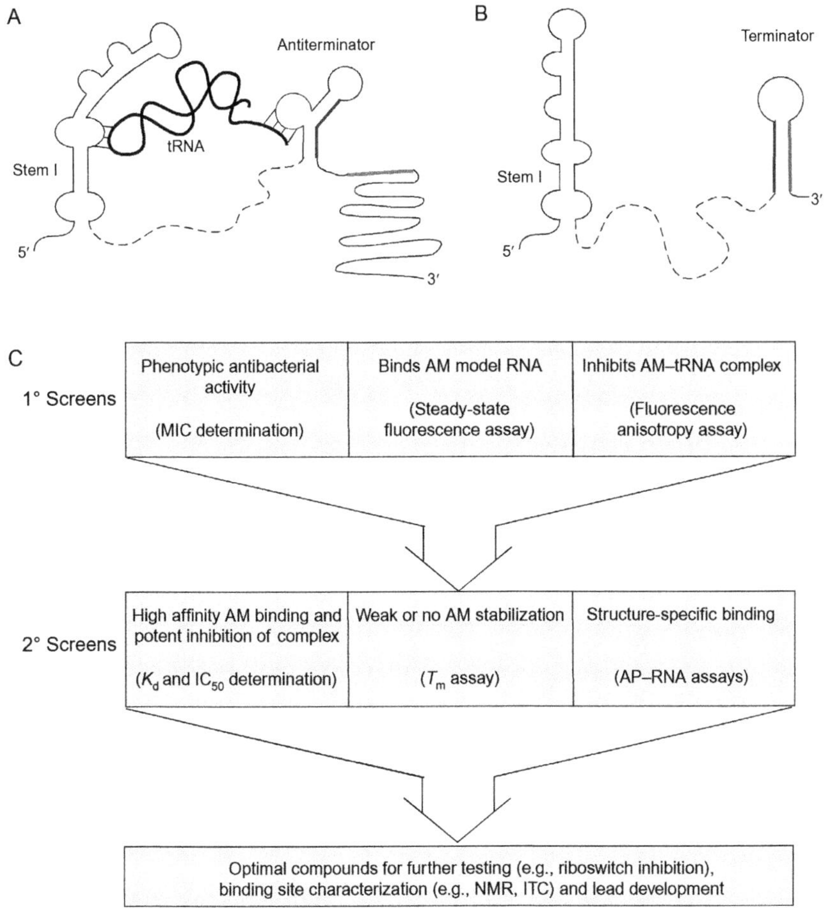

Figure 1 T box riboswitch drug discovery. (A) Transcription antitermination mechanism in the presence of uncharged tRNA, (B) Transcription termination, (C) T box antiterminator (AM)-targeted drug discovery screening cascade for identifying optimal hit compounds.

discovery (Jentzsh & Hines, 2011), given the role it serves in the riboswitch mechanism, the essential genes regulated by the T box riboswitch (Green et al., 2010) and the high sequence conservation of the antiterminator (Gutierrez-Preciado et al., 2009; Vitreschak et al., 2008). Similar criteria can be applied to the selection of valid targets in other riboswitches and, more generally, other noncoding RNA regulatory elements: (1) the RNA element serves (or affects) a key mechanistic role in the function of the RNA, (2) expression of essential genes and/or the function of essential gene products is regulated by the RNA, and (3) the primary sequence and structure of the RNA element is highly conserved (and/or is less likely to develop mutational resistance to targeted therapeutic agents).

1.2. Screening cascade for RNA-targeted drug discovery

Using information gained from functional and structural biology studies of the T box antiterminator RNA, we developed a screening cascade of functionally relevant, target-based assays that monitor T box antiterminator binding and tRNA–antiterminator RNA complex disruption. The fundamental design principles of these target-based assays are readily applicable to other noncoding RNA targets and can be complemented by phenotypic assays (e.g., antibacterial activity that correlates with the presence of essential genes regulated by the noncoding RNA). The combination of target-based and phenotypic assays helps to identify hit compounds and characterize structure-activity relationships for lead inhibitor design. This type of comprehensive drug discovery strategy is routinely used in the discovery of protein-targeted therapeutic agents (Hughes, Rees, Kalindjian, & Philpott, 2011), but has not been widely applied to targeting riboswitches and other noncoding RNA.

The screening cascade starts with primary (1°) screens that are amenable to moderate or higher throughput to allow a rapid, low cost method for identifying initial hit compounds (Fig. 1C). The steady-state fluorescence-monitored ligand-binding assay (Means et al., 2007) identifies compounds that bind antiterminator RNA by monitoring changes in the fluorescence of 5′-fluorophore-labeled RNA, which correlate with ligand-induced structural changes. This screen provides preliminary information regarding ligand RNA specificity by comparing ligand-induced changes in functional antiterminator model RNA (AM, Fig. 2A and B) with that in a reduced function variant (AM_{C11U}) and a model lacking the highly conserved seven-nucleotide bulge of the antiterminator ($AM_{control}$). This fluorescence-monitored binding assay strategy can be applied to other noncoding RNA elements, provided that ligand binding induces a sufficient structural response in the RNA element to alter the fluorescence of the end-labeled fluorophore (e.g., through changes in RNA helix-fluorophore stacking).

Fluorescence anisotropy assays are useful to identify RNA-binding ligands, but they depend on monitoring disruption of an RNA complex, thus necessitating that there be a known, functionally relevant complex which forms between the target RNA and another molecule (e.g.,

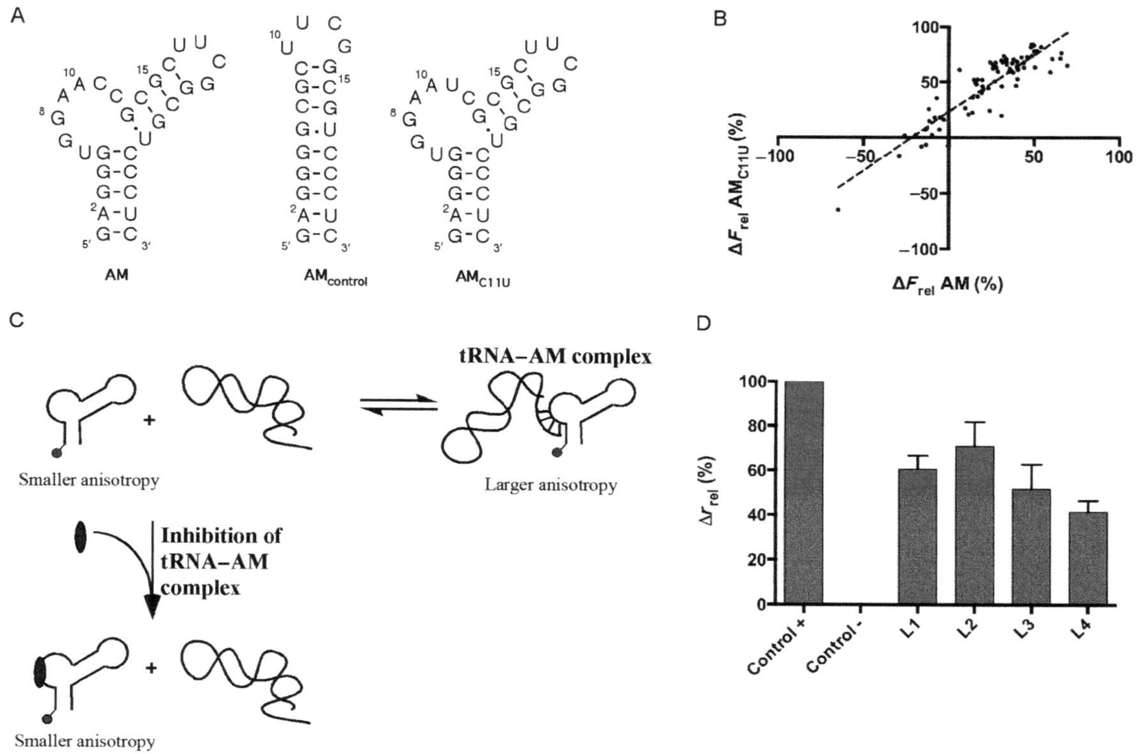

Figure 2 Fluorescence-monitored 1° screens. (A) T box antiterminator model RNAs, (B) Representative steady-state fluorescence binding screen data, (C) Fluorescence anisotropy monitored complex disruption assay, (D) Representative fluorescence anisotropy data

metabolite, protein, RNA etc.), For example, the T box riboswitch fluorescence anisotropy assay (Fig. 2C) readily monitors ligand effects on formation of tRNA–antiterminator RNA complex since the higher molecular weight complex has a larger anisotropy value than the antiterminator alone (Zhou, Acquaah-Harrison, Bergmeier, & Hines, 2011). The assay identifies antagonists (compounds that inhibit formation of the complex, Fig. 2D) and agonists (compounds that enhance complex formation). The steady-state fluorescence and anisotropy assays when analyzed together provide complementary information. For example, in the T box assays, a "hit" in the ligand-binding assay that also disrupts the complex likely does so by binding to the antiterminator rather than the tRNA.

The 1° screens are followed by secondary (2°) screening assays to confirm and validate initial hit compounds. These assays provide more detailed information about the ligand–RNA interaction ($K_{d,}$ IC_{50}) and provide preliminary information on ligand-binding site location. The design principles for these 2° assays can be readily applied to targeting other noncoding RNA by making strategic choices for fluorophore incorporation that are informed by functional and structural biology studies of the RNA element. In the T box antiterminator RNA, the fluorescent base analog, 2-aminopurine (AP), incorporated at positions 2, 9, 10, or 13 in AM (i.e., 2-AP-AM, 9-AP-AM, etc.) is used to monitor localized ligand-induced changes in antiterminator structure (Means et al., 2007, 2009). In addition, the ligand-induced change in fluorescence of 9-AP-AM compared to 9-AP-AM_{C11U} is used as a binding specificity assay (Means et al., 2006). A fluorescence-quenching monitored thermal denaturation (T_m) assay (Zhou, Acquaah-Harrison, Jack, Bergmeier, & Hines, 2011) is used to identify compounds that stabilize the antiterminator RNA element. Ligand-induced stabilization is a negative attribute for T box-targeted drug discovery since compounds that stabilize the antiterminator might act as tRNA-independent T box riboswitch agonists. In general, this fluorescence-monitored T_m assay can be readily applied to other noncoding RNA elements (regardless of whether stabilization or destabilization is the desired attribute) and allows for greater throughput than a typical UV-monitored T_m determination.

The 2° screening assays are also useful for follow-up studies on ligands with inconclusive results in the 1° assays. For example, a compound that disrupts the tRNA–antiterminator complex, but is "silent" in the 1° binding assay may still be binding the antiterminator but not in a manner that affects the fluorescence of the 5′-fluorophore-labeled RNA used in the assay. In this case, the AP-RNA assays, which are more sensitive to local environment (Means et al., 2007, 2009), may detect ligand binding. If not, then the ligand may be disrupting complex formation by binding tRNA, which could be tested by isothermal titration calorimetry (ITC).

Ultimately, RNA–ligand binding interactions need to be characterized with techniques that provide greater biochemical detail (e.g., ITC, chemo-enzymatic probing, and NMR; Orac et al., 2011; Zhou, Means, Acquaah-Harrison, Bergmeier & Hines, 2011) and assayed for riboswitch inhibition (Anupam et al., 2008). These more time-consuming and costly assays, however, are not as readily amenable to higher throughput and are reserved for compounds that are identified as hits in the 1° and 2° assays. The activity threshold used to classify a compound as a hit becomes successively more stringent as the activity of the compound library increases with successive rounds of library design optimization. This chapter details the methods for the moderate to higher throughput 1° and 2° fluorescence-monitored screening assays used to identify ligands that specifically bind T box antiterminator RNA and disrupt the tRNA–antiterminator complex.

2. GENERAL CONSIDERATIONS

2.1. Equipment

Microplate fluorometer for steady-state fluorescence and fluorescence anisotropy assays (e.g., SpectraMax M5 (Molecular Devices))
Spectrofluorometer for 2-aminopurine assays (e.g., SPEX Fluoromax-3 spectrofluorometer (JY Horiba))
Real-time quantitative PCR instrument for T_m assay (e.g., Mx3000P qPCR (Stratagene))
Microplate vortexer and spinner
Pipettes (single- and 8-channel)

2.2. Setup and optimization

Every aspect of the fluorescence-monitored assay needs to be optimized to minimize the interference from background fluorescence, scattered light, photobleaching, pH/temperature effects on the fluorophore, and ligand fluorescence.

2.2.1 Microplate setup and spectrometer settings

The outermost wells around the entire assay microplate should not be used to avoid possible artifacts at the edges. Blank control samples containing the same buffer, but lacking the fluorescently labeled RNA should be analyzed to determine the presence of scattered excitation and Raman scattering, which, for dilute or low quantum yield fluorophores, may be significant (Lakowicz, 2006). Spectrometer settings can be optimized to maximize the fluorescence signal while minimizing the effect of the Raman peak by varying the excitation and emission wavelengths and applying suitable cutoff filters. Since the Raman scatter always occurs at a constant wavenumber from the excitation (Lakowicz, 2006), optimal settings can be found by lowering the excitation to shift the wavelength of the scattered light away from that of the fluorophore's emission peak maximum, such that it does not interfere or can be removed with a cutoff filter set 5–15 nm below the emission wavelength. The emission wavelength monitored can also be increased to further separate signal from the Raman scatter. The best excitation wavelength is one that results in the maximum RFU of the fluorophore's emission peak while still having the background emission of the blank sample <1 RFU. Finally, the instrument gain needs to be the same for the measurement of the sample and any controls and blank solutions (Lakowicz, 2006) to enable meaningful background subtraction and normalization.

2.2.2 Ligand fluorescence

Possible ligand fluorescence in the presence and/or absence of the RNA target needs to be assessed and suitably subtracted if possible. For this reason, relative or normalized ligand-induced changes in fluorescence are typically reported. Ligand fluorescence may be an issue when using fluorophores with short excitation wavelengths (e.g., AP), since some functional groups present in compound libraries might also fluoresce at these wavelengths.

2.3. Pipetting and assay preparation

Excellent accuracy and precision are required to ensure assay reproducibility and sensitivity. Proper pipetting techniques are essential since very small percentage volume errors in pipetting fluorophore solutions (especially concentrated stock solutions) can result in significant errors in the assays. Techniques such as prewetting tips used for replicate pipetting and using pipettes that "lock" on the pipette tip (e.g., ClipTip pipettes (Thermo Scientific)) help minimize pipetting errors.

All solutions should be mixed gently by careful repeated pipette suction and dispensing (drawing in only liquid and no air) or by using a microplate vortex mixer (1 min). This gentle mixing step minimizes the introduction of oxygen from the air into the reaction solution. Dissolved oxygen can quench the observed fluorescence (Lakowicz, 2006). Prior to analysis, the loaded 384-well assay plate is sealed with a microplate sealing film and spun in a PCR plate spinner (15 s) to remove trapped air bubbles at the bottoms of the wells. Techniques for pipetting, mixing and microplate vortexing and spinning should be optimized by preparing replicate experiments (e.g., 24–36 wells of 100 n*M* fluorophore-labeled RNA in buffer) and then acquiring and evaluating the fluorescence data to ensure that there is excellent precision. In addition to optimizing the experimental setup technique, the assay itself can be optimized for high-throughput screening. The quality of a high-throughput assay can be quantified using the Z'-factor (Zhang, Chung, & Oldenburg, 1999); however, trade-offs need to be considered. For example, in the anisotropy assay, a balance is needed between optimizing the Z'-factor and optimizing conditions favorable for detecting competitive inhibitors of the tRNA–antiterminator complex (Zhou, Acquaah-Harrison, Bergmeier, & Hines, 2011). The T box fluorescence-based screening assays described here have been optimized to balance these trade-offs along with time and cost considerations.

2.4. Materials and reagents

All materials and reagents are molecular biology grade, RNase- and DNase-free (if available). Pipette tips are filter tips. Buffer solutions are prepared using ultrapurified water and sterilized by autoclaving for 25 min. Sterile techniques are used throughout to avoid RNase contamination. Proper personal protective equipment (e.g., gloves, goggles, etc.), handling and disposal procedures should be used at all times for work with harmful or reactive chemicals (e.g., DMSO stock solutions of compounds with unknown toxicity, etc.).

Unlabeled RNA (e.g., *B. subtilis* $tRNA^{Tyr}$(A73U); Gerdeman, Henkin, & Hines, 2002) can be purchased as a custom-synthesized RNA or synthesized by *in vitro* transcription (for tRNA, see Fauzi, Jack, & Hines, 2005). Fluorophore-labeled RNA can be purchased (e.g., Thermo Scientific, Trilink, and IDT). In addition to the excitation and emission characteristics of the fluorophore, other factors should be considered. The fluorophore signal should be insensitive to small pH changes, especially for the steady-state experiments; thus, tetramethyl rhodamine (TAMRA) is a better choice than fluorescein, which has a strong pH-dependent absorption and emission profile (Lakowicz, 2006). For the thermal denaturation assays, the fluorophore should be as insensitive to temperature as possible (e.g., HEX and ROX Marras, 2006), and it should stabilize the RNA structure as little as possible (Zhou, Acquaah-Harrison, Jack, Bergmeier, & Hines, 2011). For the fluorophore used in the anisotropy assay, the effect of linker length needs to be considered (Zhou, Acquaah-Harrison, Bergmeier, & Hines, 2011). Too short of a linker may result in the fluorophore interfering with the complex formation. Too long or flexible of a linker may allow too great a rotational freedom of the fluorophore (Unruh, Gokulrangan, Lushington, Johnson, & Wilson, 2005), thus limiting the maximum possible anisotropy change observed upon complex formation. In addition, the fluorophore should be attached to the smaller molecule involved in the complex (lowest molecular weight) to maximize the change in anisotropy upon complex formation. For the T box antiterminator–tRNA disruption assay, having a fluorophore attached to the antiterminator model RNA via an amino hexyl linker is optimal.

Dialyze all RNAs prior to use against dialysis buffer using a suitable MWCO membrane (3.5 kDa for T box antiterminator model RNA (MW = 9 kDa)) and a microdialysis system. Prior to preparing the RNA solutions for the assay, the RNA stock solutions should be renatured in Dilution Buffer (i.e., low salt, no Mg^{2+}) by heating to 90 °C for 1.5 min followed by slow cooling to room temperature (~15 min). This renaturation procedure is suitable for the model RNAs and tRNA described in these protocols, but it should be appropriately optimized if other RNAs are used. All fluorophore-labeled RNA solutions should be protected from light whenever possible (e.g., covered with aluminum foil) to avoid photobleaching of the fluorophore.

RNA concentrations can be determined after dialysis by measuring the UV absorbance of the stock solution. Most vendors of synthetic RNA supply the predicted extinction coefficient with the RNA, but it can also be calculated directly from nearest neighbor principles (Cavaluzzi & Borer, 2004; Puglisi & Tinoco, 1989). Measurement of the RNA in water (no salt) and at a denaturing temperature (e.g., ~80–85 °C) will help to minimize underestimating the concentration as a result of hypochromic effects in structured RNA. Comparing the absorbance at 260 nm versus 280 nm can be used to determine if there are protein and/or solvent contaminants. An A_{260}/A_{280} ratio of 1.8–2.0 usually indicates that the RNA is not contaminated with protein, phenol, or DNA (Sanbrook, Fritsch, & Maniatis, 1989). The concentration of fluorophore-labeled RNA is determined by measuring the absorbance at 260 nm, 280 nm, and $\lambda_{\text{dye max}}$ nm. The corrected absorbance A_{corr} is used to calculate the RNA concentration: $A_{260\text{corr}} = A_{260} - (A_{\text{dye max}} \times CF_{260})$ and $A_{280\text{corr}} = A_{280} - (A_{\text{dye max}} \times CF_{280})$, where $CF = A_{260\ (\text{or}\ 280)}/A_{\text{dye max}}$

for the free fluorophore (van der Rijke, Heetebrij, Talman, Tanke, & Raap, 2003). For example, for 5′-TAMRA-RNA $A_{260corr} = A_{260} - (A_{554} \times 0.27)$.

For compounds that have limited water solubility, DMSO can be used to help solubilize concentrated solutions, provided that the final assay concentration is $\leq 5\%$ (v/v) DMSO. This amount of DMSO does not interfere with the antiterminator RNA structure or antiterminator–tRNA complex formation (Zhou, Acquaah-Harrison, Bergmeier, & Hines, 2011; Zhou, Acquaah-Harrison, Jack, Bergmeier, & Hines, 2011). Higher concentrations of DMSO ($\geq 10\%$), however, are known to destabilize RNA structure and should be avoided (Strauss, Kelly, & Sinsheimer, 1968). Even low amounts of DMSO ($\leq 5\%$) can alter RNA structures (Lee, Vogt, McBrairty, & Al-Hashimi, 2013). Consequently, if DMSO cannot be avoided, then suitable control experiments must be run to fully document the extent to which DMSO affects the function of the RNA and to confirm that it does not interfere with the ligand screening assays.

3. EXAMPLE PROTOCOLS

3.1. 1° Screening assays

The 1° screening assays are used to efficiently screen compound libraries to identify potential RNA-targeted hit compounds.

3.1.1 Steady-state fluorescence assay: Ligand–antiterminator RNA binding assay

The steady-state fluorescence assay can be used as a rapid screen to identify potential hit compounds that bind target RNA.

3.1.1.1 Materials and reagents

RNA Dialysis/Storage Buffer: 10 m*M* MOPS, pH 6.5, and 0.01 m*M* EDTA

2.5 × Reaction Buffer: 125 m*M* MOPS, pH 6.5, 125 m*M* NaCl, 37.5 m*M* $MgCl_2$, and 0.025 m*M* EDTA

Dilution Buffer: 10 m*M* MOPS, pH 6.5

Ligand Solvent: H_2O or up to 1:1 DMSO:H_2O (v/v) depending on ligand solubility,

Ligand Stock: 1 m*M* ligand in Ligand Solvent

RNA stock: 200 n*M* 5′-TAMRA-RNA (TAMRA linked via aminohexyl linker) in Dilution Buffer. For the examples given below, model RNAs used are AM, $AM_{control}$, and AM_{C11U}.

96-well V-bottom microplates (for Reagent and Mixing plates)

384-well round bottom black polystyrene microplate, low volume, non-binding surface (for Assay plate).

3.1.1.2 Reagent plate preparation

Prepare a Reagent Plate in a 96-well V-bottom microplate to facilitate assembly of the reaction mixtures. Each row contains 8 wells of solution.

Row 1: 140 μL 2.5 × Reaction Buffer

Row 2: 10 μL H_2O

Row 3: 10 μL Ligand Solvent

Rows 5, 7, and 9: 10 μL of Ligand Stock, each well containing a different ligand for a total of 24 ligands

Row 11: 160 μL RNA stock

Rows 6, 8, and 10 are intentionally left blank between the rows of ligand-containing mixtures and the fluorescently labeled RNA containing wells to avoid any possible inadvertent cross-contamination.

3.1.1.3 Mixing plate preparation

Reaction mixtures are prepared in duplicate rows on a 96-well V-bottom microplate (Mixing Plate) by pipetting appropriate amounts from the relevant wells of the Reagent Plate followed by mixing. Mixing Plate rows are assembled using a multichannel pipette as follows (8 wells per row):

Rows 1 and 2 (Blank): 12 μL of 2.5 × Reaction Buffer, 3 μL H_2O, and 15 μL RNA stock

Rows 3 and 4 (Background): 12 μL of 2.5 × Reaction Buffer, 3 μL Ligand Solvent, and 15 μL RNA stock

Rows 5 and 6 (Ligand-Binding Reactions for ligands 1–8): 12 μL of 2.5 × Reaction Buffer, 3 μL Ligand Stock, and 15 μL RNA stock, with a different ligand stock in each well

Rows 7 and 8 (Ligand-Binding Reactions for ligands 9–16): 12 μL of 2.5 × Reaction Buffer, 3 μL Ligand Stock, and 15 μL RNA stock, with a different ligand stock in each well

Rows 9 and 10 (Ligand-Binding Reactions for ligands 17–24): 12 μL of 2.5 × Reaction Buffer, 3 μL Ligand Stock, and 15 μL RNA stock, with a different ligand stock in each well

A similar Mixing Plate is set up with Ligand Control reactions in place of the rows containing Ligand-Binding Reactions. Each row of the Ligand Control reactions in this Mixing Plate is assembled using: 12 μL of 2.5 × Reaction Buffer, 3 μL Ligand Stock, 15 μL Dilution Buffer with a different ligand stock in each well (for a total of 8 per row).

3.1.1.4 Assay plate preparation and data acquisition

Using an 8-channel pipette, transfer 20 μL of each well in a row (e.g., Row 1) from the Mixing Plate to the 384-well Assay Plate. This results in, for example, Row B in the 384-well Assay Plate having solution in every other column (i.e., Columns 2, 4, 6, etc.). Then transfer 20 μL of each well from the respective replicate row of the Mixing Plate to the interleaved, empty wells in Row B (i.e., Columns 3, 5, 7, etc.) to obtain duplicate reaction mixtures in adjacent columns (e.g., wells B2 and B3 contain replicate reactions). The specific arrangement described above is not absolutely required and can be readily altered to accommodate spectrometer-specific data output organization to simplify subsequent data analysis.

The spectrometer parameters are set for top read fluorescence. For 5′-TAMRA-RNA, the excitation wavelength = 554 nm and the cutoff = 590 nm. An emission scan can be acquired (e.g., 570–630 nm and 5 nm steps) or a single emission wavelength, (600 nm). Data are acquired at multiple time points (e.g., 10, 60, 90, and 120 min). Excess dissolved oxygen is more apt to dissipate over time, thus improving reproducibility. However, ligands with poor water solubility may begin to precipitate and/or cause RNA aggregation with time, both of which will affect the observed fluorescence. In addition, some ligands may be slow to bind the RNA (possibly resulting in an alternate RNA fold). Consequently, the time point analyzed may be dependent on the types of compounds being assayed. To minimize evaporation, the plate should be sealed with a microplate sealing film between measurements or an optically transparent sealing film used throughout.

3.1.1.5 Analysis

The fluorescence of the Blank and Background reactions are compared to ensure that there are no significant changes in the fluorescence of the 5′-TAMRA-RNA as a consequence of the Ligand Solvent. The Ligand Control reaction is then analyzed to determine if the ligand fluoresces under the conditions used. If so, the effect of this fluorescence (and possible change upon RNA binding) needs to be taken into consideration.

The Ligand-Binding reaction data are analyzed by calculating F_{rel} for each ligand using $F_{rel} = [(F - F_0)/F_0] \times 100$, where F is the fluorescence at 600 nm in the presence of ligand and F_0 is the average fluorescence (600 nm) of the Background replicates. The F_{rel} is calculated for each ligand replicate and the replicate error determined. A significant difference in the F_{rel} for ligand binding to AM compared to binding to $AM_{control}$ indicates the ligand likely binds in the bulge region of the antiterminator. A significant difference in the F_{rel} for ligand binding to AM compared to AM_{C11U} indicates the ligand likely binds the bulge nucleotides in a very structure-specific manner since there are known structural differences in the bulge region of AM_{C11U} compared to AM (Gerdeman, Henkin, & Hines, 2003).

3.1.2 *Fluorescence anisotropy assay: Ligand-induced tRNA–antiterminator complex disruption assay*

The fluorescence anisotropy assay is used to rapidly screen for hit compounds that disrupt the formation of a functionally relevant RNA-molecular complex.

The assay is prepared using Reagent, Mixing, and Assay plates similar to the steady-state fluorescence protocol (Section 3.1.1). This assay can also be run using MOPS as the buffer instead of NaH_2PO_4 to avoid potential complications from magnesium phosphate salts forming.

3.1.2.1 Materials and reagents

RNA Dialysis/Storage Buffer: 10 m*M* NaH_2PO_4, pH 6.5, 0.01 m*M* EDTA

2.5× Binding Buffer: 125 m*M* NaH_2PO_4 pH 6.5, 125 m*M* NaCl, 37.5 m*M* $MgCl_2$, and 0.025 EDTA

Dilution Buffer: 10 m*M* NaH_2PO_4, pH 6.5

Ligand Solvent: H_2O or up to 1:1 DMSO:H_2O (v/v) depending on ligand solubility

Ligand Stock: 1 m*M* ligand in Ligand Solvent

TAMRA-AM Stock: 400 n*M* 5′-TAMRA-AM (TAMRA attached via aminohexyl linker) in Dilution Buffer

tRNA stock: 10 μ*M* *B. subtilis* $tRNA^{Tyr}$(A73U) in Dilution Buffer

96-well V-bottom microplates (for Reagent and Mixing plates)

384-well round bottom black polystyrene microplate, low volume, nonbinding surface (for Assay plate)

3.1.2.2 Reagent plate preparation

Prepare a Reagent Plate. Each row contains 8 wells of solution.

Row 1: 100 μL of 2.5 × Binding Buffer

Row 2: 15 μL Ligand Solvent

Row 3: 120 μL TAMRA-AM stock

Row 4: 120 μL tRNA stock

Rows 5, 7, and 9: 9 μL of Ligand Stock, each well containing a different ligand for a total of 24 ligands

3.1.2.3 Mixing plate preparation

Reaction mixtures are prepared in the 96-well Mixing Plate by pipetting appropriate amounts from the relevant wells of the Reagent Plate followed by mixing. Mixing Plate rows are assembled as follows (8 wells per row):

Rows 1 and 2 (Background): 12 μL of 2.5 × Reaction Buffer, 3 μL Ligand Solvent, and 15 μL Dilution Buffer

Rows 3 and 4 (Positive Control): 12 μL of 2.5 × Reaction Buffer, 3 μL Ligand Solvent, 7.5 μL tRNA stock, and 7.5 μL TAMRA-AM stock

Rows 5 and 6 (Negative Control): 12 μL of 2.5 × Reaction Buffer, 3 μL Ligand Solvent, 7.5 μL Dilution Buffer, and 7.5 μL TAMRA-AM stock

Rows 7 and 8 (Ligand Disruption Reactions 1–8): 12 μL of 2.5 × Reaction Buffer, 3 μL Ligand Stock, 7.5 μL tRNA stock, and 7.5 μL TAMRA-AM stock

Rows 9 and 10 (Ligand Disruption Reactions 9–16): 12 μL of 2.5 × Reaction Buffer, 3 μL Ligand Stock, 7.5 μL tRNA stock, and 7.5 μL TAMRA-AM stock

Rows 11 and 12 (Ligand Disruption Reactions 17–24): 12 μL of 2.5 × Reaction Buffer, 3 μL Ligand Stock, 7.5 μL tRNA stock, and 7.5 μL TAMRA-AM stock

3.1.2.4 Assay plate preparation and data acquisition

The Assay Plate is prepared by pipetting 20 μL for each control, background and ligand disruption reaction mixture and replicates into the 384-well plate in the same manner as described in Section 3.1.1. The covered assay plate is incubated at 25 °C for 10 min in the fluorometer. The microplate sealing film is then removed and anisotropy measurements are acquired using the fluorescence polarization anisotropy instrument settings with excitation = 545 nm, emission = 590 nm, cutoff = 590 nm, PMT high, and *G* factor = 0.947. The *G* factor is an instrument-dependent correction factor and should be determined for a given instrument by following the manufacturer's *G* factor determination protocol using the uncomplexed fluorophore (e.g., tetramethylrhodamine-NHS ester (Glen Research, Inc.)).

3.1.2.5 Data analysis

The ligand effect on tRNA–antiterminator complex formation is evaluated by plotting the ligand-induced change in relative anisotropy of the complex, $\Delta r_{rel} = (r - r_-)/(r_+ - r_-)$. Alternatively, the ligand-induced inhibition of the complex is calculated, $\Delta r_{inhib} = [(r - r_+)/(r_+ - r_-)] \times 100$. For both methods, r_+ is the average anisotropy of Positive controls, r_- is the average anisotropy of Negative controls, and r is the anisotropy of the Ligand Disruption Reaction. Ligands that result in >50% inhibition ($\Delta r_{inhib} > -50$) are likely effective inhibitors of the AM–tRNA complex formation (Zhou, Acquaah-Harrison, Bergmeier, & Hines, 2011).

3.2. 2° Confirmation and characterization screening assays

The 2° screening assays are used to confirm that the hit compounds initially identified in the 1° screens actually bind and disrupt the function of the target RNA. In addition to this hit validation, the 2° assays also provide further characterization of the RNA-binding interactions in a moderate throughput format.

3.2.1 5′-TAMRA-RNA binding isotherms: K$_d$ determination

The steady-state fluorescence 1° screening assay protocol (Section 3.1.1) can be used as a follow-up 2° assay to determine the binding affinity of hits identified in the preliminary screen. The only difference is that the wells in the Ligand-Binding reaction rows in the Mixing Plate contain a serial dilution of a single ligand (e.g., ranging from 0–400 μ*M*). The binding isotherm of $|F_{rel}|$ versus [ligand] is then fit to a binding model to determine the K_d. For example, the single site binding model equation is $|F_{rel}| = (F_{max} \cdot [\text{ligand}])/(K_d + [\text{ligand}])$.

3.2.2 Fluorescence anisotropy disruption assay: IC$_{50}$ determination

The fluorescence anisotropy 1° screening assay protocol (Section 3.1.2) can be used as a follow-up 2° assay to determine the IC_{50} of compounds that inhibit the complex. The only difference is that the wells in the Ligand Disruption Reactions contain a serial dilution of a single ligand (spanning several log units). The plot of Δr_{rel} versus log[ligand] is then fit to a dose–response curve to determine the IC_{50}.

3.2.3 Steady-state AP-labeled RNA fluorescence: Binding site localization and K_d determination

Monitoring the steady-state fluorescence of aminopurine-labeled RNA (AP-labeled RNA) can be used to characterize the ligand–RNA interaction, both in terms of identifying the binding site (and/or ligand-induced structural changes) and binding affinity.

3.2.3.1 Data acquisition and analysis

For microplate assays we observe significant background signal contributing to the observed fluorescence of the aminopurine-labeled T box antiterminator model RNA 10-AP-AM (0.5 μ*M*) in black 384-well polystyrene plates (~60%) and polypropylene plates (~30%) if we use acquisition settings reported by others (Chirayil, Chirayil, & Luebke, 2009). However, by optimizing the fluorometer settings to maximize the signal-to-noise ratio, and by using polypropylene plates and higher concentrations of AP-labeled RNA (e.g., 1 μ*M* 10-AP-AM), we have had some success with 384-well microplate assays (background RFU < 1, emission RFU = 12 for 40 μL binding reaction mixtures with excitation = 333 nm; emission = 430 nm; cutoff = 420 nm; and PMT = high).

The most reliable results for the AP-labeled RNA 2° assays, however, have been obtained using a spectrofluorometer (rather than a microplate reader). The fluorescence spectra are obtained at 20 °C in a square 5 mm quartz fluorescence cuvette. The binding reaction containing the AP-labeled RNA is excited at 310 nm, and the emission spectrum is obtained over the range of 330–600 nm (excitation and emission slit widths = 5 nm, integration time = 1 s). A spectral scan is obtained, rather than data from only a single wavelength, since we have observed ligand-induced changes in the AP emission λ_{max} concomitant with changes in fluorescence intensity (Means et al., 2006).

The cuvette contains 400 μL of Reaction Buffer (50 m*M* NaH_2PO_4, pH 6.5, 0.01 m*M* EDTA, 50 m*M* NaCl, and 5 or 15 m*M* $MgCl_2$). Single concentration ligand binding assays are determined by measuring the normalized ligand-induced change in fluorescence ΔF_{norm}, where $\Delta F_{norm} = (F_{ligand+APAM} - F_{ligand+AM})/F_{APAM}$ at 370 nm. The reaction mixture for measuring $F_{ligand+APAM}$ contains AP-labeled AM RNA (100 n*M*) in Reaction Buffer and ligand (100 μ*M*; $[DMSO]_{final} \leq 5\%$). The ligand control mixture (for measuring $F_{ligand+AM}$) is an identical solution containing 100 n*M* unlabeled AM (instead of the fluorescent AP-labeled AM). The fluorescence control mixture (for measuring F_{APAM}) contains

100 nM AP-labeled RNA in Reaction Buffer with addition of the same amount of the ligand solvent that was used in the $F_{\text{ligand+APAM}}$ mixture. A comparison of the ligand's ΔF_{norm} when the AP is incorporated at different locations throughout the RNA (e.g., 2-, 9-, 10-, and 13-AP-AM) can be used to begin to localize the ligand binding site and/or ligand-induced structural changes.

3.2.3.2 K_d determination

Ligand–RNA K_d values can also be determined by monitoring ligand-induced changes in the fluorescence of aminopurine-labeled RNA. Successive 2 μL additions of ligand stock (followed by gentle stirring with a plastic stirring rod and 5 min equilibration) are added to AP-labeled AM RNA (100 nM in Reaction Buffer) up to a total of 20 μL (overall ~5% dilution and ≤5% DMSO (v/v)). A comparable series of background spectra are obtained by adding ligand stock to Reaction Buffer containing 100 nM unlabeled AM and subtracted from the relevant spectrum obtained from the labeled-RNA series to remove possible ligand fluorescence, water Raman peak interference and dilution/DMSO addition effects. The background-subtracted spectra are analyzed to determine the F_{rel} at 370 nm and then the K_d is determined as described in Section 3.2.1.

3.2.4 Fluorescence-monitored thermal denaturation (T_m) assay

The fluorescence-monitored thermal denaturation assay is used to screen for potential ligand-induced changes in RNA stability. This information helps to confirm hit compound binding and provides information regarding a possible structural mechanism of action for a compound that also disrupts the RNA's function.

3.2.4.1 Reaction mixture preparation

The fluorescence T_m screening uses the fluorophore-quencher labeled RNA 5′-ROX-AM-3′-BHQ2 (or 5′-HEX-AM-3′-DABCYL). Each reaction is mixed in a 96-well V-bottom microplate and then transferred into two PCR tubes (25 μL for each; optically clear tubes and caps) for the duplicate T_m measurements. Reaction mixtures contain 30 μL of 20 mM NaH_2PO_4, pH 6.5, 0.02 mM EDTA buffer, 24 μL of 250 nM 5′-ROX-AM-BHQ2 and 6 μL of 1 mM Ligand Stock (dissolved in H_2O or up to 1:1 DMSO:H_2O (v/v) as the ligand solvent) for a final concentration of 100 nM of 5′-ROX-AM-BHQ2, 100 μM ligand in 10 mM NaH_2PO_4, pH 6.5, 0.01 mM EDTA buffer, 5% DMSO. Control reactions (with no ligand) are prepared in the same manner except 6 μL Ligand Solvent is added instead of ligand stock.

3.2.4.2 Data acquisition and analysis

The PCR tubes are incubated in the qPCR instrument at 30 °C for 15 min. The instrument is programed to increase the temperature by 1 °C, maintained for 3 min at that temperature and then the fluorescence measured (excitation = 585 nm and emission = 610 nm) for a total of 61 cycles to reach the final temperature of 90 °C. The first derivative (dF/dT) is calculated (using the instrument's software or a data analysis software program) and the T_m is determined from the maximum of the dF/dT versus T plot. Ligand-induced temperature changes (ΔT_m) are calculated using the equation $\Delta T_m = T_m - T_{m0}$, where T_{m0} is the RNA melting temperature in the absence of ligand and T_m is the melting temperature in the presence of the ligand. Depending on ligand effects, there may be more than one transition observed.

4. CONCLUSIONS

Success in RNA-targeted drug discovery is greatly facilitated by the use of a well-designed screening cascade. As illustrated in this chapter, efficient primary and secondary screening assays can be developed by combining knowledge from functional and structural biology studies with strategic choices of fluorescence-based assay methods. The cascade of assays presented here monitor ligand–RNA binding, ligand-induced RNA complex disruption and ligand-induced changes in RNA stability. While the assays described have been developed for T box riboswitch-targeted drug discovery, the fluorescence methods and screening cascade design principles can be applied to drug discovery efforts targeted toward other medicinally relevant noncoding RNA.

ACKNOWLEDGMENTS

We thank the National Institutes of Health (GM61048, GM073188) for support of this work.

REFERENCES

Anupam, R., Bergmeier, S. C., Green, N. J., Grundy, F. J., Henkin, T. M., Means, J. A., et al. (2008). 4,5-disubstituted oxazolidinones: High affinity molecular effectors of RNA function. *Bioorganic & Medicinal Chemistry Letters, 18*, 3541–3544.

Boucher, H. W., Talbot, G. H., Bradley, J. S., Edwards, J. E., Gilbert, D., Rice, L. B., et al. (2009). Bad bugs, no drugs: No ESKAPE! an update from the infectious diseases society of America. *Clinical Infectious Diseases*, *48*, 1–12.

Cavaluzzi, M. J., & Borer, P. N. (2004). Revised UV extinction coefficients for nucleoside-5'-monophosphates and unpaired DNA and RNA. *Nucleic Acids Research*, *32*, e13.

CDC (2013). *Antibiotic resistance threats in the United States: U.S. Department of health & human services*. Atlanta, GA: Centers for Disease Control and Prevention.

Chirayil, S., Chirayil, R., & Luebke, K. J. (2009). Discovering ligands for a microRNA precuror with peptoid microarrays. *Nucleic Acids Research*, *37*(16), 5486–5497.

Fauzi, H., Jack, K. D., & Hines, J. V. (2005). In vitro selection to identify determinants in tRNA for *Bacillus subtilis tyrS* T box antiterminator mRNA binding. *Nucleic Acids Research*, *8*, 2595–2602.

Gerdeman, M. S., Henkin, T. M., & Hines, J. V. (2002). *In vitro* structure-function studies of the *Bacillus subtilis tyrS* mRNA antiterminator: Evidence for factor independent tRNA acceptor stem binding specificity. *Nucleic Acids Research*, *30*(4), 1065–1072.

Gerdeman, M. S., Henkin, T. M., & Hines, J. V. (2003). Solution structure of the *B. subtilis* T box antiterminator RNA: Seven-nucleotide bulge characterized by stacking and flexibility. *Journal of Molecular Biology*, *326*, 189–201.

Green, N. J., Grundy, F. J., & Henkin, T. M. (2010). The T box mechanism: tRNA as a regulatory molecule. *FEBS Letters*, *584*(2), 318–324.

Grigg, J. C., & Ke, A. (2013). Sequence, structure, and stacking: Specifics of tRNA anchoring to the T box riboswitch. *RNA Biology*, *10*(12), 1761–1764.

Grundy, F. J., Moir, T. R., Haldeman, M. T., & Henkin, T. M. (2002). Sequence requirements for terminators and antiterminators in the T box transcription antitermination system: Disparity between conservation and functional requirements. *Nucleic Acids Research*, *30*(7), 1646–1655.

Guan, L., & Disney, M. D. (2012). Recent advances in developing small molecules targeting RNA. *ACS Chemical Biology*, *7*, 73–86.

Gutierrez-Preciado, A., Henkin, T. M., Grundy, F. J., Yanofsky, C., & Merino, E. (2009). Biochemical features and functional implications of the RNA-based T box regulatory mechanism. *Microbiology and Molecular Biology Reviews*, *73*, 36–61.

Hughes, J. P., Rees, S., Kalindjian, S. B., & Philpott, K. L. (2011). Principles of early drug discovery. *British Journal of Pharmacology*, *162*, 1239–1249.

Jentzsh, F., & Hines, J. V. (2011). Interfacing medicinal chemistry with structural bioinformatics: Implications for T box riboswitch RNA drug discovery. *BMC Bioinformatics*, *13*(Suppl. 2), S5–S10, GLBIO 2011 Special Issue.

Lakowicz, J. R. (2006). *Principles of fluorescence spectroscopy* (3rd ed.). New York, NY: Springer Science & Business Media.

Lee, J., Vogt, C. E., McBrairty, M., & Al-Hashimi, H. M. (2013). Influence of dimethylsulfoxide on RNA structure and ligand-binding. *Analytical Chemistry*, *85*(20), 9692–9698.

Marras, S. A. E. (2006). Selection of fluorophore and quencher pairs for fluorescent nucleic acid hybridization probes. *Methods in Molecular Biology*, *335*, 3–16.

Means, J. A., Katz, S. J., Nayek, A., Anupam, R., Hines, J. V., & Bergmeier, S. C. (2006). Structure activity studies of oxazolidinone analogs as RNA-binding agents. *Bioorganic & Medicinal Chemistry Letters*, *16*(13), 3600–3604.

Means, J. A., Simson, C. M., Zhou, S., Rachford, A. A., Rack, J., & Hines, J. V. (2009). Fluorescence probing of T box antiterminator RNA: Insights into riboswitch discernment of the tRNA discriminator base. *Biochemical and Biophysical Research Communications*, *389*, 616–621.

Means, J. A., Wolf, S., Agyeman, A., Burton, J. S., Simson, C. M., & Hines, J. V. (2007). T box riboswitch antiterminator affinity modulated by tRNA structural elements. *Chemical Biology & Drug Design*, *69*, 139–145.

Orac, C. M., Zhou, S., Means, J. A., Boehme, D., Bergmeier, S. C., & Hines, J. V. (2011). Synthesis and stereospecificity of 4,5-disubstituted oxazolidinone ligands binding to T-box riboswitch RNA. *Journal of Medicinal Chemistry, 54*, 6786–6795.

Puglisi, J. D., & Tinoco, I., Jr. (1989). Absorbance melting curves of RNA. *Methods in Enzymology, 180*, 304–325.

Sanbrook, J., Fritsch, E. F., & Maniatis, T. (1989). *Molecular cloning: A laboratory manual* (2nd ed.). Cold Spring Harbor: Cold Spring Harbor Lab. Press.

Smith, A. M., Fuchs, R. T., Grundy, F. J., & Henkin, T. M. (2010). Riboswitch RNAs: Regulation of gene expression by direct monitoring of a physiological signal. *RNA Biology*, 7(1), 104–110.

Strauss, J. H., Jr., Kelly, R. B., & Sinsheimer, R. L. (1968). Denaturation of RNA with dimethyl sulfoxide. *Biopolymers, 6*, 793–801.

Unruh, J. R., Gokulrangan, G., Lushington, G. H., Johnson, C. K., & Wilson, G. S. (2005). Orientational dynamics and dye-DNA interactions in a dye-labeled DNA aptamer. *Biophysical Journal, 88*, 3455–3465.

van der Rijke, F. M., Heetebrij, R. J., Talman, E. G., Tanke, H. J., & Raap, A. K. (2003). Fluorescence properties, thermal duplex stability, and kinetics of formation of cyanin platinum DNAs. *Analytical Biochemistry, 321*, 71–78.

Vitreschak, A. G., Mironov, A. A., Lyubetsky, V. A., & Gelfand, M. S. (2008). Comparative genomic analysis of T-box regulatory systems in bacteria. *RNA, 14*, 717–735.

WHO (2014). *Antimicrobial resistance: Global report on surveillance.* Geneva, Switzerland: WHO Library, World Health Organization.

Zhang, J. H., Chung, T. D. Y., & Oldenburg, K. R. (1999). A simple statistical parameter for use in evaluation and validation of high throughput screening assays. *Journal of Biomolecular Screening*, *4*(2), 67–73.

Zhou, S., Acquaah-Harrison, G., Bergmeier, S. C., & Hines, J. V. (2011). Anisotropy studies of tRNA - T box antiterminator RNA complex in the presence of 1,4-disubstituted 1,2,3-triazoles. *Bioorganic & Medicinal Chemistry Letters, 21*, 7059–7063.

Zhou, S., Acquaah-Harrison, G., Jack, K. D., Bergmeier, S. C., & Hines, J. V. (2011). Ligand-induced changes in T box antiterminator RNA stability. *Chemical Biology & Drug Design, 79*, 202–208.

Zhou, S., Means, J. A., Acquaah-Harrison, G., Bergmeier, S. C., & Hines, J. V. (2011). Characterization of a 1,4-disubstituted 1,2,3-triazole binding to T box antiterminator RNA. *Bioorganic & Medicinal Chemistry, 20*, 1298–1302.

CHAPTER NINETEEN

Monitoring Ribosomal Frameshifting as a Platform to Screen Anti-Riboswitch Drug Candidates

Chien-Hung Yu, René C.L. Olsthoorn[1]

Department of Molecular Genetics, Leiden Institute of Chemistry, Leiden University, Leiden, The Netherlands
[1]Corresponding author: e-mail address: olsthoor@chem.leidenuniv.nl

Contents

Abstract

Riboswitches are regions within mRNAs that can regulate downstream expression of genes through metabolite-induced alteration of their secondary structures. Due to the significant association of bacterial essential or virulence genes, bacterial riboswitches have become promising targets for development of putative antibacterial drugs. However, most of the screening systems to date are based on *in vitro* or bacterial systems, lacking the possibility to preobserve the adverse effects to the host's translation machinery. This chapter describes a novel screening method based on monitoring the riboswitch-induced −1 ribosomal frameshifting (−1 FS) efficiency in a mammalian cell-free lysate system using $preQ_1$ class-I ($preQ_1$-I) riboswitches as model target.

Methods in Enzymology, Volume 550
ISSN 0076-6879
http://dx.doi.org/10.1016/bs.mie.2014.10.041

1. INTRODUCTION

Riboswitches are gene regulation elements generally located in the 5′-untranslated regions (5′-UTR) of bacterial mRNAs to control gene expression by forming two mutually exclusive structures elicited upon binding of small metabolites (Roth & Breaker, 2009). Specifically, the ligand-binding domain (aptamer) acts as a molecular sensor, switching the secondary structures along with following variable sequences (expression platform) when the metabolite exceeds a threshold level, resulting in gene regulation at the transcription or translation level. Although riboswitches are found in all three domains of life, the majority of examples are identified across bacterial species. These bacterial riboswitches are mainly located upstream of the biosynthesis or transporter genes of corresponding metabolites, offering the possibility to develop antibiotics against bacterial diseases. Since the selectivity of riboswitches is entirely programmed in the metabolite-sensing domain, great efforts have been made to determine the high-resolution metabolite-bound RNA structures in order to facilitate anti-riboswitch drug designs (Serganov & Patel, 2012). Interestingly, among these available riboswitches aptamer structures, some adopt hairpin-type (H-type) pseudoknot structures that display a wide range of highly specific functions in a variety of biological processes (Brierley, Pennell, & Gilbert, 2007), such as −1 FS, highlighting the convergent role of this important structural motif.

−1 FS is a translational recoding mechanism by which translating ribosomes slip one nucleotide (nt) into the 5′-direction (−1 reading frame) on the mRNA and generate an alternative protein (Farabaugh, 2000). It is well known that two *cis*-acting RNA elements are the main signals to induce −1 FS: (i) a heptameric nucleotide sequence called the slippery sequence, where the ribosome changes reading frame with consensus X XXZ ZZN (where X and Z are any nucleotide, N≠Z, and spaces denote the initial reading frame) and (ii) a stimulatory RNA structure, such as a hairpin or a pseudoknot, downstream of the slip site. The length of the spacer between slip site and downstream structure, generally 6–9 nts, is also crucial for efficient −1 FS. The appropriate spacer length presumably serves to fine-tune the tension generated by the downstream RNA structure, thereby eliciting the appropriate fraction of frameshifting.

Due to the high similarity between H-type pseudoknotted riboswitch aptamers [preQ1-I from *Fusobacterium nucleatum* (*F. nucleatum*) (Roth et al., 2007) as the described example] and frameshift-inducing pseudoknot structures, we introduced a novel method to investigate ligand-binding ability by monitoring the −1 FS efficiency (Fig.1; Yu, Luo, Iwata-Reuyl, & Olsthoorn, 2013). This method can be used to study other ligand-sensing riboswitches and has been successfully applied to a class II *S*-adenosyl-L-methionine riboswitch as well (C.H. Yu & R.C.L. Olsthoorn, unpublished data).

The method described here offers advantages: (i) the possibility for high-throughput screening when fused to adequate reporters, (ii) simultaneously monitoring potential adverse effects on the eukaryotic translation machinery by using a rabbit reticulocyte cell-free lysate, and (iii) the riboswitch aptamers are embedded within a large mRNA and thereby better mimic bacterial polycistronic mRNAs.

2. MATERIALS

2.1. Nucleobases

1. Nucleobases of adenine (Ade) (A8626), guanine (Gua) (G11950), 2,6-diaminopurine (DAPu) (247847), xathine (Xan) (X0626), hypoxanthine (Hpx) (H9377), and 2,4-diaminopyrimidine (DAPy) (4682131) were from Sigma-Aldrich. PreQ_1 compound was a generous gift from Dr. Iwata-Reuyl, Portland State University, Oregon, USA
2. Dimethyl sulfoxide (DMSO) (Sigma-Aldrich)
3. Potassium hydroxide (KOH) (Sigma-Aldrich)

2.2. Plasmid DNA template

1. The frameshifting reporter (pSF208) (Olsthoorn, Reumerman, Hilbers, Pleij, & Heus, 2010) contains an abridged influenza virus A/PR8/34 PB1 gene with inserted U_3A_3C slippery sequence followed by *Spe*I and *Nco*I sites for cloning preQ_1 riboswitch aptamers through complementary oligonucleotides (Note 1).
2. Restriction enzymes *Spe*I and *Nco*I (Thermo Scientific).
3. Synthesized pair of stabilized *F. nucleatum* preQ_1 riboswitch aptamer (stab-Fnu) oligonucleotides: 5′-CTAGTTGACGCGGTGCTAGCAA AACCCGCGTTAAACAAACTAGACTTCATG-3′ (Note 2) (Sigma-Aldrich).
4. Gel extraction spin column (Machery-Nagel).
5. T4 DNA Ligase (Thermo Scientific).
6. *E. coli* XL-1 Blue competent cells.

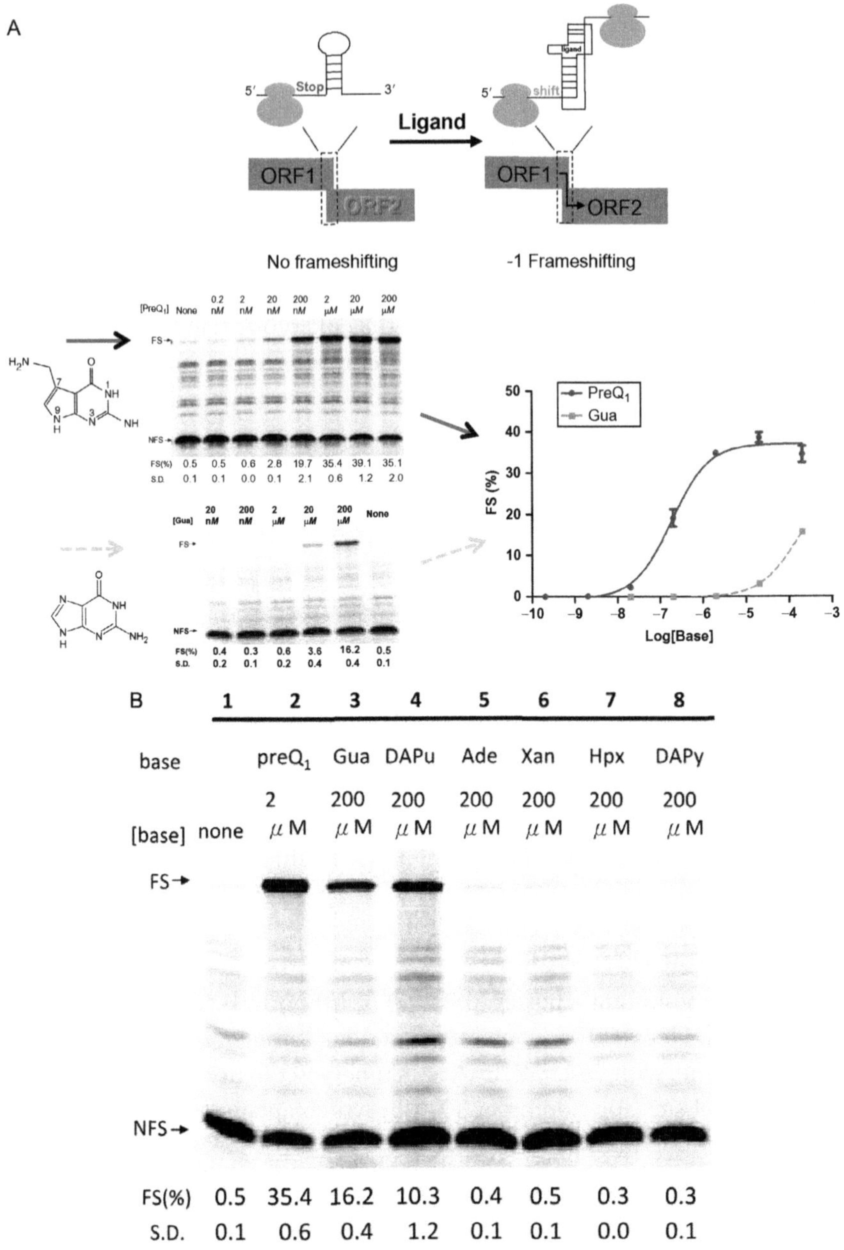

Figure 1 (A) Conceptual illustration of the evaluation of ligand-preQ$_1$-I aptamer binding affinities by monitoring −1 FS efficiency. The cartoon describes the principle of ligand-induced −1 FS. Given examples are two preQ$_1$ aptamer ligands: preQ$_1$ (optimal) and guanine (Gua, suboptimal). −1 FS efficiencies induced by either preQ1 or guanine at the indicated concentrations are quantified by the appearance of −1 FS protein products (FS) on the SDS-PAGE (see Note 11 for details). The dose (various ligand concentration, *X*-axis)-responsive (−1 FS efficiency, *Y*-axis) curves can be drawn to compare the ligand-binding affinity. (B) The SDS-PAGE shows the level of −1 FS efficiency induced by various preQ1 analogs at the indicated concentrations.

2.3. *In vitro* transcription

1. *BamH*I for template linearization (Thermo Scientific).
2. RiboMAX Large-Scale RNA Production System-SP6 (Promega).
3. Phenol–chloroform–isoamyl alcohol (25:24:1).
4. Micro-Bio-Spin P-6 gel column, tris buffer (Bio-Rad).

2.4. Cell-free translation

1. Rabbit reticulocyte lysate (RRL) System, nuclease Treated (Promega).
2. EasyTag L-[^{35}S]-Methionine, >1000 Ci/mmol (PerkinElmer).
3. 2 × Laemmli sample buffer [120 m*M* Tris–HCl (pH 6.8), 20% glycerol, 4% sodium dodecyl sulfate (SDS), 0.02% bromophenol] stored at room temperature.
4. 2-Mercaptoethanol (Sigma-Aldrich).
5. 30% acrylamide/bis solution (29:1) (Bio-Rad).
6. Ammonium persulfate.
7. Tetramethylethylenediamine (Bio-Rad).
8. SDS polyacrylamide gel (SDS-PAGE).
   - 8.1. stacking gel (5% acrylamide/bis, 0.13 *M* Tris–HCl pH 6.8, 0.1% SDS).
   - 8.2. separation gel (13% acrylamide/bis, 0.375 *M* Tris–HCl pH 8.8, 0.1% SDS).
9. SDS-PAGE running buffer (25 m*M* Tris–HCl pH 8.3, 250 m*M* glycine, 0.1% SDS).
10. Whatman filter paper (3 mm).
11. Saran wrap.

2.5. Equipment

1. Spectrophotometer [e.g., Varian Cary 5000 UV–vis-NIR (Agilent)].
2. NanoDrop 1000 (Thermo Scientific).
3. Gel Dryer [e.g., Model 583 (Bio-Rad)].
4. PhosphorImager screen (Molecular Dynamics)
5. PhosphorImage scanner [Personal Molecular Imager FX System (Bio-Rad) or Typhoon 9400 (GE Healthcare)].
6. Quantity-One software (Bio-Rad demo version suffices).
7. SDS-PAGE gel running system (e.g., Mini-Protean 3 Bio-Rad).
8. Electrophoretic power supply.
9. Waterbath.
10. Heating block.

11. Micropipettes.
12. Freezer.
13. Eppendorf centrifuge.

3. METHODS

3.1. Preparation of nucleobases stock

1. Dissolve $PreQ_1$, Xan, Hpx, and DAPy at 200 m*M* in DMSO as stocks, and store at −20 °C (Note 3).
2. Dissolve Gua and Ade at 200 m*M* in 0.15 N KOH (Note 4).
3. Dissolve DAPu at 200 m*M* in RNase-free water.
4. Check suitably diluted (0.1 < absorption < 1) nucleobase samples by spectrophotometer and verify the concentration by Beer's law.
5. For working solution, dilute all above nucleobase stocks to 2 m*M* by RNase-free water, followed by adding 1 μl into translation reaction mixtures (Step 1 in Section 3.4) in total volume of 10 μl. Serial dilutions are applied to make desired concentrations of $preQ_1$ solutions.

3.2. Preparation of DNA template for *in vitro* transcription

1. Digest the parental pSF208 by *Spe*I and *Nco*I at 37 °C for 1 h.
2. In the meantime, mix stab-Fnu oligonucleotides pair at final concentration of 10 μ*M* of each strand, followed by heating up the mixtures to 95 °C for 1 min. Then, the mixtures were allowed to cool to room temperature for annealing.
3. Purify the digested pSF208 plasmid by gel extraction spin column after migration on 1% agarose gel.
4. Ligate the *Spe*I–*Nco*I digested pSF208 with annealed stab-Fnu fragments at room temperature for 30 min followed by transformation into *E. coli* XL-1 blue cells.
5. Isolate plasmids from selected clones and verify the sequence by automated dideoxy sequencing with chain terminator dyes.

3.3. *In vitro* transcription

1. Linearize 10 μg of stab-Fnu as determined by NanoDrop by *BamH*I to create the template for run-off transcription.
2. Purify the linearized plasmid by phenol–chloroform extraction with successive ethanol precipitation or by one of the commercially available spin columns.

3. Dissolve the linearized plasmid in RNAse-free water to make a working solution of 500 ng/μl.
4. Transcribe the DNA template with the RiboMAX SP6 Kit according to the manufacturer's protocol. Typically, incubate 10 μl of the reaction solution containing 2 μl of the plasmids at 37 °C for 2 h (Note 5).
5. Add 0.5 μl DNase RQ1 to the transcription mixtures with further incubation at 37 °C for 15 min.
6. Check the quality and quantity of the transcribed mRNA by comparing the migration and brightness with the control RNA with similar length and known quantify after ethidium bromide-stained agarose gel electrophoresis (Notes 6 and 7).
7. Dilute the mRNA to 15 ng/μl as working solution for *in vitro* translation.
8. Aliquot the mRNA to small volumes and store them at −20 °C.

3.4. Cell-free translation

1. Label eppendorf tubes and add 1 μl of designated nucleobases working solution into reaction tubes and leave them on ice.
2. Prepare a translation reaction master mix on ice by adding 4 μl nuclease-treated RRL, 2 μl mRNA working solution, 0.25 μl amino acids mix without methionine, 0.25 μl ^{35}S-methionine, and 2.5 μl RNase-free water for each sample.
3. Add 9 μl translation mix into each nucleobase-containing tube (Note 12).
4. Mix the solution briefly and incubate in a 28 °C water bath for 1 h.

3.5. Monitoring −1 FS

1. Stop the translation reaction by adding 10 μl cold 2× Laemmli sample buffer followed by immediately boiling the sample for 3 min.
2. Separate the translation products by 13% SDS-poly acrylamide gel electrophoresis (SDS-PAGE) at a constant voltage of 200 V.
3. Stop running when bromophenol blue dye front reaches the bottom of the separating gel (Note 8).
4. Dry the gel on 3 mm Whatman paper (cover the gel with Saran wrap) by gel dryer, applying PAGE gel program cycle at 80 °C for 2 h.
5. Expose the dried gel (remove Saran wrap) to a phosphorimager screen (Note 9).
6. Scan the phosphorimager screen at 50 or 100 μm resolution.
7. Quantify the in-frame and frameshifted bands by Quantity-One (Note 10).
8. Calculate the −1 FS efficiency induced by designated ligands (Note 11).

4. NOTES

1. For potential high-throughput screening, the −1 FS cassette, including slippery sequence, spacer, and target riboswitch aptamer, can be inserted into suitable reporter (for example, dual-luciferase reporter).
2. The stabilized version of the *F. nucleatum* aptamer is created by mutating the A–U pair to G–C pair within the first stem region. This mutant, without losing ligand specificity, provides a better platform for ligand-binding assays due to its higher sensitivity to preQ1 compound. Although this way of improving aptamer-ligand-binding may be applied to other targets, the orientation of the G–C pair should be determined experimentally. We observed that a G–C flipped mutant shows lower $preQ_1$ binding, probably due to a nonoptimal stem-loop tertiary interaction. The spacer length, i.e., the sequence between slippery sequence and riboswitch aptamer, is another tunable factor for optimal −1 FS efficiency.
3. These compounds are quite stable in DMSO. The working solution (Step 5 in Section 3.1) can be stored at 4 °C for several months without noticeable decay.
4. Gua and Ade stocks should be kept at −20 °C for no longer than two weeks. Significant decay is observed afterward. Alternatively, stocks can be made freshly.
5. Since noncapped and nonpolyadenylated mRNAs can be translated in RRL, we do not add cap analog into our transcription mixtures, and also no polyA-tail is transcribed cotranscriptionally or added post-transcriptionally. As a consequence, some non-specific translational products with negligible intensity are observed on gel due to occasional usage of alternative start codons.
6. To prepare control mRNA, the DNase-treated transcription mixtures are first extracted by phenol–chloroform solution, followed by purification of the mRNA through P-6 gel filtration column to remove DNA and unincorporated ribonucleotides. Then, the eluted control mRNA is quantified by NanoDrop.
7. Using nonpurified mRNAs can better mimic the molecular crowding conditions as well as the complexity in the cytoplasm.
8. Since the gel will be dried on the filter paper without staining, it is not necessary to fix the gel to prevent diffusion. We did fix the gels by 45%

methanol, 45% water, 10% glacial acetic acid, and find no noticeable difference to the unfixed ones.

9. In our experimental condition, a 16-h exposure time (overnight) is enough to obtain clear signals.
10. Typically, the intensities of target bands are determined by using the function of "Volume Rect Tool" in Quantity-One. For background subtraction, an identical rectangular volume is selected immediately above the corresponding target band area as the defined background.
11. The -1 FS efficiency representing the riboswitch aptamer-ligand affinity is calculated by: [intensity of the -1 FS band (FS)] − (intensity of the background of the -1 FS band)/{[intensity of the in-frame band (NFS)] − (intensity of the background of the in-frame band) + [intensity of the -1 FS band (FS)] − (intensity of the background of the -1 FS band)} × 100, after correcting for the number of incorporated methionines.
12. To save on RRL, the reaction can also be carried out in 5 μl volumes but this requires good pipetting skills.

REFERENCES

Brierley, I., Pennell, S., & Gilbert, R. J. (2007). Viral RNA pseudoknots: Versatile motifs in gene expression and replication. *Nature Reviews Microbiology*, *5*, 598–610.

Farabaugh, P. J. (2000). Translational frameshifting: Implications for the mechanism of translational frame maintenance. *Progress in Nucleic Acid Research and Molecular Biology*, *13*, 87–96.

Olsthoorn, R. C. L., Reumerman, R., Hilbers, C. W., Pleij, C. W. A., & Heus, H. A. (2010). Functional analysis of the SRV-1 RNA frameshifting pseudoknot. *Nucleic Acids Research*, *38*, 7665–7672.

Roth, A., & Breaker, R. R. (2009). The structural and functional diversity of metabolite-binding riboswitches. *Annual Review of Biochemistry*, *78*, 305–334.

Roth, A., Winkler, W. C., Regulski, E. E., Lee, B. W., Lim, J., Jona, I., et al. (2007). A riboswitch selective for the queuosine precursor preQ1 contains an unusually small aptamer domain. *Nature Structural & Molecular Biology*, *14*, 308–317.

Serganov, A., & Patel, D. J. (2012). Molecular recognition and function of riboswitches. *Current Opinion in Structural Biology*, *22*, 279–286.

Yu, C. H., Luo, J., Iwata-Reuyl, D., & Olsthoorn, R. C. L. (2013). Exploiting $preQ_1$ riboswitches to regulate ribosomal frameshifting. *ACS Chemical Biology*, *8*, 733–740.

AUTHOR INDEX

Note: Page numbers followed by "*f*" indicate figures, "*t*" indicate tables and "*np*" indicate footnotes.

A

B

C

D

E

F

H

L

M

N

S

T

X

Y

SUBJECT INDEX

Note: Page numbers followed by "*f*" indicate figures and "*t*" indicate tables.

C

D

S

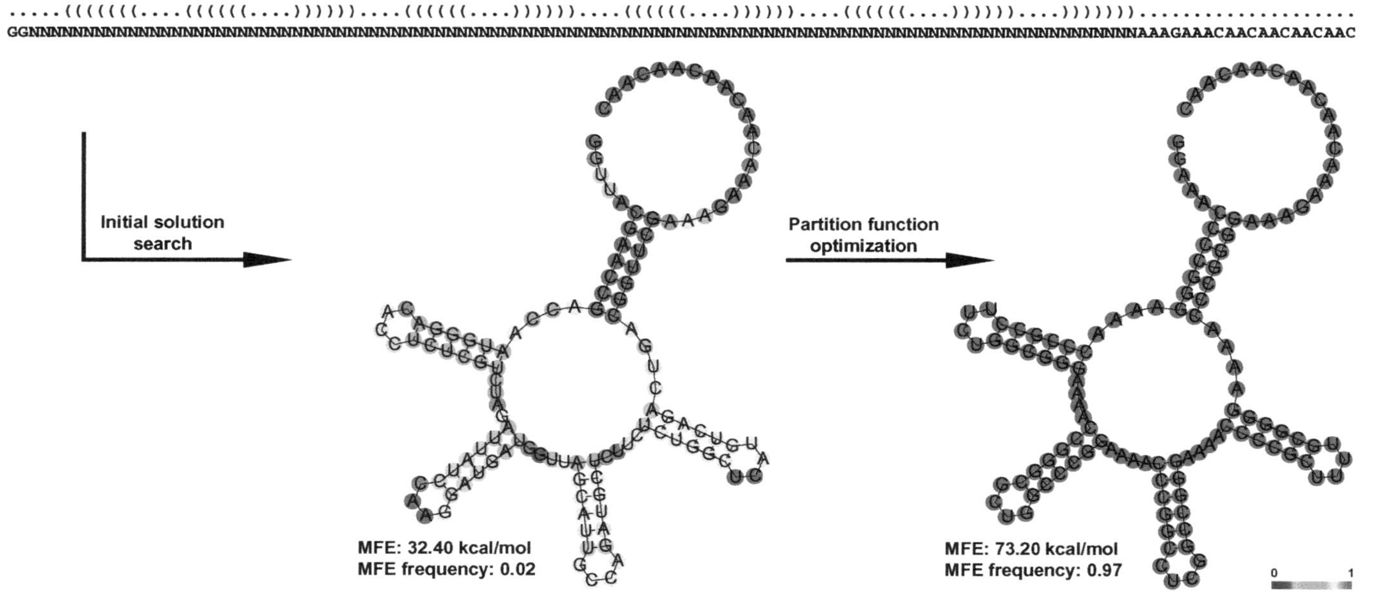

Sven Findeiß *et al.*, Figure 2 Example output of `RNAinverse` solving one task of the `EteRna` lab competition (Lee et al., 2014). The secondary structure (in dot–bracket notation) and the sequence constraint (nucleotides A, C, G, U, and N for any of those) strictly define the design target. `RNAinverse` initially searches for a sequence that adapts the target structure as ground state. Using the partition function option, it further optimizes the initial solution such that the energy is minimized and that the ground state dominates the structure ensemble. The color code (gray shading in the printed version) indicates base pair probability. Note that structural probing experiments of this and other examples have shown that the prediction and the structure of the synthesized RNA molecule can differ significantly (Lee et al., 2014).

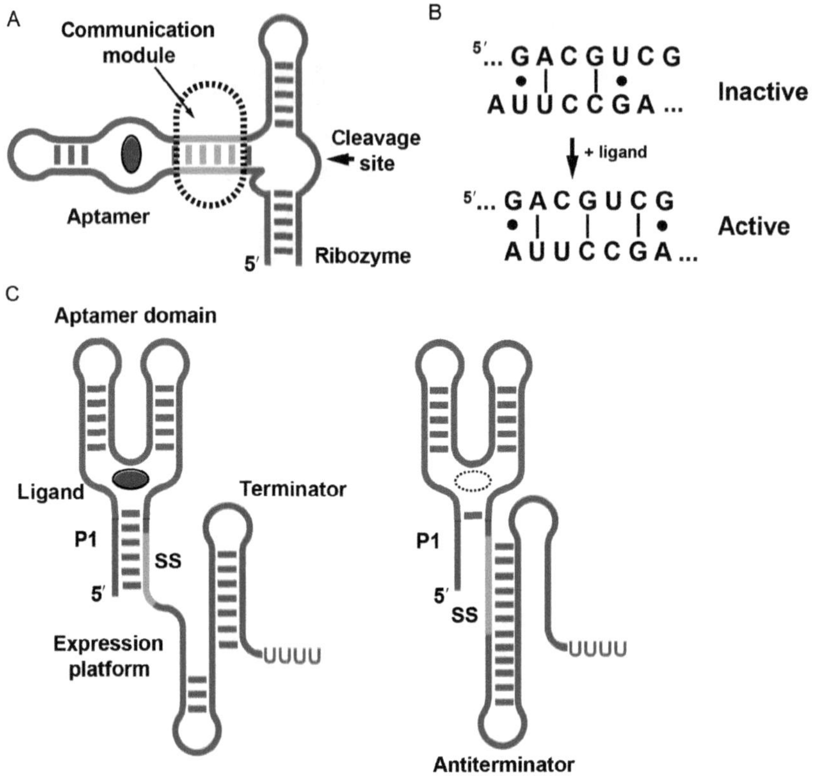

Jeremiah J. Trausch and Robert T. Batey, Figure 1 Cartoon representation of two RNA devices. (A) Domain organization of an engineered allosteric ribozyme (aptazyme) highlighting the aptamer (red), ribozyme (blue), and communication module (green). Binding of an effector ligand to the aptamer (magenta) activates the ribozyme via a structural change in the communication module that in turn organizes the active site for strand scission. (B) Sequence of a representative communication module (Soukup & Breaker, 1999) showing the two alternative structures adopted. (C) Domain organization of a natural riboswitch that regulates transcriptional termination. Binding of ligand to the aptamer domain (red) directs the secondary structural switch of the expression platform (blue). A sequence element that is found in one of the two mutually exclusive helices (of which the P1 helix is one) is referred to as the "switching sequence" (green).

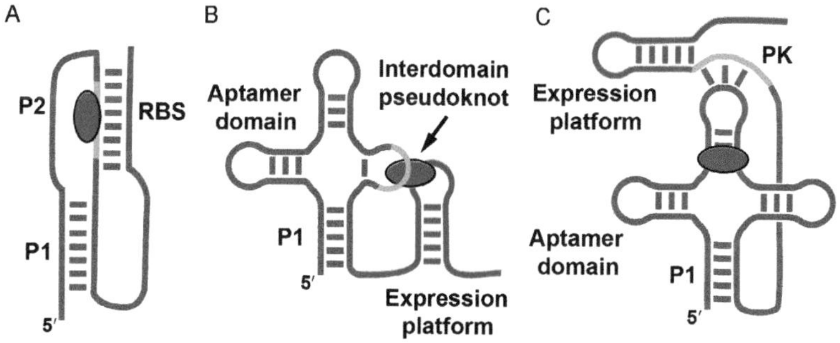

Jeremiah J. Trausch and Robert T. Batey, Figure 2 Domain organizations of natural riboswitches unsuitable for development of modular expression platforms. Riboswitch architectures in which the ligand-dependent structural switch is incorporated into a pseudoknot motif are generally unsuitable for modular aptamers or expression platforms without significant redesign. Examples include the (A) SAM-II, (B) cobalamin, and (C) SAM-IV riboswitches. Note that the switching sequence (green) in each case is part of a pseudoknot.

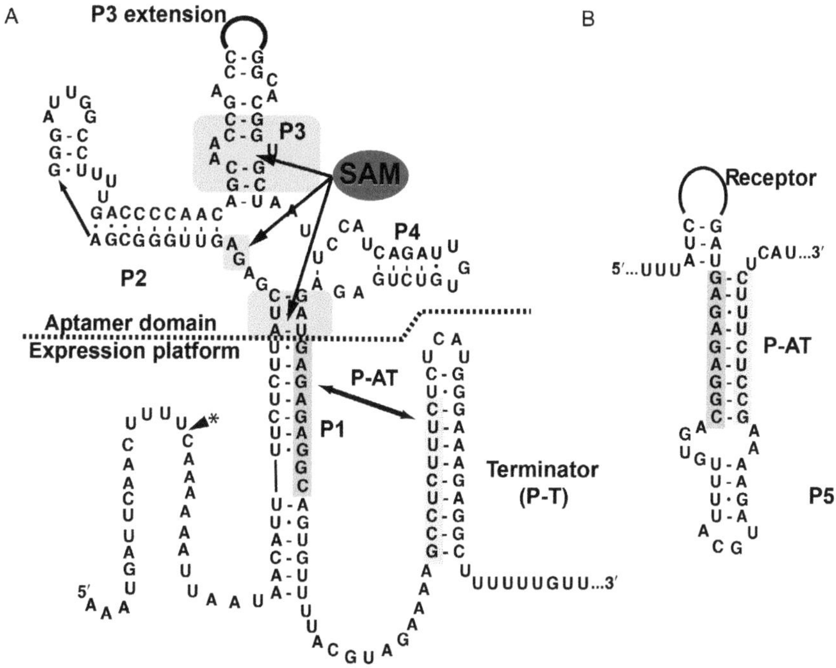

Jeremiah J. Trausch and Robert T. Batey, Figure 3 Example riboswitch with separable domains. (A) The secondary structure of the *S*-adenosylmethionine bound "OFF" state of the *B. subtilis metE* SAM-I riboswitch. The regions of the RNA that are in direct contact with SAM are highlighted in orange; of particular note are two essential A–U pairs proximal to the four-way junction. The switching sequence of the expression platform is highlighted in green and its complementary sequence in blue. (B) Secondary structure of the alternative antiterminator helix (P-AT).

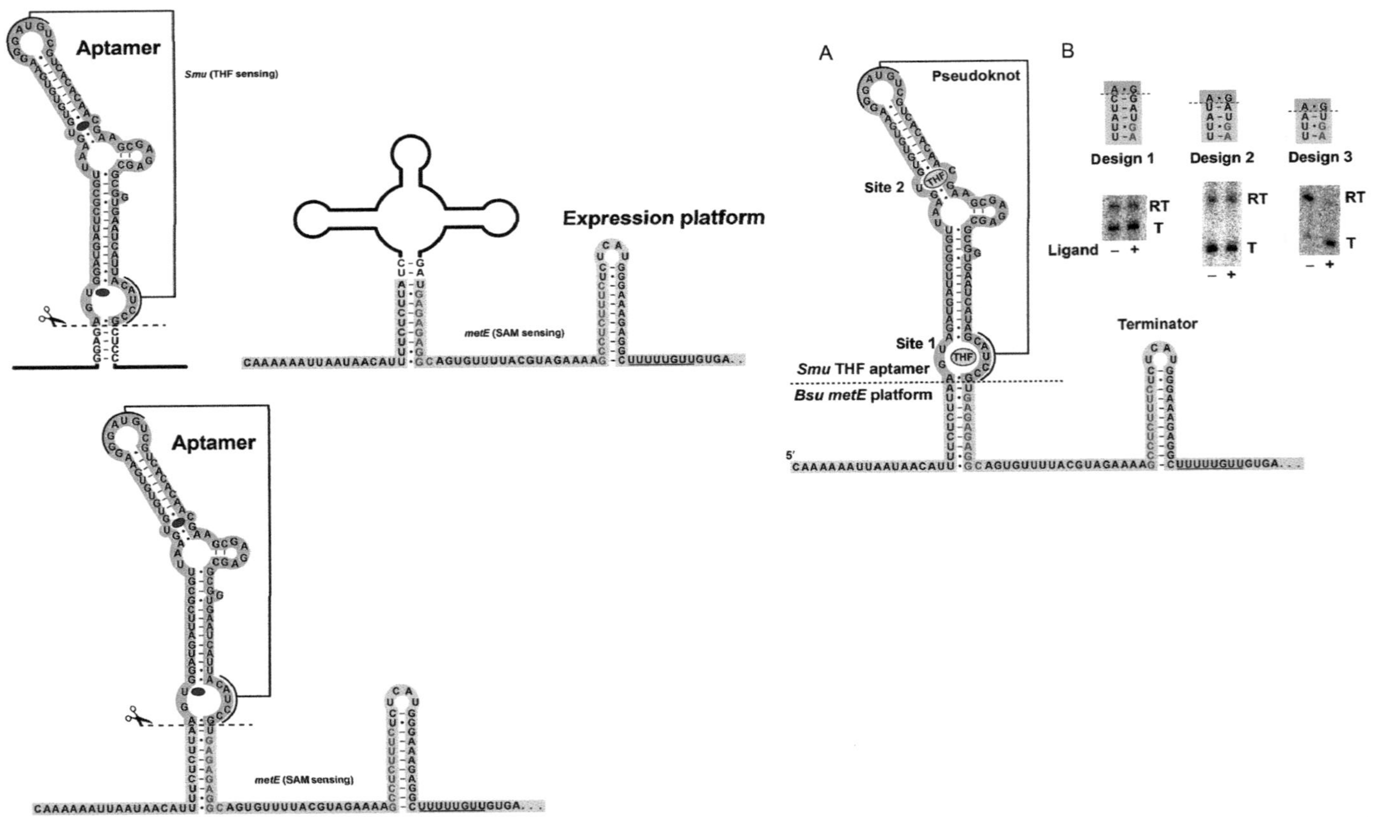

Jeremiah J. Trausch and Robert T. Batey, Figure 4 Structure and activity of a designed chimeric riboswitch. (A) The chimeric riboswitch formed by the fusion of the *Streptococcus mutants folT* THF aptamer domain (orange) and the *Bacillus subtilis metE* SAM-I class riboswitch expression platform (cyan). The secondary structure is of "design 3" variant in the ligand-bound "OFF" state. Note that the aptamer domain is not completely wild type; P4 has been shortened and capped with a tetraloop, corresponding to the sequence used for structural analysis (PDB ID 4LVV). (B) Secondary structure of the three design variants in the P1 stem around the aptamer/expression platform fusion site (dashed line). Below each is the products of an *in vitro* transcription assay in the absence and presence of ligand; "RT" denotes the read-through product ("ON" state) and "T" denotes terminated form ("OFF" state). While "design 1" and "design 2" show little or no ligand-dependent switching

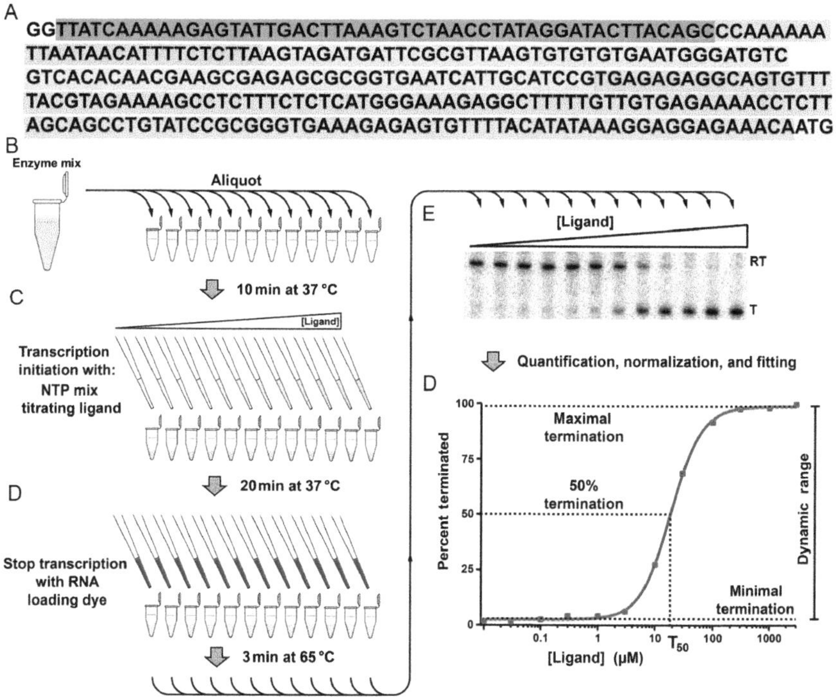

Jeremiah J. Trausch and Robert T. Batey, Figure 5 Workflow for single-turnover transcription assay. (A) Annotated sequence of the DNA template for the *Smu folT/Bsu metE* "design 3" riboswitch. The T7A1 promoter is highlighted in red, the *Smu folT* aptamer in orange, the *Bsu metE* expression platform in cyan, and the initiating ATG codon in yellow. (B) The enzyme mix is aliquoted into thin-walled PCR tubes and incubated in a thermocycler. (C) The reaction is initiated by the addition of NTP mix. Because this solution contains ligand, a separate NTP mix must be made for every concentration tested. (D) The reaction is terminated by the addition of loading dye and a raise in temperature to 65 °C prior to separation of the products by 6% (29:1) denaturing polyacrylamide gel electrophoresis. (E) Example of a gel using the chimeric *Smu folT/Bsu metE* riboswitch showing the read-through (RT) from the terminated (T) products as a function of ligand concentration. (F) Fit of data to a two-state-binding isotherm to calculate the maximum and minimum percent terminated as well as the dynamic range and T_{50}.

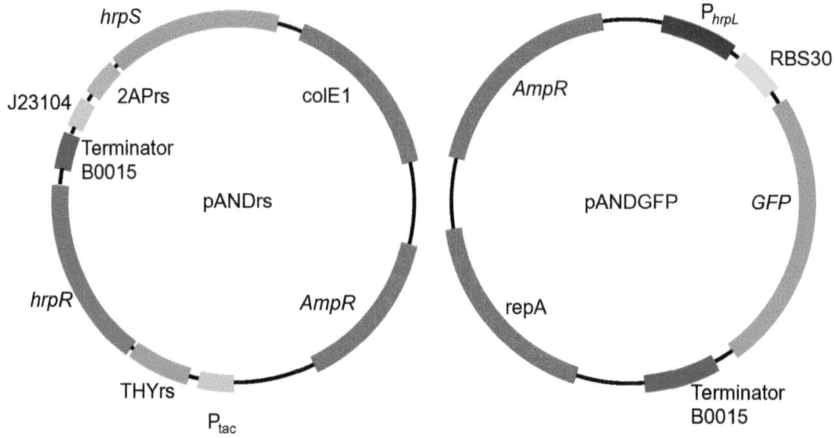

Michael S. Goodson *et al.*, Figure 2 Plasmid maps of genetic integration circuit. *AmpR:* ampicillin resistance; P_{tac}: tac promoter; THYrs: theophylline riboswitch; J23104: strong constitutive promoter; 2APrs: 2-aminopurine riboswitch; colE1: origin of replication; P_{hrpL}: *hrpL* promoter; RBS30; strong ribosome-binding site; repA: origin of replication.

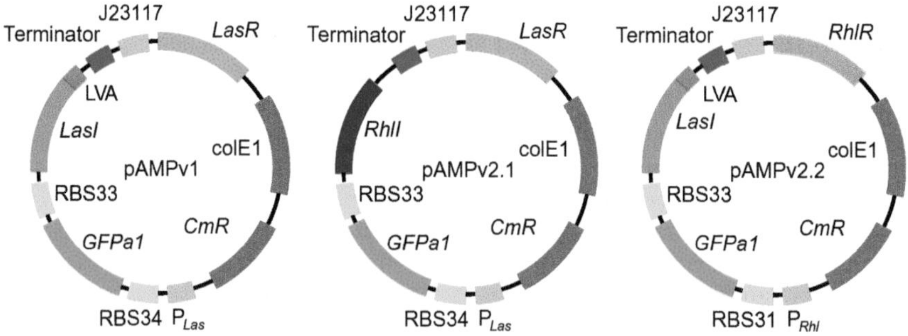

Michael S. Goodson *et al.*, Figure 4 Plasmid maps of amplification circuits. *CmR*: chloramphenicol resistance; P_{Las}: *Las* promoter; RBS34: strong ribosome-binding site; RBS33: weak ribosome-binding site; LVA: protein degradation tag; J23117: weak constitutive promoter; P_{Rhl}: *Rhl* promoter.

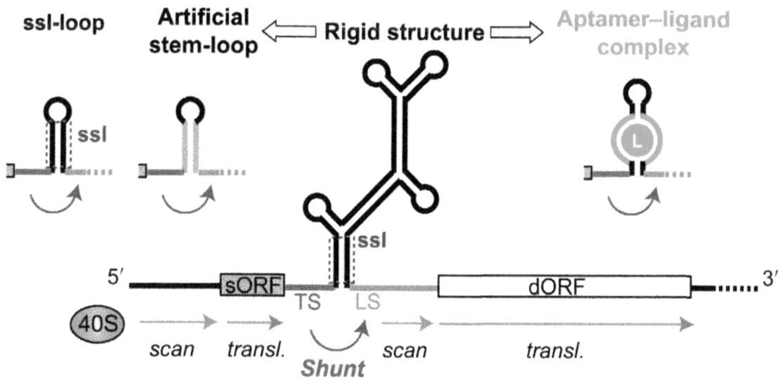

Atsushi Ogawa, Figure 1 Schematic diagram of ribosomal shunt on the CaMV 35S mRNA. Thick lines and boxes represent noncoding regions and ORFs, respectively. The ribosomal movements are shown by single arrows and italics under the mRNA. The 60S ribosomal subunit is not shown here to simplify the diagram. The intergenic rigid structure required for shunting is replaceable with an ssl-loop (far left), a stable artificial stem-loop (second left), or an aptamer–ligand complex (right). The circle L represents the ligand.

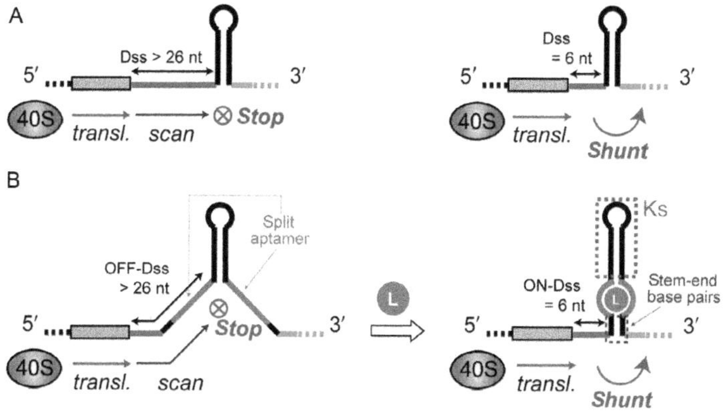

Atsushi Ogawa, Figure 3 Design of ribosomal shunt-modulating riboswitches by using the mechanism for Dss-dependently switching the mode of ribosomal movement after sORF translation. (A) The effect of Dss on the ribosomal movement. The 40S ribosome that has finished translation of sORF scans TS and stops at the rigid stem when Dss is longer than 26 nt (left), while it efficiently shunts over the rigid stem when Dss is properly short (~6 nt) (right). (B) General design of ribosomal shunt-modulating riboswitches with a split aptamer (left, the OFF state; right, the ON state).

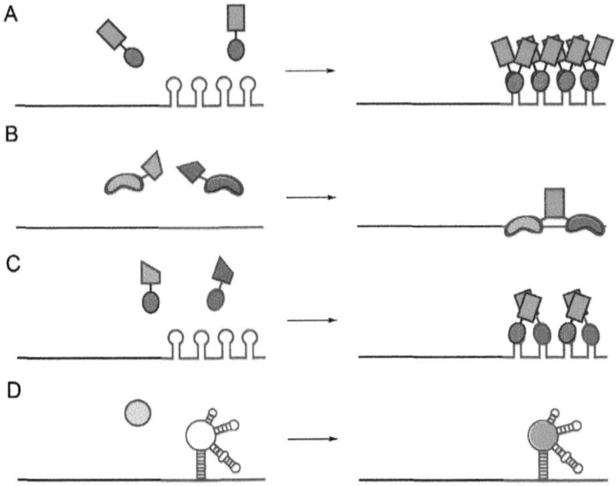

Rita L. Strack and Samie R. Jaffrey, Figure 1 Genetically encoded strategies for labeling RNA in living cells. (A) The MS2-GFP imaging system involves labeling an mRNA of interest (black line) with up to 24 copies of the MS2 RNA hairpin in the 3′ UTR (gray line). When this tagged RNA is coexpressed with the MS2 coat protein fused to GFP, these hairpins are bound by MS2-GFP as dimers, leading to tagging by up to 48 MS2-GFP molecules. (B) The PUM-HD split-FP imaging system involves engineering two PUM-HDs to distinct 8-nt regions in an RNA of interest. Each PUM-HD is then fused to half of a split fluorescent protein. When both PUM-HDs are bound to the target RNA, fluorescence complementation occurs and the RNA is labeled with a single fluorescent protein. (C) The MS2/PP7 split-FP imaging system involves tagging an RNA with both the MS2 RNA hairpin (gray hairpin) and the PP7 RNA hairpin (blue hairpin) in alternation. MS2 and PP7 coat proteins are each tagged with complementary halves of an FP. When both are bound, complementation occurs, leading to labeling of RNAs. (D) Imaging an RNA of interest by Spinach tagging involves fusing Spinach to either the 5′ or 3′ end of an RNA and incubating cells with a dye that is nonfluorescent in solution. Only when the dye is bound to the tagged RNA does the Spinach–dye complex become fluorescent, specifically labeling the tagged RNA.

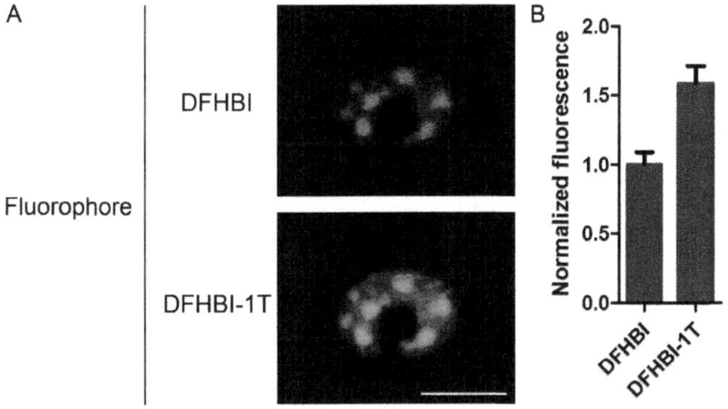

Rita L. Strack and Samie R. Jaffrey, Figure 3 Spinach2–DFHBI-1T is brighter than Spinach2–DFHBI in live cells. (A) COS-7 cell expressing CGG_{60}-Spinach2. The cell was imaged using a widefield microscope with EGFP filter sets and a 100-ms exposure time. Cells were first incubated with 20 μ*M* DFHBI and imaged. Media was then exchanged with media lacking dye for 30 min to remove DFHBI. This media was then supplemented with 20 μ*M* DFHBI-1T for 30 min. Spinach2–DFHBI-1T images were collected following this incubation. (B) Quantification of green fluorescence signal from Spinach2–DFHBI and Spinach2–DFHBI-1T in living cells. Scale bar, 20 μm. *Images were used with permission from Song et al., 2014.*

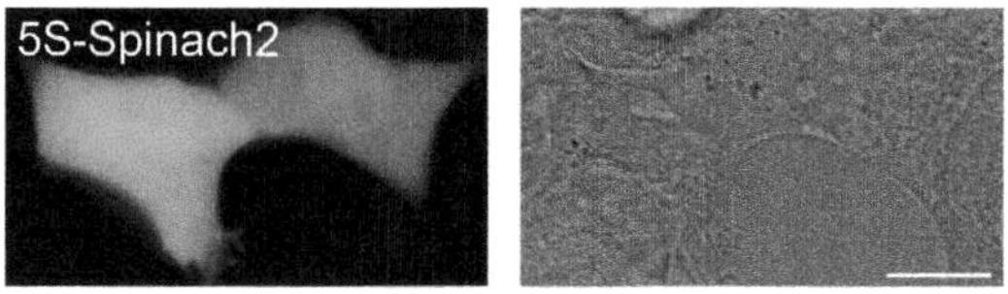

Rita L. Strack and Samie R. Jaffrey, Figure 4 HEK-293T cells expressing 5S-Spinach2. Cells were incubated with 20 μ*M* DFHBI for 30 min prior to imaging. Shown are green fluorescence (left) and differential interference contrast (DIC, right) images. Fluorescence image was collected by widefield microscopy with EGFP filter sets with a 1-s exposure time. Scale bar, 10 μ*M*.

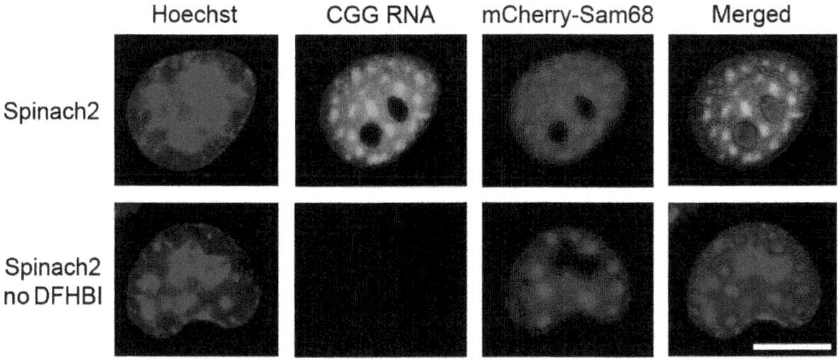

Rita L. Strack and Samie R. Jaffrey, Figure 5 CGG_{60}-Spinach2 colocalizes with mCherry-Sam68. Shown are images of COS-7 nuclei containing CGG_{60} aggregates labeled with Spinach2–DFHBI or mCherry-Sam68. Spinach signal was collected by widefield microscopy with EGFP filter sets with a 100-ms exposure time. mCherry signal was collected using a Texas Red filter set and 200-ms exposure time. CGG_{60}-Spinach2 and mCherry-Sam68 are shown with and without DFHBI. Scale bar, 20 μm.

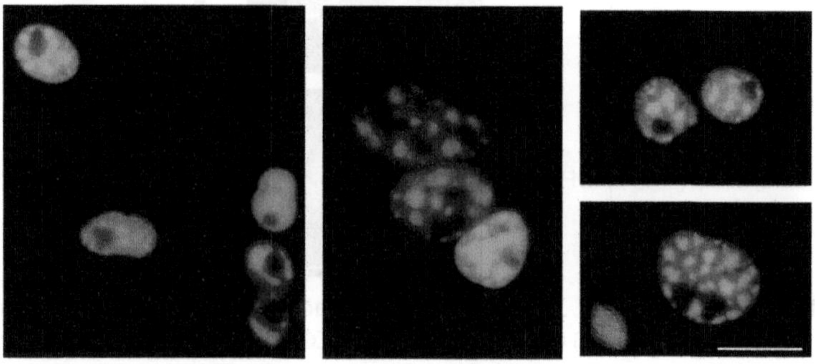

Rita L. Strack and Samie R. Jaffrey, Figure 6 Examples of COS-7 nuclei with CGG_{60}-Spinach2 aggregates. Aggregates are highly heterogeneous and range in both size and number. Some representatives are shown. Images were collected by widefield microscopy with EGFP filter sets with 50- to 200-ms exposure times.

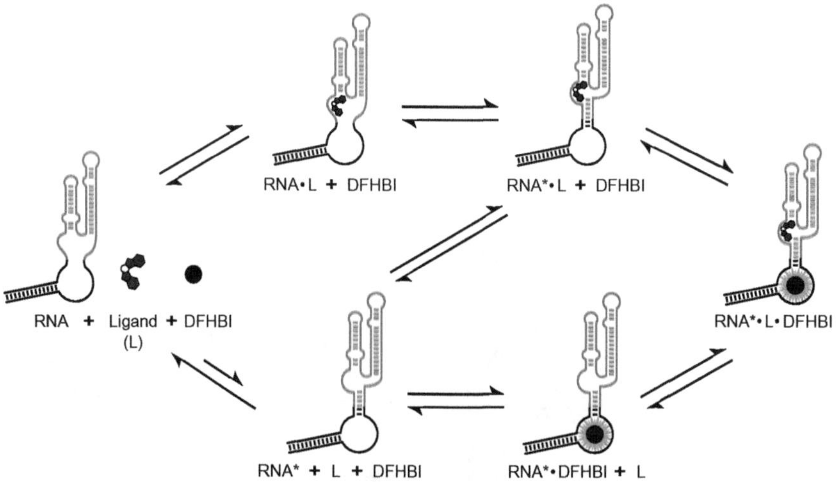

Colleen A. Kellenberger and Ming C. Hammond, Figure 4 Association of RNA, ligand, and DFHBI to form the fluorescent ternary complex. Before the stable ternary complex forms, ligand binds to the RNA, RNA presumably undergoes a conformational change to RNA*, and DFHBI binds to RNA*. Predicted pathways of complex formation are depicted.

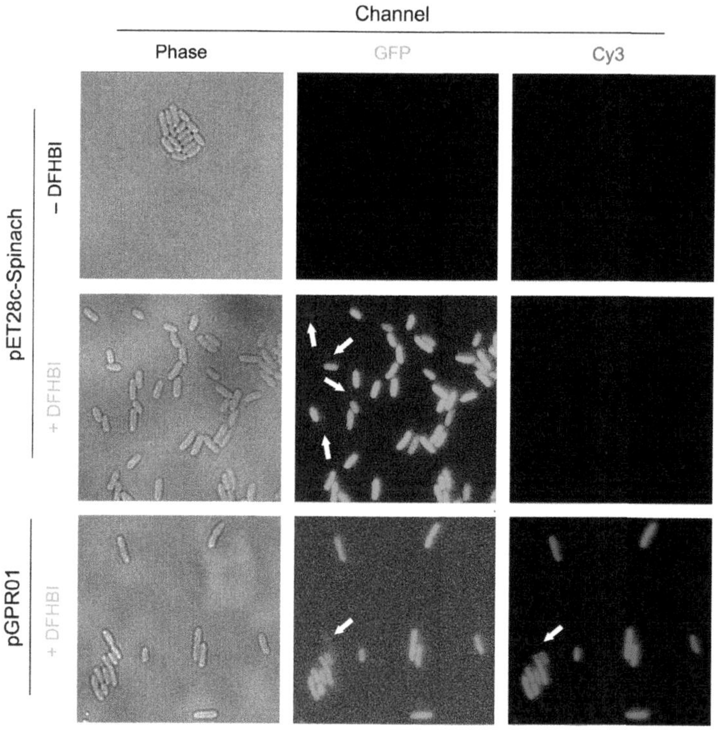

Georgios Pothoulakis and Tom Ellis, Figure 2 Live-cell fluorescence (GFP/Green and Cy3/Red) and phase images of BL21(DE3) *E. coli* cells carrying either the pET28c-Spinach or pGPR01 plasmid in the presence or absence of 200 μ*M* DFHBI following IPTG induction. OFF-population cells not showing any fluorescence are highlighted with arrows.

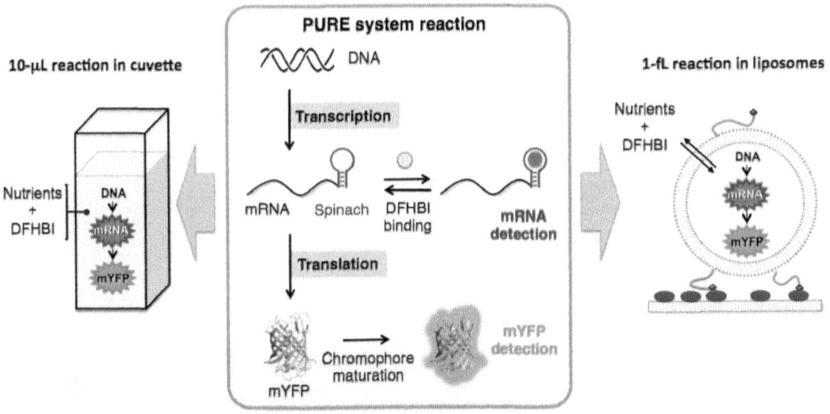

Pauline van Nies *et al.*, Figure 1 Schematic overview of our two-reporter assay for mRNA and protein levels in bulk (microtubes or cuvettes) and in liposome-confined gene expression reactions. The Spinach technology consisting of an RNA aptamer sequence that binds and turns the chromophore DFHBI into a fluorescent state was used to monitor transcription activity.

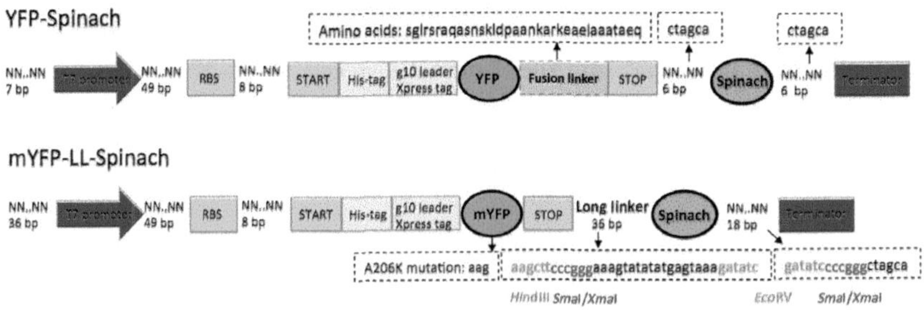

Pauline van Nies *et al.*, Figure 3 Schematic representation of the two DNA constructs primarily used in this study. Their main features and regulatory elements are depicted. The new construct mYFP-LL-Spinach has been designed based on our previously described YFP-Spinach gene (van Nies et al., 2013). The abbreviation RBS stands for ribosome binding site.

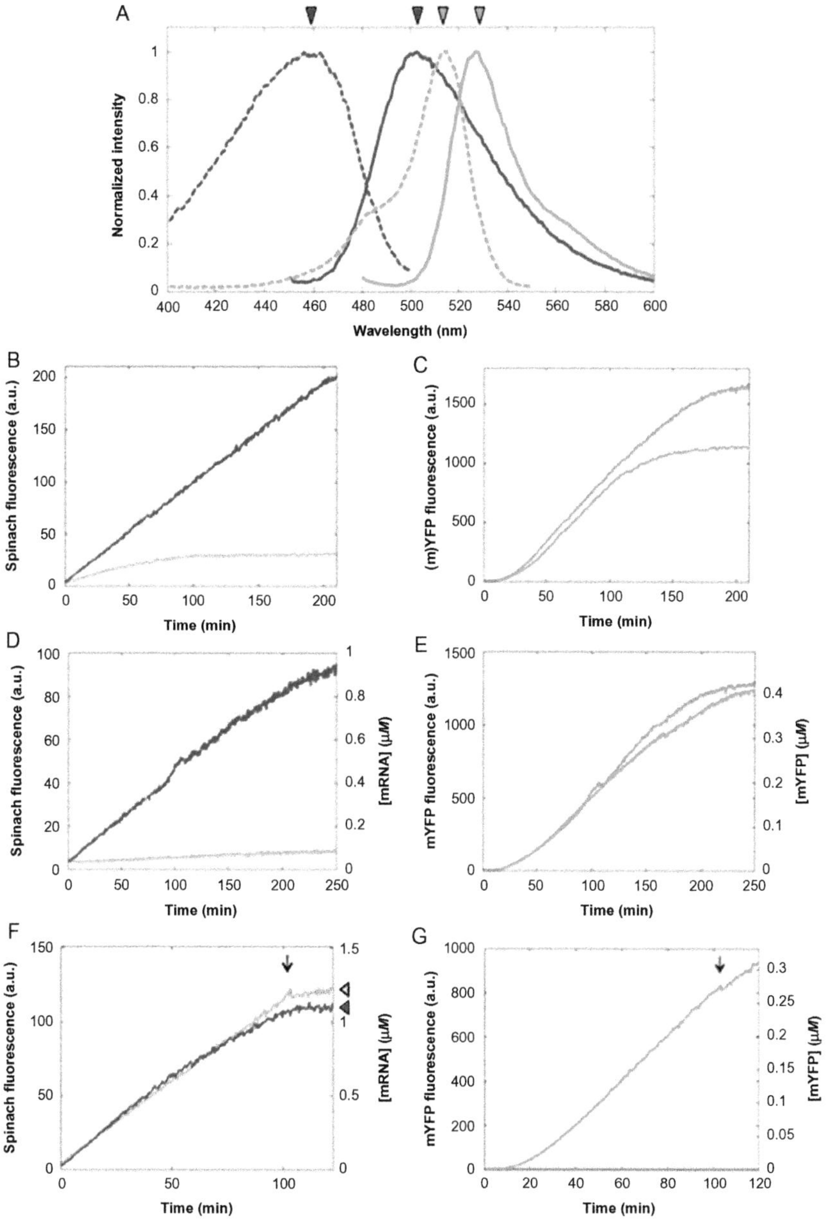

Pauline van Nies *et al.*, Figure 4 See legend on the next page.

Pauline van Nies *et al.*, Figure 4 (A) Fluorescence excitation (dashed lines) and emission (solid lines) spectra of LL-Spinach (blue) and mYFP (green) measured in the PURE*frex* expressing the mYFP-LL-Spinach gene. The LL-Spinach spectra were measured in the PURE*frex* ΔR, that is devoid of ribosome, in the presence of 20 μ*M* DFHBI. The mYFP spectra were collected in the PURE*frex* without DFHBI. The arrowheads depict the excitation and emission wavelengths used for kinetics measurements. (B) Fluorescence intensity profiles of Spinach produced from the mYFP-LL-Spinach (dark blue) or YFP-Spinach (light blue) construct. (C) Apparent kinetics of mYFP (dark green) and YFP (light green) synthesis monitored simultaneously as in (B). (B and C) DNA concentration for both genes was 7.4 n*M*. (D) Plots of LL-Spinach fluorescence versus time using the mYFP-LL-Spinach (dark blue) or mYFP (light blue) construct. (E) Apparent kinetics of mYFP produced from the mYFP-LL-Spinach (dark green) or mYFP (light green) construct and monitored simultaneously as in (D). (D and E) DNA concentration for both genes was 0.74 n*M*. (F) Progression of Spinach fluorescence versus time in a PURE*frex* ΔR (dark blue) or PURE*frex* (light blue) reaction starting from 11.7 n*M* of the mYFP-LL-Spinach DNA. The arrowheads on the right axis point to the final intensity values used for calculating the conversion factor between fluorescence a.u. and mRNA concentration. (G) Apparent kinetics of mYFP synthesis in a PURE*frex* ΔR (dark green) or PURE*frex* (light green) reaction monitored simultaneously as in (F). (F and G) The arrow at around 100 min indicates the addition of DNaseI to stop transcription. (D and F) Concentrations of mRNA were calculated using a conversion factor of 10 n*M*/a.u. (E and G) Concentrations of mYFP were calculated using a conversion factor of 0.33 n*M*/a.u.

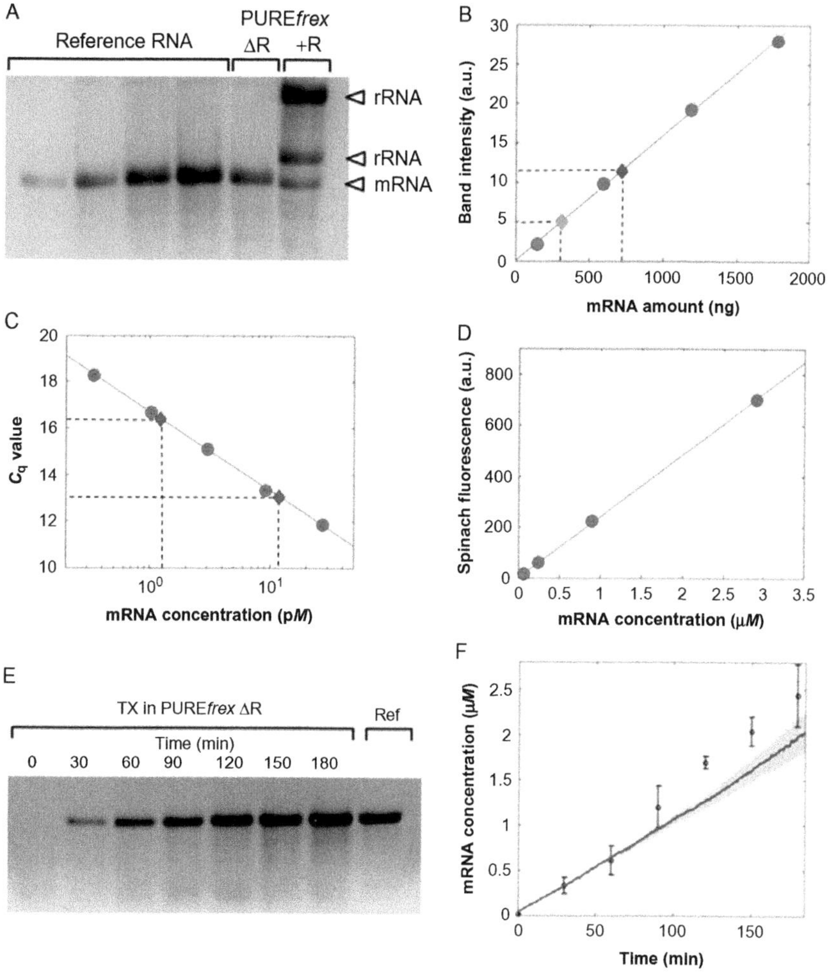

Pauline van Nies *et al.*, Figure 6 Quantification of mRNA synthesis in PURE system reactions starting from 7.4 n*M* of the mYFP-LL-Spinach DNA template. (A) RNA samples loaded on an agarose gel. The band intensities of mRNA produced in a PURE*frex* reaction (Fig. 4F) with (+R) or without ribosome (ΔR) were compared to reference RNA samples of known concentrations. The 1.5 and 2.9-kb ribosomal RNA bands are visible in the +R reaction condition. (B) Calibration curve plotted as the measured band intensities of reference RNA samples versus their predetermined concentrations. The mRNA band intensities of the PURE*frex* samples shown in (A) were appended on the calibration curve, after which the amount of synthesized transcript can be determined. (C) Reference RNA samples were analyzed by RT-qPCR and their C_q values plotted as a function of concentration. The obtained standard curve has a typical equation of $y = -1.479\ \ln(x) + 16.723$; $R^2 = 0.998$. The measured C_q values of diluted samples from PURE*frex* ΔR reactions were appended on the calibration curve and their concentrations were determined. Two samples of 10-fold different dilution factors are displayed. (D) Calibration curve consisting of Spinach fluorescence intensities measured for

(Continued)

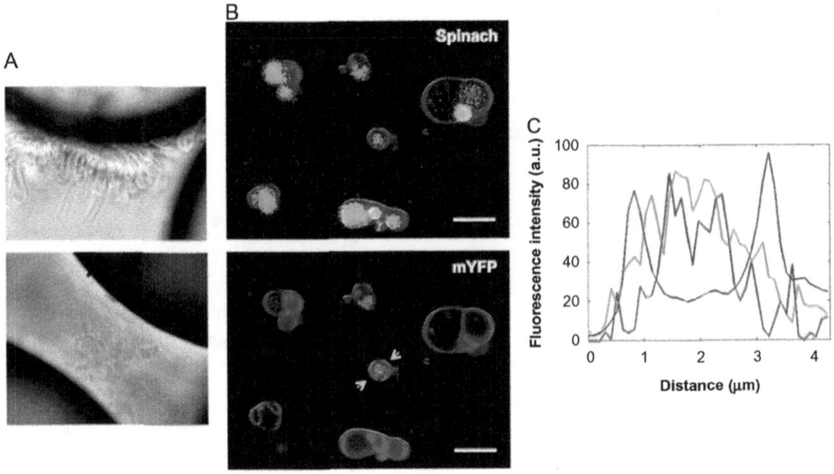

Pauline van Nies *et al.*, Figure 9 Fluorescence imaging of liposomes. (A) Phase contrast micrographs of lipid film swelling from glass bead surfaces (top). The tethered tubular liposomes eventually detach from the glass surface and remain trapped within the bead cavities (bottom) until the bead stack is gently disassembled for liposome harvesting. (B) Fluorescence confocal images of surface-tethered liposomes (dioleoyl phospholipids, composition 2) postexpression of the mYFP-LL-Spinach gene. The vesicle membrane (red) is localized using TRITC-labeled phospholipids. Scale bar is 5 μm. (C) Fluorescence intensity profiles of the TRITC, Spinach and mYFP signals measured along the line defined between the two arrows in (B). Color coding is the same as in (B).

Figure 6—Cont'd different concentrations of reference RNA solutions. The slope gives a conversion factor of 4.1 n*M*/a.u. (E) Time series gel analysis of mRNA produced in a PURE*frex* ΔR reaction. The band intensities were compared with that of a reference RNA (right-most lane). (F) The dynamics of transcription was reconstructed by plotting the concentrations of mRNA as determined in (E) at different time points. Error bars indicate SEM, $n=3$. For comparison, the apparent kinetics obtained by monitoring the Spinach fluorescence in real time is overlaid. The blue curve is the mean of three independent measurements and the gray-shaded area denotes the min and max deviation.

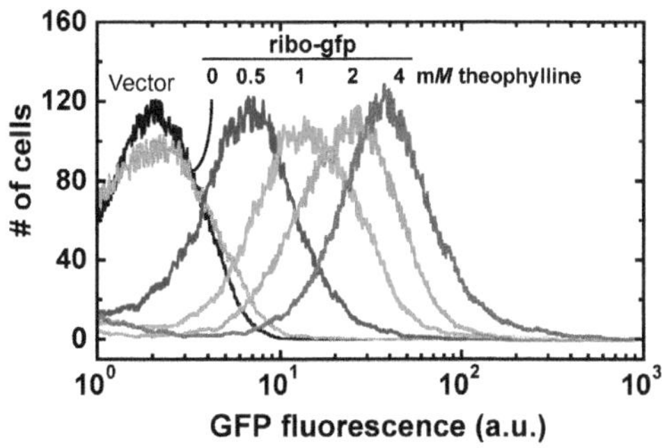

Erik R. Van Vlack and Jessica C. Seeliger, Figure 4 Flow cytometry of theophylline-dependent GFP expression shows that the E* riboswitch is titratable. *Adapted from Seeliger et al. (2012).*

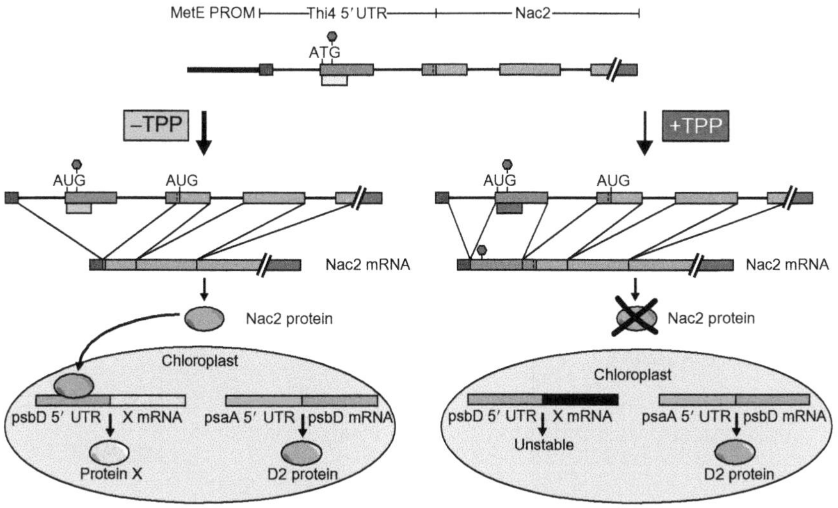

Silvia Ramundo and Jean-David Rochaix, Figure 1 Reversible vitamin-mediated repression of chloroplast gene expression. Scheme of the vitamin-mediated repressible chloroplast gene expression system. The *Nac2* gene, which is specifically required for the accumulation of the chloroplast *psbD* mRNA, is fused to the *Thi4* 5′UTR containing the TPP-responsive riboswitch and the *MetE* upstream region is used as promoter. Addition of thiamine causes alternative splicing in the riboswitch region, which results in translation termination due to the inclusion of a stop codon (red flag) (Croft et al., 2007). The yellow box below the second exon indicates the genomic location of the TPP riboswitch. The change of the box color represents in a schematic way the conformational change of the riboswitch upon binding of TPP (green: no TPP binding; red: TPP binding). Because Nac2 acts specifically on the *psbD* 5′UTR, it is possible to render the expression of any chloroplast gene dependent on Nac2 by fusing its coding sequence to the *psbD* 5′UTR. To allow for phototrophic growth in the presence of the vitamins, the *psbD* 5′UTR of the *psbD* gene was replaced by the *psaA* 5′UTR, thus making *psbD* expression independent of Nac2. *Reproduced from Ramundo, Rahire, et al. (2014) with permission (Copyright American Society of Plant Biologists).*

A

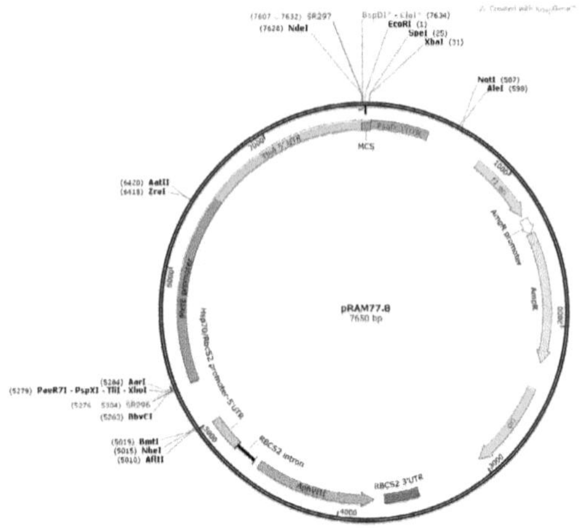

B

	DNA template	Primer Fw	Primer Rv
(psbD 5′) (amplicon 1)	Chloroplast genome	SR 186	SR185
(Rps12 - flanking region for HR) (amplicon 2)	Chloroplastgeonome	SR184	SR183
(psbD 5′-Rps12 – flanking region for HR) (OE-PCR amplicon)	Template 1 + Template 2	SR186	Sr183

C

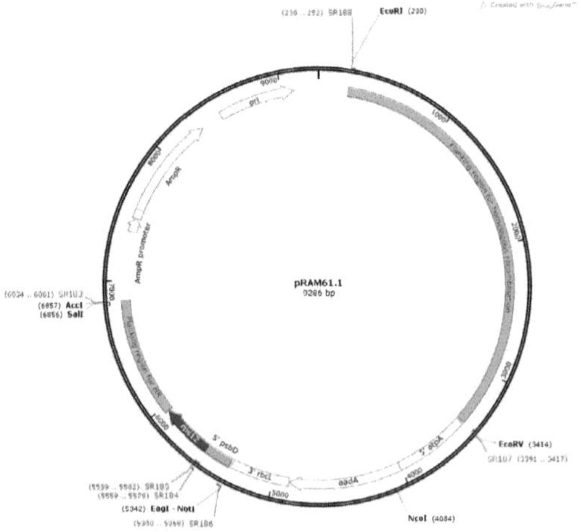

Silvia Ramundo and Jean-David Rochaix, Figure 2 See legend on the opposite page.

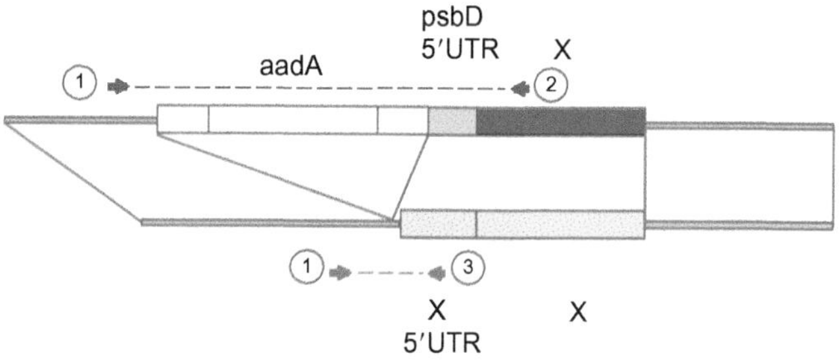

Silvia Ramundo and Jean-David Rochaix, Figure 3 Test for homoplasmicity. The upper line shows the construct with the *aadA* cassette and the *psbD* 5′UTR-geneX inserted into the chloroplast genome through homologous recombination. The lower line represents the locus of gene X in the wild-type chloroplast genome. Primers 1 and 2 amplify a fragment specific for the inserted *psbD* 5′UTR-X construct (red-dashed) whereas primers 1 and 3 amplify a wild-type fragment (blue-dashed). A strain homoplasmic for the *psbD* 5′UTR construct will only amplify the red PCR product, whereas a heteroplasmic strain will amplify both the blue and the red PCR product. Since there are about 80 copies of the chloroplast genome per *Chlamydomonas* cell, to prove that the transformed strain is homoplasmic, it is highly recommended to perform a PCR with 80 times less wild-type DNA and ensure that a single copy of the wild-type gene is detectable with PCR under the same conditions.

Silvia Ramundo and Jean-David Rochaix, Figure 2 (A) Map of pRAM77.8, the vector containing the *MetE* promoter and *Thi4* 5′UTR derived from pSL18; the illustrated primers in magenta SR296 (ccgctcgagTACTTCGTGCAGGTGTCTTA) and SR297 (ccgctcgagTACTTCGTGCAGGTGTCTTA) were used to amplify the *MetE* promoter and the *Thi4* 5′UTR from pRAM23.1; some relevant and unique restriction sites are indicated with a bold text string; MCS, Multicloning site. (B) Example of overlap extension polymerase chain reaction (OE-PCR) to fuse the *psbD* 5′UTR/promoter to the plastid gene *rps12*. (C) Map of pRAM61.1, the plasmid used to transform the A31 strain to generate a repressible expression system for the plastid gene *rps12*. In this case, the *aadA* gene was cloned in the same orientation as the *psbD5′-rps12* gene although no read-through transcription was observed. Some relevant and unique restriction sites are indicated with a bold text string. The sequence of the illustrated primers in magenta (Ramundo, Rahire, et al., 2014) and the map of the pUC-atpX-AAD plasmid are available (Goldschmidt-Clermont, 1991).

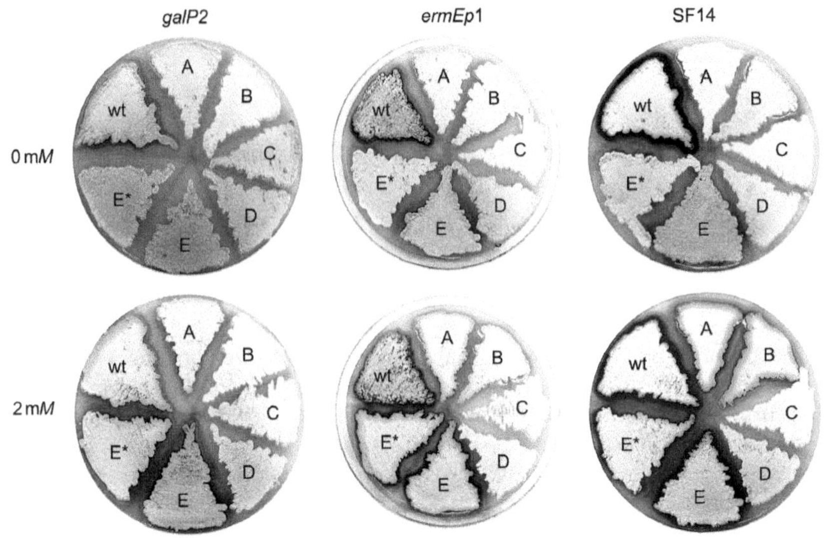

Martin M. Rudolph *et al.*, Figure 3 Analysis of different theophylline-dependent riboswitches in *S. coelicolor*. X-Gluc staining of *S. coelicolor* strains growing on MS agar expressing *galP2-/ermEp1-*/SF14-driven *gusA* under control of the theophylline riboswitches A, B, C, D, E, E*, or without riboswitch (wt). The medium was supplemented with 2 m*M* theophylline (theo) for induction.

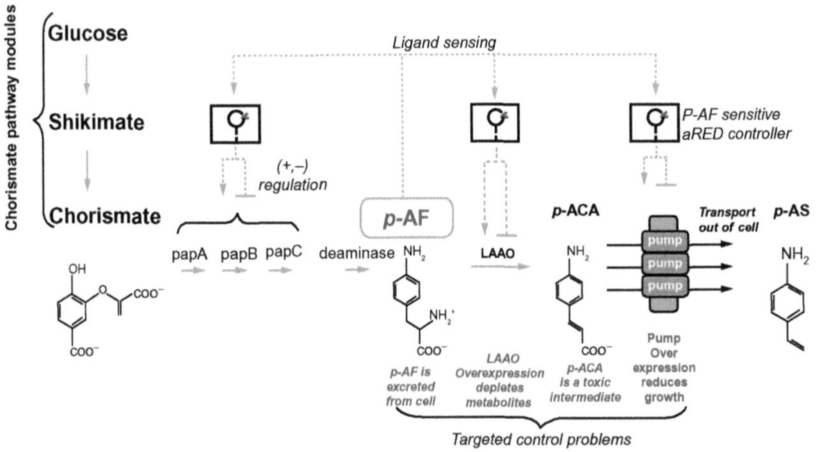

David Sparkman-Yager *et al.*, Figure 1 Schematic of the metabolic pathway for the production of *p*-aminostyrene, from glucose. Targeted control problems for the pathway are indicated. Dashed blue arrows represent the possible control points for a *p*-AF-sensing aptazyme.

A

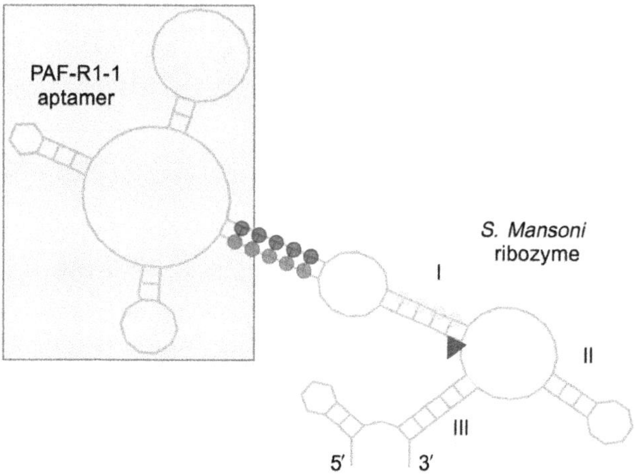

B

1	2
5′ GGGCGAAAGCCGGCGCGUCCUGGAUUCCAC NNNNNAUGUCCCUACCAUACGGGAUUGCCCAGCUUC GGCUGCCAUGCCGGCCAAACGGUAACCGGCCUACG NNNNNGGUACANNNNGCUGAUGAGUCCCAAAUAGGA CGAAACGCGCU 3′	5′ GGGCGAAAGCCGGCGCGUCCUGGAUUCCACC AAGCAUGUCCCUACCAUACGGGAUUGCCCAGCU UCGGCUGCCAUGCCGGCCAAACGGUAACCGGC CUACGGGAGGGGUACACAACGCUGAUGAGUCC CAAAUAGGACGAAACGCGCU 3′

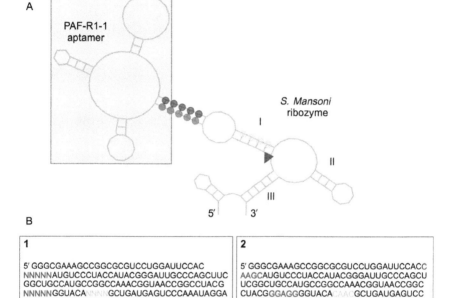

David Sparkman-Yager *et al.*, Figure 3 Structure and sequences of a *p*-AF-sensing aptazyme design pool. (A) Secondary structure diagram for the active state of the designed aptazymes. The artificial *p*-AF-R1-1 aptamer (gray box) is fused to the natural *S. mansoni* hammerhead self-cleaving ribozyme by a random linker (Carothers et al., 2010; Yen et al., 2004). (B) 1. Sequence of the aptazyme *in vitro* selection pool. Colored letters represent randomized nucleotides (N's) in selection pool. 2. Sequence of a functional aptazyme (*p*-AF-2) identified from the aforementioned *in vitro* selection pool.

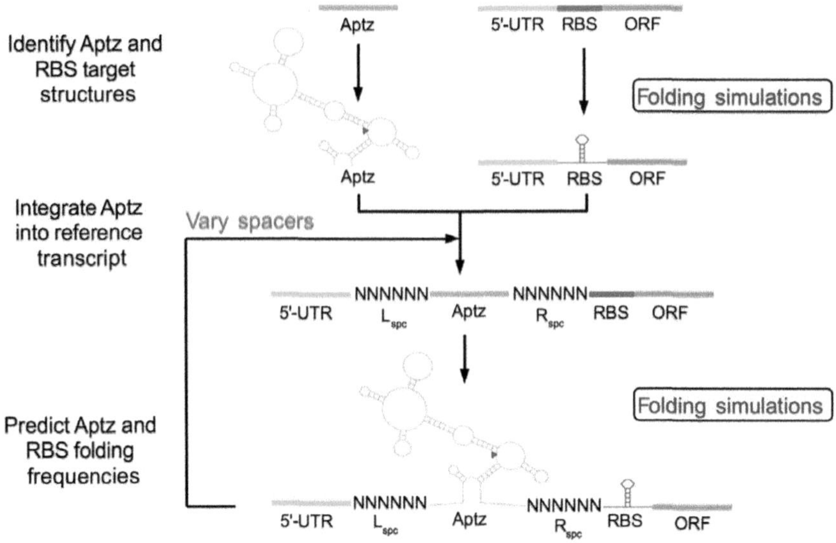

David Sparkman-Yager *et al.*, Figure 4 *In silico* workflow for the design of genetic devices with predictable outputs. First, the native secondary structures for the aptazyme and ribosome binding site are determined through stochastic folding simulations. Second, the aptazyme sequence is inserted upstream of the RBS, flanked by random spacer hexamers. Next, the generated sequences are folded and evaluated on the structural agreement of their component parts with their target structures. Spacer sequences are varied until the desired folding frequencies are achieved.

T7 promoter 5′UTR Lspc
pAF-2 Aptazyme Rspc **RBS**
5′ end of DsRed (RFP)

5′ GTCTAATACGACTCACTATAGGGACGACGACAGGCACCCGAACTCNNNNNNGGGCGA
AAGCCGGCGCGTCCTGGATTCCACCAAGCATGTCCCTACCATACGGGATTGCCCAGCTT
CGGCTGCCATGCCGGCCAAACGGTAACCGGCCTACGGGAGGGGTACACAACGCTGATG
AGTCCCAAATAGGACGAAACGCGCTNNNNNN**GAAAGAGGAGAAATACTAG**ATGGCTTCC
TCCGAAGACGTTATCAAAGAGTTCATGCGTTTCAAAGTTCGTATGGAAGGTTCCGTTAACG
GTCACGAGTTCGAAATCGAAGGTGAAGGTGAAGGTCGTCCGTACGAAGGTACCCAGACC
GCTAAACTGAAAGTTACCAAAGGTGGTCCGCTGCCGTTCGCTTGGGACATCCTGTCCCC
GCAGTTCCAGTACGGTTCCAAAGCTTACGTTAAACACCCGGCTGACATCCCGGACTACCT
GAAACTGTCCTTCCCGGAAGGTTTCAAATGGGAACGTGTTATGAAC...3′

David Sparkman-Yager *et al.*, Figure 5 Sequence of a *p*-AF biosensor used for transcript design example. The *p*-AF-2 aptazyme is inserted downstream of the T7 promoter, and upstream of the RBS and coding sequence for RFP. The RBS target structure is evaluated within the given context, without the presence of the aptazyme or spacer sequences.

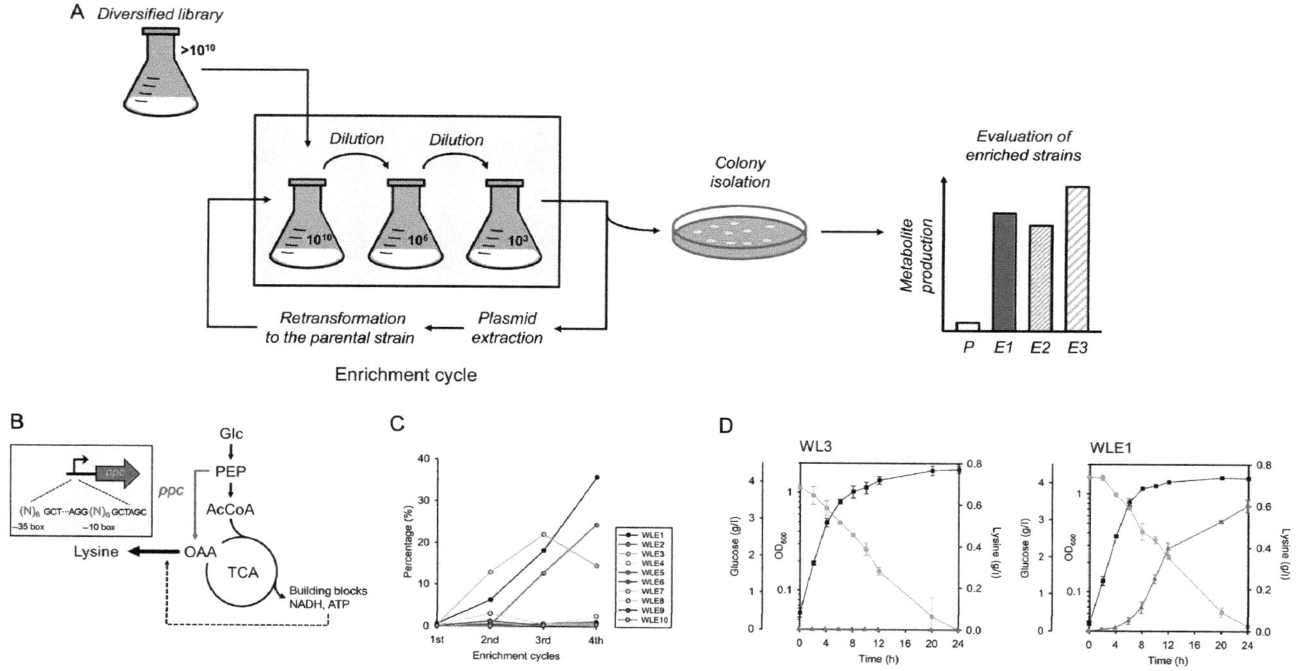

Sungho Jang *et al.*, Figure 3 See legend on the next page.

Sungho Jang *et al.*, Figure 3 Application of Riboselector for pathway engineering. (A) Program for optimization of metabolic pathway by means of Riboselector is illustrated here. Phenotypically diversified library is cultured in the presence of the Riboselector and corresponding selection pressure. Repetitive dilution should shift the microbial population to metabolite high-producing strains. Single colonies are isolated after enrichment cycles, and each variant is tested for metabolite production by flask culture and proper quantification method. (B) Schematic metabolic pathway toward lysine in *E. coli*. Expression level of *ppc* was diversified to construct the library of lysine producing *E. coli*. variants with different productivity by randomizing sequences of −35 and −10 boxes of promoter (BBa_J23100) that drives transcription of *ppc* gene encoded in plasmid. Glc, glucose; PEP, phosphoenolpyruvate; AcCoA, acetyl-CoA; OAA, oxaloacetate. (C) Population of the library was analyzed using next-generation sequencing of promoter region of *ppc*. The *y*-axis represents the percentage of each variant in the total population. The *x*-axis represents the enrichment cycle depicted in (A), and each enrichment cycle corresponds to three serial cultures and one plasmid preparation. (D) Comparison of physiology between the parental strain (WL3) and one of the enriched strains (WLRE1) that occupied the population up to 40% after four enrichment cycles. The left *y* offset and right *y*-axis represent concentration (g/l) of glucose (green circles) and lysine (red triangles), respectively. The left *y*-axis represents optical density (black rectangles) at 600 nm in log scale. The *x*-axis represents the culture time (h). The experiments were replicated twice. *Panels (B–D): Adapted from Yang et al. (2013) with permission.*

Edwards Brothers Malloy
Ann Arbor MI. USA
January 13, 2015